Hemp

Industrial Production and Uses

CW00550702

Hemp

Industrial Production and Uses

Edited by

Pierre Bouloc

La Chanvriere de L' Aube (LCDA), France

Serge Allegret

La Chanvriere de L' Aube (LCDA), France

and

Laurent Arnaud

Ecole Nationale des Travaux Public de l' Etat (ENTPE), France

Development Editor

David P. West, PhD

Plant Breeder, USA

Translated from orginal French text by

Glen Cousquer, BVM&S, BSc, CertZooMed, MRCVS

Independent Translator, UK

www.cabi.org

CABI is a trading name of CAB International

CABI
Nosworthy Way
Wallingford
Oxfordshire OX10 8DE
UK

CABI
745 Atlantic Avenue
8th Floor
Boston, MA 02111
USA

Tel: +44 (0)1491 832111
Fax: +44 (0)1491 833508
E-mail: info@cabi.org
Website: www.cabi.org

Tel: +1 617 682 9015
E-mail: cabi-nao@cabi.org

A catalogue record for this book is available from the British Library, London, UK.

Library of Congress Cataloging-in-Publication Data

Chanvre industriel. English
Hemp: industrial production and uses / edited by Pierre Bouloc, Serge Allegret, and Laurent Arnaud; translator: Glen Cousquer.
    p. cm.
  Translation of: Le chanvre industriel: production et utilisations.
  Includes bibliographical references and index.
  ISBN 978-1-84593-792-8 (pbk: alk. paper) – ISBN 978-1-84593-793-5 (hardback: alk. paper)
    1. Hemp. 2. Hemp–Utilization. 3. Hemp industry. I. Bouloc, Pierre. II. Allegret, Serge. III. Arnaud, Laurent, 1965-

  SB255C4713 2012
  633.5′71–dc23

2012004199

ISBN-13: 978 1 84593 792 8 (pbk)
      978 1 84593 793 5 (hbk)

Commissioning editor: Sarah Hulbert
Editorial assistants: Alexandra Lainsbury and Gwenan Spearing

Typeset by SPi, Pondicherry, India
Printed and bound in the UK by CPi Group (UK) Ltd, Croydon, CRO 4YY

First printed in 2013. Reprinted in 2019, 2020.

Contents

Contributors

Serge Allegret, La Chanvriere de L'Aube (LCDA), France
Laurent Arnaud, Ecole Nationale des Travaux Public de l'Etat (ENTPE), France
Olivier Beherec, FNPC, France
Dr Janoš Berenji, Institute of Field and Vegetable Crops, in Novi Sad, Serbia
Sylvestre Bertucelli, FNPC, France
Pierre Bouloc, La Chanvriere de L'Aube (LCDA), France
Bernard Boyeux, Consultant, France
Bernard Brochier, formerly Centre Technique du Papier, France
Michael Carus, European Industrial Hemp Association (EIHA), c/o nova-Institut GmbH, Chemiepark Knapsack, Industriestrasse, 50354 Hürth, Germany
Nicolas Cerruti, Institut Technique du Chanvre (ITC), France
Gabriel Cescutti, European Patent Office, Verrijn Stuartlaan, 2288 ER Rijswijk, The Netherlands
Brigitte Chabbert, INRA, France
Catherine Dejean, Hospital Pharmacist, Centre hospitalier Henri Laborit, Poitiers, France, and Faculty of Medicine and Pharmacy, University of Poitiers, France
François Desanlis, Consultant, France and South Africa
Holger Fischer, Faserinstitut Bremen e.V. – FIBRE –, Am Biologischen Garten 2, 28359 Bremen, Germany
Gilbert Fournier, Laboratory of Pharmacology of the Faculty of Pharmacology at the University of Paris-Sud, France
Yves Hustache, Consultant, France
Tanya Jobling, Consultant, Australia
Bernard Kurek, INRA, France
Sandrine Legros, Institut Technique du Chanvre (ITC), France
Gero Leson, Consultant, Leson & Associates, Berkeley, California, USA
Gérard Mougin, AFT Plasturgie – Agro Fibres Technologies Plasturgie, Fontaine les Dijon, France
Jörg Müssig, Hochschule Bremen, University of Applied Sciences, Faculty 5, Biomimetics/ Biological Materials, Neustadtswall 30, 28199 Bremen, Germany
Sébastien Picault, Institut Technique du Chanvre (ITC), France
Denis Richard, Hospital Pharmacist and Head of Service, Centre hospitalier Henri Laborit, Poitiers, France, and Faculty of Medicine and Pharmacy, University of Poitiers, France

Vladimir Sikora, Institute of Field and Vegetable Crops, Novi Sad, Serbia
Hayo van der Werf, INRA, Rennes, France
Nick Veltre, Consultant, Vietnam
Philip Warner, Consultant, CEO ECOFIBRE, Australia

David P. West, Plant Breeder, USA
Glen Cousquer, Independent Translator, UK

1 Hemp: A Plant with a Worldwide Distribution

Pierre Bouloc
La Chanvriere de L'Aube (LCDA), France

1.1 Where is Industrial Hemp Grown?

Hemp is thought to have originated in the Yunnan Province of China and its utilitarian form, *Cannabis sativa*, has spread progressively across the globe. In general, it demonstrates a preference for the temperate zones situated between the 25th and 55th parallels on either side of the equator.

This great versatility is not unusual, for there are many plants, including wheat, that flourish in both the northern and southern hemisphere. Historical records showed that hemp was widely cultivated across most of the planet by people who recognized its great utility.

1.2 Modern-day Industrial Hemp Production

Table 1.1 provides a comprehensive list of all those countries producing industrial hemp, even if their output is small. It bears testimony to the ongoing importance of this crop.

The information provided for the 25 European countries is particularly reliable, as the figures are derived from an industry that is strictly controlled and closely monitored.

The subsidies available to support hemp production further ensure the accuracy of these reported figures.

The production of certain other countries, such as the Baltic states and the ten countries who began production in 2003, remains small. We know, however (Chapter 2), that these Baltic states and northern Russia were largely responsible, during the 17th and 18th centuries, for the supply of hemp to the navies of France, Britain and Holland, to name the largest three.

All the new members of the European Union, historically having produced significant hemp crops, have requested authorization to continue production.

Among the countries of Eastern Europe are Russia, Serbia and Romania. Historically, they have devoted significant areas to the production of hemp. Today, their production is modest, although sooner or later financial backing will be found to fund the cultivation of this crop. It should not be forgotten that, despite various financial problems, countries such as Hungary, Poland, Serbia and Ukraine maintain hemp research centres that rival France's Fédération Nationale des Producteurs de Chanvre (FNPC) in Le Mans.

In Asia, modest production is reported from countries like Korea and Japan. This reflects

Table 1.1.

Countries Cultivating hemp in 2004			
Europe 25[3]	Area of hemp in 2004 (ha)	Other European countries	Area of hemp in 2004 (ha)
Germany	1730	Romania	2000
Austria	397	Russia	2500
Belgium	0	Serbia	200
Denmark	0	Ukraine	1000
Spain	678	**TOTAL**	
		Other European countries	**5700**
Estonia	PM	**ASIA[4] (estimations)**	
Finland	6	China	65000
France	8 427	North Korea	13000
Great Britain	1 640	South Korea	224
Holland	27	Japan	20
Hungary	539	Turkey	700
Ireland	22	**TOTAL Asia**	**78944**
Italy	885	**Australia**	
Latria	PM	Australia	**250**
Lithuania	PM	**North America**	
Poland	81	Canada	**5500**
Slavakia	PM	**South America**	
Sweden	PM	Chile (FAO)	4300
Czech Republic	500	**Africa**	
TOTAL		South Africa	PM
Europe 25	**14932**	**TOTAL WORLD**	**105756**

3. Data from the EU.
4. Data from FAO and estimations
* PM = not recorded

the political alignment of these two countries with the repressive restrictions of the USA. However, the price of oil, together with current environmental concerns, may well see production increase in these countries, particularly following recent moves in favour of the use of natural fibres.

The figures for China have been estimated, as it has been impossible to obtain a reliable estimate. Chinese production is undeniable, however, for:

1. Every year the Chinese export to Europe 5000–6000 t of hemp seed for use as animal (bird) feed. The Chinese also supply neighbouring markets with organic hemp seed, Japan in particular. This allows us to estimate, taking into account yields, a surface area of some 10,000 ha under cultivation.
2. China exports a wide range of clothing, textile and decorative items made from hemp

to various countries in the developed world. The existence of dioecious hemp, from which the best fibres for weaving are obtained, can therefore be assumed. The large quantities of exported hemp textiles point to an extensive programme of cultivation.
3. We now know that five large companies cultivate, for their own use, some 50,000 ha. Their crops are produced using modern technologies and are destined for the textile industry.
4. If we add the surface area under cultivation for internal demands, a total area of 65,000 ha under cultivation can be proposed as the absolute minimum.

In Australia, the pioneers of Ecofibre have come a long way in 15 years to re-establish hemp as a legitimate crop. They have lobbied successfully for changes in the law, undertaken seed trials, developed appropriate local farming and harvesting techniques, as well as developing markets, farming agreements and financing plans.

In the USA, the industrial production of hemp is still not permitted by law. That said, the country of prohibition, in which hemp production has been outlawed since 1950, is slowly opening itself to the potential uses of hemp seed. This seed comes from neighbouring Canada. Here, production has been legal since 1998 and is growing rapidly, judging by the production figures for Ontario, Manitoba and Saskatchewan. If we take into account the experiments being undertaken in Quebec, New Brunswick and the Prairies, approximately 12,000 ha were under production in 2009. All these areas were for seed production destined for human consumption. Chapter 16 of this book provides a scientific description of the properties of this seed. The benefits associated with hemp seed are currently being investigated under the aegis of the Canadian Hemp Trade Alliance (CHTA). Further growth in North American hemp production is to be anticipated.

In Africa, there are few figures to report, as production is negligible. The development of hemp production in South Africa remains a possibility, as various pioneers have initiated projects.

1.3   Future Perspectives

Interest is growing across the world, and especially in Europe, in industrial hemp and in natural hemp fibre(s) in particular. The preceding account bears testimony to this.

A growing demand for natural products, our increasing interest in and concern for the environment, together with rising fuel and raw material costs provide, among other factors, a unique opportunity for hemp production.

At the same time, selection techniques, cultivation practices and industrial equipment have all advanced. The most important development, however, concerns our improved ability to use and work with natural fibres. Over the next 10 years, we can expect to see a significant growth in raw production and in the uses found for this remarkable plant and its various constituents.

This book will help all those who seek to develop and diversify their production.

2 The History of Hemp

Serge Allegret
La Chanvriere de L'Aube (LCDA), France

2.1 Introduction

Industrial hemp makes for an interesting and intriguing book title, for the uninitiated will be largely ignorant of this plant and its industrial role.

The contemporary significance of this subject may not, at first, be obvious to the reader. Few, other than a small number of professionals and botanists, will have set eyes on this plant of some 3 m in height, growing thickly in fields like millions of tall, green pencils, for that is about how thin they are in optimal culture.

Who is aware that, in 2009, of the 29 million ha (Mha) under cultivation in France, 12,000 ha were devoted to the cultivation of hemp? Only chance will bring people into contact with this plant. Only our grandparents and great grandparents will be able to draw on their memories of the hemp plant and provide explanations to enlighten younger generations.

And yet hemp has accompanied humans from their earliest industrial endeavours. It is now several thousand years since humans transformed hemp into serviceable products, and for a long time, hemp was one of our most important commodities, and was recognized as such. And then, little by little, it fell into disuse, becoming a statistical non-entity and eventually just a word in our dictionaries.

After a period of great utility, bordering on the indispensable, it fell victim to various industrial discoveries. Coal-fired steam engines saw it disappear from the shipbuilding industry, where it had previously supplied both sails and rope, while synthetic fibres replaced it as a material for use in the textiles industry.

Towards the end of the 20th century, humans suddenly became conscious of their insensitive brutality towards the natural world. They could no longer ignore the widespread and overwhelming levels of pollution and the fact that natural resources were not infinite, as previously thought, but were disappearing little by little.

The late but necessary reaction of humans to this situation has been to institute measures designed to save the environment. There has been a growing realization that what has been used up has gone forever.

It is the ambition of the authors of this book to detail all the uses of this ancient plant and to show that, following on from its distinguished history, the hemp plant can expect a great future. In presenting the history of hemp, this book will show its principal uses and demonstrate its historical importance.

Drawing on various sources, the book will explain how hemp has accompanied humans in their day-to-day life and how it has supported them in their battles.

In the past, hemp has known both highs and lows, and it is possible that, like the phoenix from the ashes, it is now making a comeback.

The book will demonstrate and present the history of this plant, while at the same time fully acknowledging that this is not an exhaustive account.

2.2 The Various Forms of Cannabis

'Hemp? ... Hemp did you say? What? As in the stuff rope, string, bags and plumbing oakum (or tow) is sometimes made of?' These might be the answers volunteered in response to a researcher posing questions about 'hemp' to passers-by. If asked about 'cannabis', or worse, 'marijuana', however, the same people would be much more forthcoming for they would have heard of this 'drug' and be able to talk about it in one way or another.

Marijuana is not to be our subject, however, for there is within the genus *Cannabis* variation so great that botanists cannot stop arguing about whether to make a separate species (Small, 1979). Thus, one often finds sources that allocate the drug types of cannabis to the species *Cannabis indica*, and the others – including what we are today calling 'industrial hemp' (though formerly just 'hemp') – to the species *C. sativa* ('sativa' being the specific applied to plants commonly found in agriculture, L., 'cultivated'). That division has now been rejected.

In 1753, Carl von Linně was the first to classify *Cannabis* using his new system of binomial nomenclature. Since that time, the genus has been placed in Moraceae (Mulberry), then Urticaceae (nettles), before graduating to its own family, Cannabaceae, which it now shares with *Humulus* (hops) and, just recently, *Celtis* (hackberry). Botanists may revisit these associations yet again once modern DNA-based tools are applied, so it may be wise to regard them merely as suggestions.

Clarke (1999) has reviewed the current view of the species-level classification: three different systems each with its adherents. Since all Cannabis is interfertile (i.e. there is no 'species barrier' or sexual incompatibility

between types), a practical taxonomy is currently in vogue by which varieties are classified by their chemical profile, a so-called 'chemotaxonomy' (Hillig, 2004, 2005). The key to this taxonomy is the genetic presence of alleles for the contrasting cannabinoids, tetrahydrocannabinol (THC) and cannabidiol (CBD) (covered later). Drug varieties of Cannabis overproduce THC, whereas 'industrial' (meaning fibre and oil) varieties have very little THC, and also much less of the resin that bears it.

Whether high in THC and copious in resin production, or low in THC with stems made to yield a long, strong fibre, the Cannabis that is found across the world bears the mark of the ancient breeder. With the probable exception of the degenerate 'ruderalis' type – which may or may not be a vestige of the wild ancestor – Cannabis of every type (no less than all our other domesticated crops) has the traits sought and enhanced by humans.

Out of this vast variation, our focus here is on the 'sativa' types bred for something other than their resin. Historically, that has meant the stem (or 'bast') fibre. In recent times, increasingly, varieties selected for the seed and its nutritional profile are emerging, though traditionally the seed was a secondary product of fibre production, used only in a few cultures of Eastern Europe and Russia.

The naming conventions of western botanical science aside, the trail of this plant is found in local names, its centrality reflected in the proliferation of appellations.

In France, names for local *Cannabis* varieties are preserved in regional dialects:

*   Aube et Haute Saône: *cheneville*
*   Berry: *chaude*
*   Bresse: *chenève*
*   Forez: *chinève*
*   Franche-Comté: *chenove*
*   Languedoc: *carbe*, etc.
*   Limousin: *chanabal*
*   Mâconnais: *chernière*
*   Meuse: *chenevoux*
*   Normandy: *cambre*
*   Picardy: *canve*
*   Poitou: *chenebeau*
*   Provence: *cannabal*, *cannebière*
*   Région toulousaine: *carbenal*

- Rouergue: *canabou*
- Saintonge: *charve* or *cherve*
- Savoie: *stenève*
- Vivarais: *chanalier*
- Walloon: *chenne*

To this we must also add the names used to describe hemp seed: *chènevis, chenève, chenevardou, chenèvard, cheneveux*.

Not only can we unearth these names from different regions of France and periods of history, but also hemp has given its name to a number of French villages.

A study of the toponyms suggested that hemp cultivation was widespread:

- Cambe, Canabal, Canabièra and Chanbier (Correze and Lot)
- Canabols (Rouergue)
- Chanabert (Ardéche)
- Chanavard (Loire)
- Chanavas (Vaucluse and Savoie)
- Chènevière (Val de Marne and Haute Marne)
- Chenôve (Burgundy)

This list of rural place names that all refer to hemp provided evidence that hemp was known and cultivated throughout most of France.

The roots of *cannabis* can be found in classical Latin *kannab*; in Arabic *kannab*; in Hebrew *kanneb*; and in Assyrian *quanabu*.

Though the suggestion has been resisted, for obvious reasons, linguistic analysis indicated the *kaneh-bosm* referred to five times in the Old Testament was, in fact, cannabis (Benet, 1936). Bennett (2010) has pursued the linguistic and cultural evidence that cannabis was also the principal ingredient in the recipe for the Vedic *soma* and Zoroastrian *haoma*.

While we know from Herodotus that the horse-riding Scythian warriors were well acquainted with the psychoactive properties of cannabis, we are left to speculate whether divergence had already occurred between their recreational cannabis and the fibre with which they formed their bridles. It is possible that it had. We perceive in this crop the artifice of ancient breeders, lost under desert sands.

2.3 Retracing Hemp's Traces into Antiquity

2.3.1 The history of hemp: a challenge?

What resources can we draw on to detect the presence of hemp?

Palynology, the study of pollen in archaeological samples, gives us a good idea of the location and time of hemp's arrival in Europe (although palynological techniques are not without caveats: pollen of *Cannabis* and *Humulus* are difficult to tell apart, microscopically).

Cannabis pollen grains dating to 3450 BC were recovered from a site in northern Italy; samples from central and northern Germany, Scandinavia, England and France have been dated between 2900 and 1700 BC; and in central Germany, pollen indicated a continuous presence of hemp from 2000 to 530 BC. The tip of the arrow went from stone to bronze to iron, but hemp remained the fibre with which it was tied.

The detection of hemp pollen in an archaeological dig implies that hemp existed near enough the site for wind to carry the pollen in a practical radius of approximately 12 km. Cannabis pollen has been found to be mixed in variable proportions with pollen from other species, indicative of plants growing wild as opposed to under cultivation. Even where pollen has been found in high concentrations, it remains difficult to confirm that these originated from cultivated fields of hemp. Hemp plants can grow in fairly dense stands in the wild. Whether cultivated or a feral camp follower of the earliest pioneering humans, hemp would have been a botanical resource vital to even the most mundane aspects of primitive life: string.

2.3.2 Chinese origins and the role of Central Asia

In China, formal proof is readily available, either from archaeological discoveries or from the administrative documents preserved by this country.

Historically, hemp makes its appearance in China very early, c.8000 BC.[1] Pottery, dated at 6200–4000 BC, has been found depicting clothing that, on analysis, was shown to be made from hemp. Other remains of hemp, including hemp seeds, have been found in the graves of nobles (2100–1900 BC), thus demonstrating that hemp was in use during these times.

Rope dating from 5000 to 4000 BC has also been recovered, while the oldest paper made from hemp was discovered in a tomb dating from 2685 to 2138 BC.

Samples of lightly starched paper (easier to write on) were found at Hatanpo (c.200–150 BC). Similar artefacts were also found at Junguan, dating from 70 to 50 BC.

Finally, we must not omit to mention Tsaï-Lun, Chinese Minister of Agriculture, who in 105 BC commercialized paper made from hemp and the bark of mulberry.

The recovery of seeds and grain from as far away as Mongolia (2400 BC), together with the presence of hemp-derived paper and textiles in occidental China (600 BC), are particularly notable for they demonstrate that the westward migration of people from these areas towards Europe brought hemp to the shores of the Black Sea.

Similarly, the migration of people southwards from, or through, Tibet and Nepal brought hemp to India.

2.3.3 The arrival of hemp in the Middle East

Hemp also made its appearance in the Middle East, with Persia acting as a staging post between India and this region. It then spread around the Mediterranean basin. In biblical times (10th century BC), it was mentioned in the calendar of the City of Gezer (a town situated on the road to Jerusalem), where the month in which hemp was due to be pulled up was specified.[2]

The texts mention that flax and hemp can be woven together into a material, as they are both products of the earth. In no case, however, is the weaving together of wool and hemp permissible, as wool is an animal product. The proportions of each plant fibre were clearly specified, allowing variation in the upper limit for hemp.

Hemp has also been found in Syria, Egypt, Lebanon and North Africa. Archaeology thus shows the spread of hemp from the Far East to the Middle East. It is likely that this was influenced in part by the psychotropic properties of the plant. Cannabis spread across North Africa to reach the Atlantic Coast, where the use of the compressed cannabis resin known as hashish became customary.

2.3.4 Hemp in Europe

What evidence do we have of hemp from Gaul and the Gallo-Roman world?

Pollen has been found in northern France, the Eure-et-Loir, Seine Maritime, Mayenne and the Loire estuary, as well as from the Aisne, Marne and Somme. Each of these territories is characterized by its humidity and is therefore well suited to the cultivation and retting of hemp (as well as flax).

This does not mean that hemp was being cultivated, although there are very good reasons for believing that it was: in a number of these regions, hemp production has been reported throughout history, and continues to exist today. Certain texts mention hemp in the Rhone Valley. This may, however, be hemp imported from Greece, or brought through the Dardanelles from Black Sea ports, or from Riga through Gibraltar, headed for the Port of Marseille.

At Chatham, Kent, the English built the empire's navy. Now preserved as an historical site, one can tour Hemp Houses 1, 2 and 3, warehouse-sized buildings where hemp arriving from Riga, Latvia, was processed from raw baled fibre to massive anchor ropes and other riggings.

But the hempery of Marseille remains in name only: *Canebière* is a well-known alleyway that opens out into Marseille's *Vieux Port*, or old harbour. The name recognized the fact that this alleyway was the thoroughfare through which much of the imported hemp was transported.

We have good information for the Gallo-Roman period. First, we know that the Romans were familiar with hemp: one of their agronomists, a man by the name of Columelle, detailed the principles of its cultivation in his agronomic treatise.

We also know, from findings on a boat (from the first half of the 1st century AD) discovered in Marseille during archaeological digs under the *Bourse de Commerce*, that rope and caulking ('oakum') were made from hemp tow mixed with a vegetable putty. The fibres of this plant were also used in the manufacture of bird aviaries.

Other evidence of hemp rope has been recovered from the forts lining the *Limes Germanicus*. The established cultivation of this plant in Gaul from the 5th century AD was no doubt made possible by its introduction with Germanic immigrants. These people are thought, in turn, to have received hemp from Eastern Europe: the Silk Road, therefore, did not assure the supply and movement of silk alone.

And what of pieces of textile? Contrary to what one might think, it is very difficult to differentiate a piece of ancient flax cloth from one made of hemp. Such cloth, if several hundred years old, is difficult to identify with any certitude, even under the microscope. Archaeological experts have contradicted each other on many occasions in their reports and publications.

Textiles purported to be hemp have been found in Switzerland, Austria, Ireland, Greece and Turkey. These all date from a period approximately 2500 BC, while carbonized remains have been uncovered in Spain (900–700 BC). That said, we should reiterate the practical difficulties of differentiating hemp from flax.

In a village in Switzerland, archaeologists discovered a number of fragments of textile. The material was identified as wool, of course, together with some canvas. The specialist investigations were very involved but concluded that half of the material was made of flax, with the remainder predominantly made of bark from the lime tree. The inventory concluded: 'and finally we found a fine piece of cord ... made of hemp!'.[2]

Was this piece of cord made locally? Was it imported from another region? Even in prehistoric times, commerce and the movement of goods were well established and the beginnings of today's trading routes were being formed.

So, it would appear that hemp has been present and cultivated in Europe, particularly central Europe, since at least 2000–1500 BC.

Where did this hemp come from? Herodotus wrote that the Scythians and the Greeks knew of hemp. The Scythians were a people from central Asia who had contacts with China. It is therefore possible to surmise that they were able to introduce hemp into the areas in which they lived. Seeds from their territory have been recovered and dated to 4000 BC.

And what of the Europeans in all this? They were in contact with the countries of central Asia and the Middle East throughout much of their history. Travellers and merchants contributed to the spread of cannabis, as did Christian pilgrims bound for Jerusalem and the movement of people caused by the crusades.

2.4 Developments Through the Ages in Europe

2.4.1 The Middle Ages and Renaissance: Iconography and literature of hemp

Information from this period yields little evidence on hemp cultivation.

The 4th century Roman agronomist, Palladius, picked up where Columelle (of the 1st century) and his predecessors left off.[6] His protocols for the cultivation of hemp were practised throughout the Middle Ages. We also know that Charlemagne recommended the cultivation of hemp. Despite the fact that little mention is made of hemp cultivation during this period, it is not possible to imagine that it had disappeared, for – among other factors – it would still have been needed for the rigging and equipping of boats.

Further proof of the ongoing use of hemp comes from iconography: a wood carving of this period mentions that a particular person 'collected all that he could find into a hemp sack'.

Carvings and illustrations also show the clothing worn by people during this period. The nobles, the rich and other important people were all, no doubt, clothed in silk and wool. For warmth, everyone wore wool. But for lighter wear and undergarments, the rich might have silk, but the clothing of the hoi polloi was likely made of hemp or flax.

In the lexicon of textiles, *drap* in French refers to woollen cloth (cf. the *villes drapantes de Flandre*) and *toile fine* ('fine cloth') refers to flax cloth (linen). In the absence of 'fine', there is little to indicate the nature of the fibre used in the manufacture of the cloth.

Hemp was primarily the fibre of cordage, rather than textiles. But there are many exceptions. Northern Italy produced very fine hemp fabrics from crops grown at maximum density. Italian 'germplasm' (genetic varieties) later became the foundation of Hungarian hemp breeding from which textile-grade hemp developed. Despite these and other exceptions, hemp was destined most often for rope, not tablecloths.

There are many early allusions to hemp, including that shown in Fig. 2.1, and others from the papacy and poets.

A pontifical letter dated 14 January 1245 (the reign of Saint Louis), which now resides in the French National Library,[8] deals with the dark story of the Senechal de Beaucaire, who was accused of taking merchandise from a number of Genoese merchants on their way to Lagny. Responding to these merchants' entreaties, the Pope wrote to the King of France, requesting that their goods be returned. The letter is entitled 'Pontifical letter on hemp yarn'. Its author goes so far as to explain what a *bullée* of hemp yarn is: 'a vulgar piece of string'. This pejorative is apparently a temporary one (in a rough draft or the summary of the negotiated settlement), however, for later letters also speak of *bullées* of silk thread.

As for our poets:

... Car il était étroit avec son matelas peu épais
Et couvert d'un gros drap de chanvre Lancelot
tout désarmé gisait ...

Fig. 2.1. Hildegarde von Bingen (1098–1179)[3] and her confessor. The 12th century Benedictine nun left us another glimpse of hemp in medieval daily life. As a doctor, she recommended hemp for the dressing of wounds: 'When the swelling of the dragon (dracaena) appears on the leg, it is to be soaked in a juice of nettles and covered by a *Chardon de Marie* (leaved stem), before then enveloping it in hemp cloth. The swelling will not develop further'.[4]

Thus wrote Chrétien de Troyes (1130–1183) in *Le chevalier de la charrette*.[5]

> ... Nos chers frères Sachets on fait monter le prix de la mèche de chanvre, chacun d'eux semble un vacher ...

wrote Rutebeuf, a *trouvère* of the 13th century.[6]

And another of the 16th century:

> ... Sa mère qui lui fait la tache Le chanvre qu'elle attache à sa quenouille de roseaux ...

Théophile Viau[7] (1590–1626), who wrote:

> ... Ou d'une bergère
> Dont le cœur innocent eut contenté mes vœux d'un bracelet de chanvre avec ses cheveux ...

During the same later, period (1583), Olivier de Serres, a gentleman of the Vivarais, who was widely recognized by agronomists as the father of modern agronomy, wrote *Le théâtre d'Agriculture et mesurage des champs*, in which he recommended growing hemp in the garden, together with flax and other plants.

The 17th century French fabulist, La Fontaine, drew his inspiration from Aesop (4th century), who in his original *Fables* mentions 'flax' on three occasions. But in *Les Hirondelles et les Petits Oiseaux* (The Swallows and the Little Birds, 1668), La Fontaine replaces this with 'hemp' (*chanvre*):

> ... Il arriva qu'au temps que le chanvre se sème ...
> ... Quand la chènevière fut verte ...
> ... Le chanvre étant tout a fait crue ...

Meanwhile, we find the suggestion of one Rétif de la Bretonne (1740–1806) that, in certain meals, hemp – the seed, we assume – makes an excellent dessert.

In 1854, the poet Alfred de Vigny, writing to the steward of his property at Maine Giraud, remarks, 'Madame you recommended that Jeanette be told to spin the hemp that she was left'. And in 1856, he writes, 'Madame, please add this note on the hemp. She wishes that, with this year's strands...'. Then, in 1857, on the subject of strands (finely hackled hemp fibre) and tow, he says, 'Please require these to be spun more finely'.

These examples showed that hemp was not a new plant, however. Unlike maize and the potato, hemp was not a plant that had arrived suddenly from the Middle East or from the newly discovered New World, but was already well established in European culture.

2.4.2 Hemp in all its forms

Wars and the constant conflict that ravaged towns and villages throughout France in the 16th century resulted in the burning and disappearance of a great many legal archives. Sources from the 17th century onwards are better preserved. References to hemp are found in property inventories of the dead (both peasant and noble), in tithe inscriptions – *livres de raison* – and in other diverse accountancy records. Such documents are a trove of data on the place of hemp in traditional French domestic life.

Cultivation becomes more widespread

The word 'hemp' appears in various forms, as we have seen already, across many areas of France. Each example draws attention to the importance of hemp. This is further evident in administrative reports. Furthermore, each commune boasts proudly of its hemp and the profit that derives from it, while those who have no hemp bemoan its absence.[8] The 18th century historian of naval riggings and author of agricultural texts, Duhamel du Monceau, was amazed on seeing a hemp plant measuring some 12 feet high (approximately 4 m) and 3 inches (approximately 8 cm) in diameter! The hemp grown in the Auvergne region of France, by comparison, generally averaged only 3.5 feet in height, was soft, full of trash and judged unsuitable for the navy. Burgundy hemp – whitish, hard and friable – was of a much lower quality, mixed as it was with approximately 50% of hemp *de Bresse*, a hemp of decidedly inferior quality.

In the Dauphiné region, the fibre was reportedly soft, fine, and measured 4–5 feet in length. Such fibre was easily combed and was preferred by the navy.

One cannot finish this presentation without speaking of Brittany.[9] In the traditional hemp growing regions of Le Maine, Mayenne, Sarthe and Le Perche, in the heart of the Breton peninsula, towns gave their names to the canvas they produced; thus, we hear of *les olonnes de Locronan*, *les olonnes de Merdrignac* and *les noyales de Rennes*. The first of these was named for wrapping that enveloped the salt produced from the salt marshes of Saint-Gilles d'Olonne. In the same way, the town of Pouldavid gave its name to *pouldavoirs*, a canvas material sold in England (particularly Cornwall, which had a special trading relationship with Brittany). Breton canvas was also marketed in Spain, where it proved very popular with Spanish mariners and the navy, and was accepted as the new French *olonnes*. Canvas orders from Spain put Bretons in competition with the Dutch. Across the countryside, in Brittany as in Le Perche, hundreds of seasonal weavers worked to supply traders.

The records show that hemp production declined during the Locronan crisis. This can be explained partly by foreign competition (from Russia in particular), with the arrival of cheaper fibre of comparable quality. But it was also a result of Louis XIV's programme during the Seven Years War of building factories and workshops, especially in Rennes, that concentrated raw imports and the workforce in one place. Faced with increasing competition, merchants paid weavers less and less for their canvas. These, in turn, compensated by lengthening the weft of their canvas, making fabric that tore easily when caught by the slightest gust of wind. This cycle of events inevitably led to the collapse of the industry in Brittany and the end of the culture it subtended.

Other, more mundane, evidence further demonstrates the presence of hemp throughout the French countryside. In Aquitaine,[10] from Libourne to Sainte Foy, the hemp crop was consumed locally and production might have been increased had prices been higher. In the area of Blaye, hemp was produced for local demand and there was not enough to meet commercial demand, necessitating importation from abroad.

The seeds were pressed for oil; some were exported to Bordeaux and Holland for pressing. In Marmande, 4720 ql (ql = quintal; 1 quintal = 100 kg) of surplus were sold to Bordeaux. These originated from the plains of the Garonne, where hemp had replaced tobacco. In this particular report, an increase in production was advocated in order to avoid importation, providing, of course, the demands from navies allowed for this. The records of this period are very useful in allowing us to become more aware of the significance of hemp. Advice was even given on how production could be increased.

In another example, in a parish of the Aveyron, in a lease farming contract dated 9 March 1669, the owner of the sharecropping farm is a sheet maker:

> On March 9, 1669, Gilbert Bousquet, master sheet-maker in the village of Cayssiols (in the parish of Ampiac, near Rodez) has rented to Pierre Béteille, a labourer from the village of Ruols (Paroisse de Luc), the Bousquet sharecropping farm for 7 years.

For the farm rental, the farmer promises to pay:

> ... Bousquet will be required to supply to the farmer, the following seed stock: 1 quart[11] of lentils and 5 quarts of *Canabou* (hemp) seed for sowing the *chènevières* (hemp fields). The hemp will belong to Beteille this year and the final year excepting the *paladou*, the whole to be delivered annually to Bousquet.

The statistical annals kept by the Gironde Archives[12] illustrate very well the commercial trends and the concerns over whether manufacturers might set up locally and threaten the income sources of local people in the Cantal.

Cannabis hemp:

> ... is generally cultivated in the communes (villages). In the villages and hamlets of our Eastern, Western and Northern valleys there are no inhabitants who do not grow this crop. After harvesting it, retting it, they grind it, process it and spin it. This work is all

undertaken by women. Once the yarn is ready, it is supplied to local weavers who will turn it into fine canvas (using the fibre of the female hemp plant). The male hemp plant produces a coarser yarn that will consequently produce a coarser cloth. The merchants from the Departments of the Herault, Tarn and the Aveyron visit the Spring fairs to buy the coarse canvas and deliver these to the ports of Toulon and Marseille where they are sold to make sails or sailor's shirts. Any surplus is taken to the Aude, Pyrénées Orientales and even as far as Spain where the cloth is similarly used to produce cloth for the people. This branch of production produces cheap, affordable cloth for all classes of society in the Cantal. Few would use manufactured cloth; the latter is a branch of commerce that generates hard currency, that meets other needs, and can even serve to pay part of the tax bill. It would be a great misfortune if canvas manufacturers were to become established in this country even if they were to produce the same cloth with the same yarn and sold it at the same price as locally made cloth. The reason is obvious: the manufacturer would buy the raw hemp from the inhabitants and would sell the manufactured product back to them. In this way, the local people would be required to pay the raw cost of the hemp plus labour charges and would thereby lose the price of their own yarn in the process.

Hemp in day-to-day life

CULTIVATION IN THE CHARENTE. The bulletin of the *Société Archéologique et Historique de la Charente*[13] contains a supplement written by Mme Alberte Cadet, revealing that in Angoumois and Saintonge, there are no less than 80 villages whose place names (according to the village elders) contain *chenevaux, chenevière, chainevars, chaînevars* or *chèrues*.

We have a certain Charles de Gorée, who, on 16 May 1652, wrote to the steward of his estate of Champagne-Mouton, 'In the meantime, spin a further two to three pounds of this same hemp yarn as we will not have enough with the 19 pounds requested and they must not be allowed to whiten'.

Between 1682 and 1789, hemp is regularly listed as an agricultural product subject to a levy (or tithe). Sometimes, the detail of the information is surprising. In 1765, levies were imposed to the value of 542 handfuls of hemp, but on only 156 in 1774. In Taizé (in La Charente), in 1789, it was as high as 2.5 bushels of hemp seed and 130 pounds of hemp fibre. At Vouzon, in the ecclesiastical accounts, the levy was for 100 pounds of hemp and 6 measures of hemp seed. And then, successively, hemp appeared in the 'tithed' farm. In this way, the priest was able to exclude it from taxation, for it was cultivated in prescribed gardens and enclosures.

The prices are known: a hectolitre (hl) of seed, in 1677, was more expensive than flax. The former cost 7 pounds and 2 sols, whereas the latter was valued at 5 pounds and 2 sols.

In 1765, at Angoulême, hemp seed was priced at 9 pounds and 12 sols; in 1768, at more than 11 pounds; and it reached 15 pounds and 15 sols in 1797, thus showing a steady rise in price. With the start of the 19th century, more accurate statistics become available and we learn that hemp and flax are cultivated on just over 1800 ha.

In 1818, in la Charente, the weight of hemp yarn produced attained 541,350 kg and was valued at 649,620 francs.

The hemp plant also yields seed. It is known that during this period 1 ha yields 6 hl of seed, equivalent to 1 hl of oil. For the whole region, a total of 1800 hl of oil would have been produced. Hemp cultivation was therefore very important.

The *sérançage*[14] employed 386 people, a third of whom would have been employed exclusively in cleaning and collecting the seeds of hemp. Spinning hemp into yarn also would have been the responsibility of the women, as there were no professional spinners or trade of that name.

During the 19th century, there were some 2028 weavers in la Charente. Of the 496 known weavers in the *arrondissement* of Angoulême, approximately 6.5% worked throughout the year. For the remainder, their work could be broken down as follows:

- 12 workers for 9 months of the year
- 130 workers for 4 months
- 84 workers for 3 months

When there was nothing else to do, they wove canvas.

The statistics for four *arrondissements* of the Charente tell of the production of some 646,000 m of canvas. This was composed of material of different qualities: canvas, *brin* (a fine canvas), tow, *réparane* (tow seconds), *sarpillière* (a large canvas made of tow fibres) and *recrue* (a larger and lighter tow used for packaging).

Some 100 *corderies*, or rope makers, manufacturing everything from string to rope, together with a number of bell foundries requiring 20–25 pounds per bell (in order to reinforce the moulds), absorbed this production.

Local production was inadequate to meet the needs of the department and it became necessary to import material from Normandy, Brittany and the Limousin.

Quénot, in his statistics, talks of 500,000 m, but these figures do not differentiate between flax and hemp. We know, however, that in this region flax was known as a crop that exhausted the soil. It was not recommended to plant flax in the same field more than once in 6 years. We can surmise, then, that a significant portion of Quénot's figure consists of hemp.

Products made from hemp were of significant economic importance. The peasants visited neighbouring villages to sell their weaving. Their wares included canvas, flat string and sewing thread. These wares were exported as far as Spain via La Rochelle. Other destinations included Bordeaux, as well as the merchants of Paris, Picardy and Normandy. Hemp also made up part of the remuneration of domestic servants in addition to their monetary wage.

Even hemp seed was the object of a curious and lucrative trade (generating profits of 200–300%). The Dutch bought up the seed and stored it in barrels for sale in Flanders and the UK. Windmills were used to transform the seed into oil and this was then exported back to Saintonge!

Let us now examine an inventory after death, in order to shed further light on the presence of hemp in daily life during this period. This example concerns the *Seigneur de Jarnac*[15] (Charente). As a noble, he was likely to have led a luxurious life that would have been beyond the imagination of peasants. The inventory itemized what this nobleman left behind after his death in 1668. The document is a lengthy one, extending to 474 pages. The word 'hemp' appears no less than 62 times. Among these, we can highlight:

- 165 new shrouds of six *aunes* in length (approximately 7 m) made of hemp. In old French, the word *linceul* means not so much a shroud but a bed sheet. It was customary to bury the dead in a bed sheet and so the terms were used interchangeably. It is worth noting that this stock represented some 1000 m of sheets.
- 65 dozen serviettes, some specified as being of fine cloth.
- 62 tablecloths measuring 1.5–1.75 *aunes* (approximately 2 m).

The *Seigneur de Jarnac* was the most important of the Charente noblemen, and his lifestyle reflected this. This example highlights the role played by hemp in the household linen department.

CULTIVATION IN THE AISNE. In northern France, or more specifically, in the Commune of Mauragny[16] near Laon, there are a number of interesting documents regarding hemp. A report appearing in the *Cahiers de l'Histoire* tells us that, between 1681 and 1880, records were kept of the number of weavers.

In 1812, some 1200 ha were planted. The hemp produced was short and thin and suitable only for making cleaning cloths. The crop appears to have been used locally. Brayer's statistics show that, in the *arrondissement* of Laon, the products of hemp were exported, although their destination was not specified.

These documents provide us with some details of the oil extracted from the hemp seed. The seeds were first pressed in a *tordoir*. Sales records inform us that a *tordoir* was made up of a press, a heating chamber and a mill, comprised of one stationary stone and one moving stone. This was usually installed in the cellar of the house.

The *tordeur* was recognized as a trade or profession. There were 56 working in the region and they would spend a third of the year there. Records show the names of several of these men during the period 1800–1859, after which the trade came to an end. The same document tells how this oil replaced walnut oil when the latter became scarce. The oil was also used for lighting (with hemp fibre being used as the wick material), as well as in the preparation of paints and soap.

Hemp seed was also used as a feed for farmyard animals.

The inventories after death also provide us with details of the amount of land under cultivation of hemp.

In 1760, hemp growers had properties little bigger than gardens. In the first quarter of the 19th century, the area under cultivation increased, although this was reduced by the sale of 17 hemp fields. A further sale in 1894 brought the average area of a property down to 36 m².

The end of the century saw hemp production fall. It is possible to conclude, therefore, from the reduction in the area under cultivation, that each peasant farmer was growing only what he needed for his own family.

The region of the Perche appeared to have resisted this decline and continued to produce hemp longer than other regions.

The eco-museum situated between Bellême and Nogent le Rotrou (at Ste Gauburge in Saint Cyr la Rosière) provides some interesting concrete information:[17] hemp was cultivated on small plots of land, close to the farm so that children could chase away the birds that landed on the newly seeded ground, and again, later, when the crop had matured!

Legend tells how Saint Martin, feeling sorry for hemp growers, sent them a bird of prey to chase away the thieving birds. To thank him, after each harvest, one or two male hemp plants were left in the ground. This led ornithologists to name the prettiest of these birds of prey *le Busard Saint Martin*, otherwise known as the hen harrier (*Circus cyaneus*).

A few further observations illustrate how hemp was viewed by those who farmed it:

It left the earth impeccable. There was never any grass or weeds amongst the hemp.

It was hard. If ever it was dry when the time came to pull the plants up, we would hurt our hands.

Hemp, just before it was harvested would produce a strong odour that could produce, especially amongst women, a drunken state with loss of balance.

By contrast, the retting ponds produced a pestilent odour. This led some local authorities to pass orders that required retting to be undertaken in the river, using free-flowing water.

And finally, some expressions of appreciation for hemp cloth:

There were jobbing men who produced hemp cloth: they made sheets. Our grandmothers would have rolls of cloth. There would always be some stored in the cupboard but... we used it as little as possible...

My mother had large shirts; they were as stiff as the law. I still don't know how she managed to put them on like that.

But unanimously, for all these hemp workers:

It was good, good value, but it was bad!

Or again:

It was the most profitable of all, and we also had a bonus...

It was the cultivation that was most profitable.

The works of Claude Cailly[18] record the events of the rural textile proto-industry in the Perche, in the 18th century.

This provides new insights into the subject of hemp. Towards the end of the 18th century, hemp farming increased. Hemp fields of 1 ha or less became rare, as the demand for textiles from the New World was felt. This coarse cloth was used to dress slaves. The coarser canvas was used to package cotton and coffee. With colonial expansion and the growth of the *economie de traite* (extracting the raw product and exporting it from a colony) heavily protected by the strict exclusivity that characterized France's trading practices with its colonies, the manufacture of hemp canvas soared throughout the 18th century.

During this time, the hemp industry would acquire certain characteristics of the wool industry. Technical organization and changes appeared as it sought to modernize.

Three categories can be distinguished: weaving apprentices and artisans (who depended on the merchant manufacturers); independent manufacturers (whose significance varied with the number of looms and the stock of hemp yarn); and, finally, the merchant manufacturers who engaged with the first group but also dealt with the cloth merchants from neighbouring towns. Many cloth manufacturers and around half of the weavers had a *clos à filasse* (a garden where hemp was cultivated).

All the operations up to and including spinning were conducted on farms. The women were largely responsible for this work, which represented up to 60–70% of the value of the yarn. This, together with the higher price paid for hemp compared to that paid for cereal crops, allowed the peasant farmer to have complementary resources. The stages of processing were as follows:

- The hemp was harvested by hand.
- It was retted in ponds or in rivers (this varied depending on the region and the regulations).
- Retted stalks were dried.
- Stalks were then 'broken' (separation of bast fibre from hurds).
- The recovered fibre was combed ('hackled').
- Clean fibre was spun, either with a spinning wheel or by hand.
- The thread was warped, i.e. prepared, before it was mounted on a loom.
- The thread was woven into canvas by a professional weaver.
- Laundering: the canvas was washed and dried, to give it a nice colour.

Preparation of the chains and weft thread and of the spool was subject to rules for the division of labour. As a rule, the women were responsible for warping the loom, the last stage before actual weaving (a masculine speciality) was begun.

One word on the home: the rural labourer and the merchant manufacturer differed in the amount of workspace they had. The latter would have a substantial workshop, whereas the former would work on a spinning wheel in his one-room home. The loom usually needed to be housed in a separate room, known as the *chambre à toile*. The house of the weaver was usually built over a large, and often dark, cellar, which provided the humidity required to maintain the suppleness of the yarn. The women's work was particularly tough, for they had to undertake household tasks during the day, before spooling, spinning and weaving in the evening. This was the life of the women working for merchant manufacturers. It was better for those working for a master weaver, for their spinning activities were continuous.

2.5 Hemp: A Strategically Important Material in the 17th and 18th Centuries

Why a strategically important material? As important as charcoal during the previous century and as important as oil today?

Before the advent of steam engines, sails (in conjunction with oars) and Aeolus, the Greek god of the winds, powered boats. When speaking of sails, we might just as well speak of hemp and flax, for these were used in the construction of the rope, cables, ladders and stays that made up a ship's rigging.

2.5.1 The case of France

The Royal Navy (of France) was officially created by Henry II by Royal Patent on 13 September 1547. One of the first men of state to preoccupy himself with the maritime strength of the country, and build a veritable fleet of war, was Cardinal Richelieu, named principal minister by Louis XIII in 1624. In 1626, Richelieu, by the King's grace, became *grand maître, chef et surintendant général de la navigation et commerce de la France*, that is to say, he was put in charge of her shipping. This went beyond the responsibilities of a grand admiral, for it gave Richelieu control of all of France's maritime affairs, from the arsenals to the nomination of her admirals, passing by way of her ports and shipyards. Richelieu supervised everything.

At this time, the English navy, as well as those of Spain and Holland, easily surpassed that of France. Both Denmark and Sweden could count more warships than France. But, with Richelieu in command, things changed fast! From 1640–1643 onwards, the French fleet crushed the Spanish Atlantic and Mediterranean fleets.

With the Civil War (*La Fronde*) of 1648–1653 over, Louis XIV and Colbert found themselves at the head of a large French fleet, able to measure up well against the English and Dutch, but which drew heavily on their resources. This provides us with a better understanding of the importance of hemp to the politics of Colbert, the finance minister under Louis XIV (1665–1683) and how this in turn affected hemp production.

In effect, we must draw the reader's attention to the following figures in order to understand better what we present below. An average-sized ship (Fig. 2.2) would use between 60 and 80 t of hemp as rope and 6–8 t in sails per year. This represents a tonnage of between 70 and 90 t.

It was Colbert's talent as a strategist that allowed him to recognize the need for industrial methods to help arm and equip the King's ships. He therefore modernized the existing arsenals in Brest and Toulon, while creating a new arsenal at Rochefort sur Mer (Charente Maritime). He attached a rope factory to the latter. This was to become famous and still

Fig. 2.2. A fine example of a sailing ship (photograph taken in 2000).

exists today in the form of a museum, the *Corderie Royale.*

Created in 1666 by Colbert, it came into service in 1670 and remained operational until approximately 1865. During the liberation it was destroyed by the Germans, who had occupied it throughout the 1939–1945 war. It was only reconstructed in 1970 and for the past 20 years has provided visitors with an exact idea of what industrial rope making in the 18th and 19th centuries looked like. This factory equipped some 400 boats and, most notably, kitted out the frigate *Hermione* that was to take the Marquis de La Fayette to the aid of the American insurgents in 1780.[19]

The arsenals of Brest and Marseille also boast similar rope factories, where thousands of metres of rope and sails would have been produced to meet the needs of the fleet of battle ships.

This work has set out to demonstrate the constant relation between the ports of Ponant, Brest, La Rochelle, the arsenal of Rochefort and the minister, Colbert, who was responsible for supplying hemp to these Atlantic ports. We know this from the correspondence he conducted with maritime officials. Everything was required to be produced in France, with a bare minimum being purchased abroad when unavoidable. This was the case, for example, for the resin required to tar the cables.

Wherever possible, manufacturing plants were established with the aid of foreign specialists who were only employed for the time necessary for them to pass on their knowledge.

Colbert and his son, Seignelay, were kept abreast of everything and supervised the operation in great detail month by month.

Every year, the same question presented itself: where shall we source the hemp from? The Atlantic supports were supplied from Brittany, from the Auvergne and from the valley of the Garonne. Hemp from Alsace and from the Rhone valley was called upon to supply the arsenal of Toulon. In all this, we can see Colbert's incessant work with the port officials, ensuring that they found the necessary hemp and negotiated as low a price as possible. There were problems, however. The product became rare, swindlers supplied bundles of hemp that were rotten inside or else packed with stones to make up the weight.

It would appear that a single merchant was able to secure the monopoly on hemp supply. A would-be competitor lowered his prices, but the merchant lowered his still further until he was the only one left selling hemp, thereby becoming the master of price. Colbert questioned this and, trying to rid himself of his services, conducted investigations. No dishonesty was brought to light, however, and when the hemp dried up in barren years, it was always the same man who got his minister out of a fix.

Hemp was so important that even Louis XIV interested himself in these matters. He chastised a supplier who had delivered a spoiled batch of hemp. He also imposed purchase prices and supervised and ordered that hemp be grown in specific provinces. After 20 years of effort, Colbert's policies bore fruit. Hemp deliveries became regular and the quantity consistent, while the price dropped. Occasionally, as a last resort, supplies had to be sought from the Baltic states or from Holland. Northern hemp, originating from Livonia (in northern Lithuania) or from Russia, was often of better quality and recognized as being both fine and supple. It was able to hold a heavier weight than that produced in France and had a further advantage, for it was more abundant and cheaper than French hemp.

In the opinion of the naval officers of the time, French hemp produced sails that were too heavy for warships, where they made for slow sail handling and rope work during manoeuvres. These sails were also characterized by the fact that they were not as good at catching the wind, especially when it was weak. This represented a disadvantage when compared to the sails of the Dutch and English navies, both of whom used a Dutch-made canvas of high-quality hemp. On closer inspection, it would appear that the manner in which the sails were woven made all the difference. The Dutch canvas was made from a single thread and the weaving was tighter, producing a sail that was less permeable to water, more supple and stronger than the French canvas. The French canvas would always be stiffer and heavier, but this did not prevent the French

Navy from winning a number of brilliant successes against the Dutch and Spanish.

Now let us explore the culture of hemp in 17th and 18th century Holland.

2.5.2 Hemp in Holland

The Dutch lived by the sea and, building on their mastery of the seas, they became a commercial power of note between the 16th and 18th centuries. This required a reliable supply of hemp in order to equip their commercial and fishing fleets. In addition to supplying rigging needs, the production of fishing nets also called for hemp.

Hemp production was therefore encouraged during this period, and the western part of central Holland was equipped for the purpose. Rural engineering works, hydraulic systems and commercial management systems were all developed for the purpose. In some areas, hemp even appeared to have become an irrigated monoculture for which fertilizers were used.

However, retting hemp stalks in streams polluted the water, posing a problem. The processing of hemp in the special techniques required resulted in the creation of an industry. As we saw earlier, France looked to Holland to supply its needs from time to time. After the 17th century, the increasing availability of cheap hemp from Eastern Europe led to a progressive decline in hemp production.

By the 18th century, the Dutch ships had kissed goodbye to the glory and wealth of yesteryear. Their need for hemp also disappeared. The Dutch farmers found it increasingly hard to make a living from this crop. This led to a gradual evolution away from hemp towards animal production, and ultimately to dairying.

2.5.3 The French Revolution and the Empire

The revolutionary period and that of the First Empire are marked by a series of important developments for the French Navy. The navy was relied upon to maintain links with the West

Indies and the USA. In 1794, this allowed a convoy of 200 ships carrying wheat to reach the port of Brest. The navy also prepared a landing in Ireland and organized an expedition to Egypt in May–October 1798, which consisted of several hundred boats that were later annihilated at Aboukir. In Boulogne, they also prepared for the unlikely invasion of England. All this activity translated itself into an immense demand for hemp.

The English had the same problem. When Napoleon I chose to impose a continental blockade on England in 1806, England's supply of hemp became compromised. She would, in consequence, open up the Baltic sea ports by force in order to secure supplies from the Baltic states. In order to facilitate this, England brought pressure on Russia to detach itself from Napoleon in return for subsidies.

During this time, hemp became a commodity of great strategic importance. The following testimony will therefore come as no surprise.

In 1812, Napoleon I embarked on his disastrous Russian campaign (18 June–18 December 1812). The captain of the Haute Marne region executed the orders he had received from his minister, and therefore from the Emperor. These dealt specifically with the production of hemp, as this was a material of strategic importance.

The following give some idea of the information sought:

- Price of 1 ha of hemp
- Amount of taxes per hectare
- Cost of labour and other preparations
- Price of hemp seed stock
- Cost of weeding, pulling the hemp and retting it
- Cost of hackling and scutching

- Number of people involved in the hackling and scutching (including their daily rate)
- Number of metres of canvas produced and its different varieties
- Number of weavers and spinners
- Price paid for the fibre, etc.

One can sense on reading this the same imperial precision as that in evidence in the writing of the Concordat of the Civil and Commercial Codes (1807).

In his answer[20] of 1 September 1812, the mayor of Orge (a commune of the Haute Marne), specified, in what appeared to be congratulatory terms, that flax was not grown within the territory of his commune: the importance of hemp to the Emperor and ancient artillery officer, Napoleon, was clear.

2.6 Hemp in the 19th and early 20th Century

By the 19th century, statistical records became readily available.[21] These are precise, meticulous even, and allow us to identify exactly what crops are cultivated. In particular, they allow us to follow the evolution of hemp production throughout the century and see how production is affected by the development of new technologies and techniques.

For 1890, 1891 and 1914, the tonnage totals both hemp and flax. Table 2.1 shows a reduction in hemp production, both in terms of hemp produced and land under hemp cultivation. Several explanations can be put forward for this trend, including scientific, technical and technological change.

The 19th century was to see a number of important inventions that would change the entire world dramatically, including the

Table 2.1. The raw materials of textiles (Renouard, 1909).

	1830	1840	1852	1862	1871	1889	1890	1891	1914
Hectares	175,000	176,148	123,352	100,114	96,395	53,825	51,990	51,602	12,500
Fibre (t)[a]		67,507	64,173	57,433	49,099	ND[b]	61,300	39,500	32,700

Note: [a]This series of figures is interesting but does not truly reflect the actual tonnage. In effect, 1–2 t of fibre are produced per hectare. Given the surface area under cultivation, the production of fibre certainly exceeds the figures presented above. Auto-consumption will, by its very nature, fail to appear in the accounts.

A few important advances in the mechanization of textile production

1733: first flying shuttle (invented by John Kay)
1767: first spinning jenny (invented by James Hargreaves)
1785: first mechanical loom
1794: first cotton threshing machine[22]
1801: first Jacquard mechanical loom (invented by Joseph Marie Jacquard)

Invention of the steam engine and its effect on transportation

1690: Denis Papin produces the first prototype of what would become a piston-driven steam engine
1790: first steamboat on the Delaware (Philadelphia, USA)
1803: first steamboat (built by Robert Fulton) on the Seine; refused by Napoleon, who thought the inventor a charlatan
1819: first crossing of the Atlantic by a steam-powered sail ship (the SS *Savannah*)
1829: first steam engine used in the shipyards at Rochefort by the French Royal Navy; the machine was called the Sphinx
From 1850: navigation by steam becomes a reality

In 1860, the French Merchant Navy arms 84,000 t of steam-powered ships compared to only 10,000 t in 1848.

Between 1870 and 1900, the number of sail ships falls from 15 to 7 Mt, while the number of steamboats increases from 5 to 33 Mt.

economy of hemp. Of particular note are weaving machines and steam engines.

2.6.1 The role of cotton

The invention in the late 18th century of the cotton threshing machine led to the rapid development of the cotton industry and the popularization of cotton. The machine allowed large quantities of cotton to be processed. The English cotton industry successfully developed cotton production on an industrial scale, thus replacing subsistence agriculture.

This phenomenon was most notable in both India and Egypt throughout the 19th century. Both Tsarist Russia and the Soviet Union achieved similar results in the Central Asian republics. The cotton threshing machine, together with developments in spinning and weaving, allowed cotton to take up a dominant role in the clothing industry through the creation of a supply of good-value clothing items that were comfortable to wear. The traditional clothes made of hemp and flax were to suffer heavy competition.

We should also add that cotton gives an impression of softness that hemp is unable to achieve. In French, the expression *aussi raide que la justice* is applied to hemp and translates as 'as rigid as the law'. Not only did cotton material possess these qualities of softness but also it was produced in large quantities. England developed and expanded cotton mills equipped with thousands of mechanical looms. This allowed England to become a great exporter of cottonware and to flood the European market with cheap materials. This wave of merchandise became so significant that, in 1840, a French deputy complained about this invasion and France's failure to keep up in this domain. A further factor was the general improvement in standards of living during the 19th century. People yearned for a more comfortable and easier life. This led to the development of new products at the expense of hemp.

2.6.2 The steam engine competes with hemp to power ships

In 1829, the French Royal Navy's first steamship left the shipyards of Rochefort. The arrival of both steam power and steel would, over the next 40 years, condemn sail power to history. The last great sailing ships were built in 1890.

This evolution in propulsion technology signalled the virtual disappearance of sails, and the hemp required to produce them.

The rural exodus further exacerbated this phenomenon and we see the progressive abandonment of hemp until its uses were limited to cordage and sacking. In this area, jute imported from India and Bangladesh provided stiff competition. Year by year, manufacturers and factories were established there, resulting in the progressive elimination of independent rural weavers. For farmers, hemp was no longer economically viable, especially when compared with the price of imported hemp. The peasant farmer restricted himself to the few hundred square metres required to meet his own needs. This practice lasted a long time, for in 1950, in L'Aveyron, one of the authors knew of a country lady, of over 80 years old, who still cultivated hemp in order to supply her small farm with a supply of string.

Figure 2.3, taken from a German study by the Nova Institute, eloquently illustrates the history and decline of hemp over the past 150 years, including a steep drop after World War I and the market's recovery as World War II approached.

The crash in the amount of land being farmed for hemp production, whether it be in France, Italy or Germany, was attributable to the arrival of imports and of the new technologies discussed above. However, the demand for and consumption of hemp did not stop.

Table 2.2 expresses the amount of hemp and flax (combined) divided by the total amount used in France.[23] The proportion of imported flax and hemp material increased from 35% in 1865 to 76% in 1914, representing a doubling in imports.

Where did these imports come from? The *Conseil Supérieur du Commerce* published the quantities imported between 1825 and 1837. According to their table, which does not include figures for flax, France imported 5420 t of hemp annually. The main supplier was Russia, with 2400 t. Tuscany, with 1850 t was second, while Austria, Prussia, the Hanseatic towns, Belgium and England each exported less than 100 t of

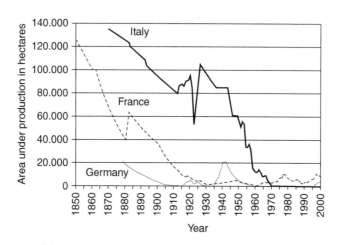

Fig. 2.3. The history and decline of hemp over the past 150 years.

Table 2.2. The amount of flax and hemp used in textiles decreases from year to year.

	1865–1874	1885–1894	1905–1914
Flax and hemp (%)	65	46.7	24

hemp to France. With imports rising to 75–80%, French hemp production clearly showed a continual decline.

So, what did France do?

2.6.3 The first production grant

The situation had not escaped the attention of the government, nor of the deputies.[24] In 1892, during the revision of customs duties, hemp was the beneficiary of a world first. The rural electorate, the agricultural companies and the agricultural fairs all demanded a raise in the customs duties for imports of flax and hemp. Rope and string makers were of a different opinion, as they were very much dependent on imports. In response to this situation, and recognizing the need to help the producers of textile plants in order to prevent their disappearance, a solution was found that would please everybody; a subsidy was introduced. It was the first time in France, and probably in the world, that such an idea had been proposed. As there was no precedent, one had to be invented. After discussions with the customs authorities, the amount of this subsidy was finalized.

Then things became more difficult. Over the centuries, the surface area under cultivation for hemp was between 2 and 5.5 times greater than that for flax. Hemp producers would therefore benefit, while the customs authorities wanted to prioritize hemp.

The next question concerned the distribution and repartition of subsidies. Was this to be given to the proprietor or by harvest on a pro rata basis? What limits should be set?

Finally, the solution settled upon was to assign subsidies according to the size of each property, while setting a minimum surface area. How was this to be controlled? That is to say, how was the surface area under cultivation to be measured? Someone had a bright idea, remembering that in each village there was one person who was familiar with agricultural measurements and the metric system and able to measure surface areas. This was the schoolmaster who, in many villages, also fulfilled the role of secretary to the mayor. He was therefore employed to undertake this work and rewarded with a well-regulated remuneration.

This initiative proved to be excellent; a contemporary report indicated that only 2 years were required to eliminate all fraud. The subsidies were set for a period of 6 years, but the surface area under cultivation continued to fall (Table 2.3).

The objective of a viable hemp agriculture was therefore not achieved and both the area under production and the amount produced continued to fall. One should question what would have happened, however, without the subsidy. What would have been the end result? In 1897, the deputies decided to renew the subsidy for a further 6 years.

At the same time, the prior 30 years had seen industrial machinery developed for one or the other textile and adapted for specific stages of fibre production (from harvesting to spinning). Several different manufacturers competed with each other. It was likely that hemp workers lacked the capital required to invest in this technology and were afraid of change. Despite the subsidy, the competition provided by imports did not predispose them to making such purchases.

The naval regulations, passed in 1898, formally specifying that sails could only be manufactured from flax or hemp might have come as consolation. It was a token

Table 2.3. Even with the subsidy in place, hemp production continued to fall.

	Surface area (ha)	Flax (tonnes)	Hemp (tonnes)
1893	41,237	26,968	11,843
1894	40,583	24,821	13,327
1895	37,216	27,289	11,962
1896	34,224	24,389	9,707
1897	33,843	23,330	9,507

consolation, however, for the previous 40 years had seen ships dispense with sails in favour of steam power.

On the eve of World War I, there remained a small number of tall ships, most notably the clippers with their 4000 m² of sail. These were largely foreign ships, few in number and did not represent a client base for the French hemp industry.

Once again, it would appear that, faced with a straightforward problem – that of a falling demand for hemp – there was a failure to ask the simple question: why?

On closer examination, it would have been obvious that cotton, together with the advent of steam power, was killing off hemp as a textile and as a means of harnessing wind power for locomotion.

We must refrain from stoning our forebears, however, for it is possible to cite a hundred examples of similar behaviour within the field of agriculture over the past 20 years.

2.7   Political Manoeuvrings and Hemp in the Modern Era

After World War I, both hemp and flax saw their importance diminish year by year. The former certainly did not appear to be used as a textile. It was, however, still used in the production of sacking, rope and string. It also held its ground in the field of paper production, for it could be transformed into fine-quality paper and was used for the production of certain special paper products.

Hemp cultivation fell into decline everywhere and even disappeared in certain countries. Only the USSR maintained its hemp production, despite the competition afforded by the Muslim Soviet republics of Central Asia. The area under cultivation fell from 870,000 ha to 250,000 ha in 20 years. Serbia, Hungary and Romania also managed to maintain a certain amount of production. As for France, in the 1930s, there remained but a few hundred hectares in production, essentially all within the department of Sarthe.

But it was in the USA, during the interwar years, that some remarkable strong-arm tactics were brought to bear on the industry.

Since the 17th century, hemp cultivation had flourished after having been introduced by immigrants (Hopkins, 1951). Of course, it should be remembered that the draft versions of the American Constitution of 1787 were written on paper made from hemp. With a large demand for hemp, the USA even imported Russian hemp. In the meantime, however, technological developments continued to appear.

In terms of paper production, the chemical processes that allowed wood pulp to be turned into paper were under development. These processes, including the use of sulfites for the extraction of lignin and the use of chlorine to whiten the pulp, allowed leafy and resinous wood to replace flax, hemp and cotton as sources of paper. And both the USA and Canada had an abundant supply of wood!

Advances in chemistry allowed formulae and procedures to be perfected that would give rise to a range of synthetic fibres, placing these products in direct competition with hemp and plant fibres in general.

Meanwhile, hemp came under attack for its pharmacological properties. The drug had no doubt always existed in the USA when, around 1910, African Americans created jazz and introduced 'marijuana'. However, the term 'marijuana' was given to this plant by the Mexicans of Pancho Villa's army (Booth, 2003).

A campaign to ban hemp was launched, arguing that the use of dope (which comes from hemp) was unhealthy and promoted laziness among the workforce. A series of advertising campaigns were begun in the name of moral standards and public health. Their objective was clear: to ban hemp.

Who waged this campaign? The newspapers of the William Hearst Group (Rosenthal, 1994) and the 'Citizen Kane' of Orson Welles, who were also the owners of large tracts of North American forests. At the time, these papers were known as the 'yellow press', for the chemicals used to transform wood pulp into paper produced a newspaper that would yellow after exposure to light.

The petrochemical giant, Du Pont de Nemours, now known as DuPont, was, at the same time, working on the development of synthetic fibres derived from the petrol industry. This group patented nylon (in 1937), Dacron (in 1950, based on a copyright of 1941) and various other plastic products such as Teflon (invented in 1938).

These new and revolutionary products needed to find their way on to the market fast and their competitors therefore had to be eliminated. Hemp was in widespread use as a component fibre in many fabrics, including the famous denim (etymologically from *de Nîmes*) used to make jeans, as well as in rope and string.

On 12 August 1930, the federal Narcotics Bureau was created to fight against illicit substances. Harry Anslinger[25] was placed at the head of this organization. It was suggested that his wife was related to Andrew Mellon, Secretary of State to the US Treasury. Mellon was the banker and proprietor of Gulf Oil. As a banker, Mellon financed Dupont de Nemours, which, between 1935 and 1937, lobbied hard for the prohibition of cannabis. This resulted in the Marijuana Tax Act, passed on 2 August 1937, which imposed heavy taxes on hemp cultivation, making it uneconomical.

From this moment, hemp was effectively banished from the USA. It made a small come-back during World War II, when the US Army needed hemp to produce tent canvas and rope. This was the occasion of a government-sponsored film, entitled 'Hemp for Victory', to promote the production of hemp.

The war over and, following the advances made by American chemists and industry in the production of nylon, the 1937 legislation was reactivated.

From 1945 onwards, hemp has effectively been banned in the USA. Finding itself a global power after its victories of World War II, the USA imposed their view of hemp on the nascent United Nations. The resulting de facto 'worldwide' ban (restricted to temperate growing regions, as hemp does not grow in tropical regions) was resisted by China, India, the USSR, Eastern Europe, Italy and France.

2.8 Conclusion

After these misadventures, hemp production went into decline.

It is difficult to conclude that we have finished writing the history of hemp. It has been a part of human history for over 8000 years. Hemp has clothed us, provided us with ropes and, in paper, a means of communication.

While some may have lost hope of hemp ever regaining its splendour of old, there is reason for hope. Hemp derivatives are well suited to a large number of uses, as described in detail elsewhere in this book. This gives us hope that hemp cultivation will take off again.

Today, we can observe encouraging developments in the hemp industry in Germany, Australia and Canada. We can also draw encouragement from the numerous projects under way in the Baltic states, Poland, the Czech Republic and Ukraine.

This plant's ability to protect the environment, most notably its role as a carbon sink and as a heavy metal trap and sewage digester, clearly opens new possibilities.

There is, therefore, cause for hope.

As historians, we know all too well that history paves the way to the future. It is our wish that this plant, with its many uses, rediscovers a future worthy of its past and that it is allowed to play its role in the protection of the environment.

Key Dates in the Chronology of Hemp

BC (before Christ)

8000: hemp is growing in the wild state in Central Asia.
2727: hemp is referenced for the first time in a Chinese pharmacopoeia.
1400: cultural and religious use of cannabis along the river Indus in Kashmir and Pakistan.
500: Buddha is said to have survived on a diet of hemp seed.
450: Herodotus describes the Scythians making clothes made of fine hemp fibre.
300: the Carthaginians and the Romans fight to gain control of the Mediterranean spice and hemp road.
105: commercial remains found in China on a paper made from hemp and blackberry fibres.

AD (after Christ)

62–117: Cai-Lun, working for the Emperor of China, perfects the manufacture of paper made largely from hemp.

600: the Germans, Francs and Vikings all use hemp fibre.

751: Battle of Talas, in Samarkand, saw the Arab Abbasid Caliphate defeat the Chinese. The vanquished Chinese prisoners passed on the secret of paper manufacture.

770: the Chinese produce the first printed book (Dharani scroll).

9th century: the introduction of cotton breaks hemp's monopoly of the clothing and yarn industries.

1109: first European document written on paper by Charif Al Idrissi – the book of King Roger II of Sicily. This was written in both Arabic and Greek for geographic purposes.

1117: according to legend, Jean Mongolfier, during the time of the Second Crusade, was taken captive by the Saracens and undertook forced labour in a Damascus paper mill. On his return to France, in 1157, he is supposed to have established a paper mill in Auvergne.

1150: The Muslims introduced hemp to Europe, leading to the first paper to be produced on the continent.

1151: a water mill was established at Xativa, near Alicante, in Spain. This mill macerated scraps of paper. Cf. Bertrand Gille, *Histoire des techniques*, Gallimard, coll. *La Pléiade*, 1978.

1154: paper first used in Italy in the form of a register written by Giovanni Scriba and dated 1154–1166. It is thought that this paper was imported from the East. No other example of paper was found in Italy until 1276 and the mention of the mills of Fabriano.

1348: a paper mill is established in the region of Troyes: the *Moulin de la Peille* is probably the oldest French mill.

1390: the first paper mill to be established in Germany is set up by Ulman Stromer, in Nuremburg. Previously, all paper used in Germany was imported from Italy.

1492: Christopher Columbus sails to the New World, a journey made possible by the 80 t of sails and rope, made of hemp, on his caravels!

1545: hemp cultivation starts in Chile, South America.

1564: Phillip of Spain orders that Spain be cultivated across his empire, which at the time extended from Argentina to Oregon.

16th and 17th centuries: the Dutch Golden Age commences, thanks to the trade in hemp.

1631: hemp is used as a unit of barter across the American colonies.

1776: draft of the Declaration of American Independence is printed on hemp paper. (Both George Washington and Thomas Jefferson cultivated hemp on their land.)

18th century: appearance of the first cotton threshing machines allowed the harvesting of cotton to become mechanized and cotton to become a competitor to hand-harvested hemp.

1807: patent deposited by Canson for *papier calque*, otherwise known as tracing paper. This was made from hemp fibre.

1850: introduction of exotic fibres. Start of the petrochemical era. Development of sulfite processing to extract lignin and of chlorine bleaching (Berthollet procedure) allow paper to be manufactured from wood pulp. Steam power replaces sail power.

1895: first use of the word 'marijuana' by Pancho Villa's partisans.

1910: African Americans introduce jazz and 'the weed' to New Orleans.

1937: the Marijuana Tax Act is enacted. Hemp cultivation is taxed so severely that it is abandoned in the USA.

1932: creation in Le Mans, France, of what was to become the Fédération Nationale des Producteurs de Chanvre (FNPC), the National Federation of Hemp Producers.

1937: nylon is patented by Dupont de Nemours.

1943–1945: 'Hemp for 'Victory', a propaganda film is produced to encourage American farmers to produce hemp to help in the war effort.

1961: *Convention Unique sur les Stupéfiants* (United Nations Single Convention on Narcotic Drugs) signed with the exception of China, India, the USSR, the Eastern block, Italy and France (http://www.incb.org/incb/convention_1961.html).

1960: hemp cultivation restarted in L'Aube, France, in order to help a paper maker, Bolloré.
1974: creation of the agricultural cooperative, *Chanvriere de l'Aube*.

1992: hemp cultivation is reauthorized in most of Europe.
1994: Hemcore grows hemp in the UK.
1997: hemp cultivation is authorized in Canada and Australia.

Notes

[1] Actes du Colloque de Pattès, Octobre 1999, sur l'archéologie des textiles. See also the Symposium de Paris, 1984: Problèmes des fibres de textiles anciens.

[2] *Les textiles aux temps bibliques* (périodiques). Dr P. Horu (1968), Bâle C.I.B.A.

[3] Hildegarde figures among the most notable and original thinkers of medieval Europe. Born of an aristocratic family of the Rhineland, she entered a convent at the age of 8 and was to stay there for the remaining 84 years of her life. Abbess, scientist and celebrity, she was erudite and noted also as a musician, prolific composer, political and religious personality and as a visionary. Her writings figured among the earliest mystical works of the Middle Ages and included a work on spirituality, *Sci Vias*, which translates as 'Know the Way'.

As the most important doctor of her day, Hildegarde von Bingen wrote a number of books that anticipated the developments to come in our understanding of the circulatory and nervous systems. The medicines she used to treat various diseases demonstrate an extensive knowledge and understanding of the pharmacology of plants. She corresponded prolifically with other great thinkers and participated in the political and religious debates of the times. Régine Pemoud, in her biography of Hildegarde, called her the 'inspirational conscience of the 12th century'.

[4] The 'swelling of the dragon' we know as phlebitis, while the *Chardon de Marie* is a wild artichoke (St Mary's Thistle, or *Silybum* – thistle family).

[5] In *Le chevalier de la charrette*, translated by A. Foulet.

[6] Text set and translated by Mick Zirk.

[7] *Œuvres poétiques*, edited by G. Subia (1999).

[8] *Nouveau manuel du cordier*. Boitard (1899), editor Roret, Paris.

[9] *Histoire de la Bretagne*. Y. Pelletier and ChUGEARD. NLF (1900) in Chapter IV.

[10] *Annales historiques de la Gironde* (1913) (T 48).

[11] Quarte: mesure de poids équivalent à environ 20 kg.

[12] *Annales historiques de la Gironde*.

[13] *Lin et chanvre d'Angoumois et Saintonge* by Mme A. Cadet. Mémoire de la société d'Archéologie et Historique de la Charente No 1-1968.

[14] Sérançage: action of cleaning the hemp before the retting process in order to remove the seeds.

[15] Inventory of the furniture and effects of the Castle of Jarnac. 1668. Published in 1890, in Angouleme. Extract from the *Bulletin Archéologique et Historique de la Charente*, annotated by E. Biais.

[16] *Les Cahiers d'Histoire: histoire de la culture du chanvre à Mauragny (Aisne)*.

[17] *Mémoire d'un Musée: Musée Départemental des Arts et Traditions Populaires du Perche*.

[18] *Cahiers Percherons. La proto industrie: Textile rural du Perche au XVIII° siècle* (cf. References) Claude Cailly (1993).

[19] This paragraph on the naval industries of the age has drawn on a number of works: Duhamel du Monceau, *Traité de la fabrique des manœuvres pour les vaisseaux ou l'Art de la corderie perfectionné*. (1769) ed. Desaint Libraire, Paris; Boudriot, J., *Le vaisseau de 74 canons*, 4 vols, Paris; *La Corderie Royale de Rochefort. Les cahiers de la Corderie*, published by Le Centre International de la Mer – Rochefort sur Mer.

[20] Original document belonging to an inhabitant of this region.

[21] For further information on this subject please refer to: *Tome III – Histoire de la France rurale*, collected works under the direction of G. Duby and Armand Wallon, Editions du Seuil, 1976.

[22] Cotton had been known of for a long time. It was reported in Mexico around 3700 BC, in Pakistan around 2700 BC and in Peru around 2500 BC. The first imports of cotton to Southern Europe, specifically Sicily and Spain, were made by the Saracens. The first cotton plantations were established in the USA in the 18th century, together with the slave trade, which ensured its cheap production. With the arrival of harvesting machines and threshing machines, cotton production took off, dealing a fatal blow to the hemp industry.

[23] *Histoire de la France rurale*, Vol IV.

[24] *Les matières premières textiles*. J.C. Fremiot.

[25] Anslinger was called the 'MacCarthy of hemp'. After his work in the USA, where he was Assistant Commissioner in the Bureau of Prohibition and subsequently First Commissioner of the Treasury's Federal Bureau of Narcotics (now the DEA), he became the international figurehead of the fight against narcotics, and cannabis in particular, as US representative to the United Nations Narcotics Commission. He resigned on 20 May 1962.

3 Physiology and Botany of Industrial Hemp

Brigitte Chabbert[1] (Part I), Bernard Kurek[1] (Part II) and Olivier Beherec[2] (Part III)
[1]*INRA, France;* [2]*FNPC, France*

3.1 Introduction

In order to do justice to the complex subjects of hemp physiology and botany, researchers Brigitte Chabbert and Bernard Kurek, of the Institut National de la Recherche Agronomique (INRA), France, present the plant's anatomy and botany, together with its chemical composition and properties.

Building on this first part, Olivier Beherec then presents and details the particularities and specificities of the plant's vegetative cycle.

3.2 Hemp Physiology

This section is presented in two parts:

- Anatomy and botany
- Constituents and chemical composition.

3.3 Part I: Anatomy and Botany

3.3.1 Classification and botanical description

Hemp was one of the first crops to be grown for the production of natural fibre. Today, this annual plant grows under a wide range of climatic conditions, but is principally found growing in Europe and Asia. The taxonomy of this dicotyledenous angiosperm is somewhat controversial. Various taxonomic approaches have been called upon. Where morphological criteria were applied, hemp was placed initially in the order *Urticacea*. Recent molecular studies have, however, produced a new phylogenic classification of the flowering plants (*Angiospermae*). This reclassification of the dicotyledons into distinct orders is based on the characterization of common molecules (such as the *Rubisco* genes). Thus, according to Chase (1998), the *Cannabaceae* family is now to be found in the order *Rosales*, which, together with six other orders, form the subclass *Rosidae*. A number of other fibre-rich plants, including flax and ramie (*Boehmeria nivea*) are also to be found in this subclass. The *Cannabaceae* family is considered to form a single genus Cannabis comprised of the species *Cannabis sativa* L., that itself regroups three subspecies (Bocsa and Karus, 1997):

1. *C. sativa* corresponds to industrial hemp and is the source of phloem fibres.
2. *C. ruderalis* is the wild form.
3. *C. indica* L. is characterized by a high content of tetrahydrocannabinol (THC), the principal cannabinoid of hemp.

More recently, a distinction has been made between the two species in the Cannabis genus. It is proposed that *C. indica* L. and *C. sativa* L. be designated two separate species. The latter would be divided into two subspecies, *C. sativa spontanea*, the wild form, and *C. sativa culta*, which collectively describes the cultivated varieties used for the industrial production of grain and fibre. This last subspecies demonstrates considerable physiological and morphological diversity according to the environment in which it is found (Bocsa and Karus, 1997). Consequently, geographical types can be recognized according to the size of the stem and whether they flower early. In addition to environmentally dependent characteristics, the industrial cultivars also demonstrate differences in fibre yield, grain production and THC concentration. The sexual dimorphism of this annual dioecious flowering herb further complicates the morphological and developmental diversity of this plant. The genetic improvement of hemp has led to the creation of a monoecious variety in which the female phenotype predominates (Arnoux *et al.*, 1969). Hemp's diversity has allowed the development and selection of cultivars according to the desired industrial end product (paper fibre, fibre for textiles or seeds). Recent studies have shown that this geographical and sexual polymorphism can be determined by molecular and protein analysis (Lucchese *et al.*, 2001; Mandolino and Carboni, 2004). Current programmes can identify molecular markers associated with the desired morphological and chemical characteristics of industrial hemp.

3.3.2 Plant morphology

C. sativa is an annual plant with a stem that can grow to a height of some 2–4 m depending on the variety. The stem rarely branches and has a diameter that averages between 1 and 3 cm. The plant's morphological characteristics can vary according to the growing conditions (sowing density, photoperiod, etc.). At low sowing densities, branching is likely to increase, whereas a high sowing density will favour the development of a tall plant with long, straight unbranched stems (Bocsa and Karus, 1997).

The root system

In comparison to the biomass above ground, the root system of hemp is relatively reduced when compared to other economically important annuals. The volume of the root system varies according to the cultivation methods used and the soil quality. It is also dependent on sexual phenotype. On average, the root volume comprises 8–9% of the total biomass. Male plants have a shorter vegetative phase than female plants, and consequently have a less developed root system. The main root can grow to a depth of 2 m, whereas the secondary roots that form the bulk of the root system reach lengths of between 10 and 60 cm.

The leaves

The leaves are composed of a petiole from which arises 7–10 lanceolate folioles of unequal size. The leaves arise following an opposite leaf pattern every 10–30 cm, although this phyllotaxy evolves towards an alternating leaf pattern as the plant flowers. This change can be identified towards the flower head. During the vegetative phase, hemp therefore demonstrates a phyllotaxy with two synchronous helices. This arrangement becomes asynchronous at the start of the flowering period, producing an alternating leaf pattern. The number of folioles is always odd and increases progressively while varying with cultivar. During its development, the plant remains green. When mature, the lower leaves on the stem may be shed, especially where the sowing density is high. (Lacombe, 1978; Bocsa and Karus, 1997).

The inflorescence

Hemp is a dioecious species made up of male and female plants. The male inflorescences form a loose cluster of flowers on the end of a central stem (rachis), with very few leaves. The female reproductive apparatus forms at the apex of the stem and the flowers are borne on racemes. Genetic selection has led to the creation of dioecious strains in which the female factor predominates. The male plants are generally weaker and start to deteriorate at the time of flowering, reducing the yield in grain and fibre. Monoecious strains of hemp exist that are exclusively female. The sexual character of monoecious strains is very much dependent on edaphic factors (photoperiod, nitrogenation), with certain conditions favouring a masculinization of the crop. In this last case, the plants demonstrate both male and female characteristics: the male flower appearing towards the middle of the stem, the apical part supporting the female inflorescence (Bocsa and Karus, 1997).

The male flowers are composed of five greeny-yellow sepals enclosing five tightly packed stamens that open to reveal little anthers (Fig. 3.1). The female flower is easily distinguished from the male. Devoid of petals, this female flower is made up of two pistils and a single seed pod. The pistils measure on average between 3 and 8 mm in length and are enclosed by a small sheath, or bract, that is then attached to the stem. The bracts bear the highest density of trichrome glands and it is these that are responsible for the production of cannabinoids in those varieties rich in THC. These structures are also found on the leaves, although at a lower concentration, together with non-glandular trichomes that do not produce cannabinoids.

The seed

This commonly used term is inappropriate from a botanical point of view. The ovule produces an indehiscent fruit (akene) in which the seed (embryo) takes up all the volume. This seed is enveloped in its own coat (testa) and outer envelope (pericarp). Hemp seeds are ovoid or spherical in shape and measure 3–5 mm in length depending on the cultivar. Each grain encloses two cotyledons rich in oil and proteins but, unlike many other plant species, these seeds contain a reduced quantity of albumen. Although hemp is not grown specifically for the purposes of oil production, the yield from certain monoecious varieties allows the industrial exploitation of hemp seeds for oil.

The stem

Regardless of the cultivar, the general morphology of the stem remains similar. The internode distance between branches can vary. The surface

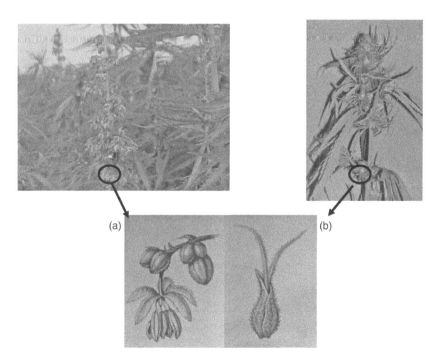

Fig. 3.1. Male and female inflorescences and the corresponding details of their respective flowers. (a) Male flower; (b) female flower.

of the stem is grooved and its diameter increases from the plant's apex to its base. Morphological characteristics such as height and diameter are, however, very dependent on the species, environment and stage of development of the plant. Male plants, for example, are between 10 and 15% taller than their female homologues, whose stems are wider. Fine stem diameters are promoted by a low sowing density, which also is propitious for the development of tall plants (Bocsa and Karus, 1997).

3.3.3 Anatomy and origins of stem tissues

The principal products of hemp are derived from the stem, for it is a source of fibres with remarkable physico-chemical properties. The anatomy of the stem tissue will therefore be presented in detail. The stem is produced as a result of the sequential addition of internode sections (the section of stem between the insertion of two successive sets of leaves).

The constituent tissues of the stem – their nature and origin

The stem of the hemp plant, as with other annual dicotyledons, demonstrates two tissue zones: a central woody area (pith) and an external epidermis, or bark. Together, these perform a number of special functions (Bowes, 1998):

- The epidermis forms a protective barrier around the stem and modulates exchanges between the plant and its environment.
- The parenchymal tissues responsible for metabolic processes such as photosynthesis are poorly represented in the stem of mature plants.
- The supporting tissues that provide the plant and its organs (collenchyma, sclerenchyma) with its rigidity and suppleness are found in the stem's outer layer.
- The tissues responsible for the movement of sap are of two types: xylem tissues move water and soluble mineral nutrients from the roots through the plant, whereas phloem tissues transport organic nutrients from the site of photosynthesis (leaves) to non-photosynthetic tissues around the plant.

The formation of the stem tissues starts with the generation (division and differentiation) of cells derived from the meristem (Esau, 1977; Robert and Catesson, 2000) (Fig. 3.2). The meristematic tissues are composed of undifferentiated cells that support the production of new cells capable of acquiring a range of specific functions (protection, metabolism, support, sap transport). Two types of meristem are recognized. The primary meristem is located at the apices of the stem and is responsible for the primary growth of the internode as well as the formation of leaves. This growth arises through division and growth of the different parenchymal, conducting and protective tissues. Primary growth essentially allows organs to lengthen and results in the generation of primary tissues. Lateral growth (widening) of the stem ('secondary growth') comes from the laterally located secondary meristematic tissue (cambium). The vascular cambium produces secondary xylem towards the inside of the stem and secondary phloem towards the outside of the stem. The cambium arises from the procambium, which itself is derived from the primary meristematic tissues that give rise to the primary xylem and primary phloem. Cambial activity is thus responsible for the production of secondary xylem and is a key component of the stem of the hemp plant.

The xylem which supports the weight of the plant is responsible for the transport of minerals. These functions are carried out by the fibrous walls and lumen respectively (Esau, 1977). The anatomy of the xylem of the hemp plant is very similar to that found in other dicotyledenous annual plants such as lucerne, although the cellular distribution seen in flax shows some variation (Day *et al.*, 2005). The bark is made up primarily of long fibres (sclerenchyma) associated with phloem tissue, parenchyma and surface epidermis. Two types of fibre can be distinguished, forming two rings around the central xylem (Esau, 1977) (Fig. 3.2):

1. The primary fibres arise from primary phloem cells, derived from the primary meristem.
2. The so-called secondary fibres are generated from the cambium.

The primary fibres can be distinguished from the woody (pith) fibre by their length

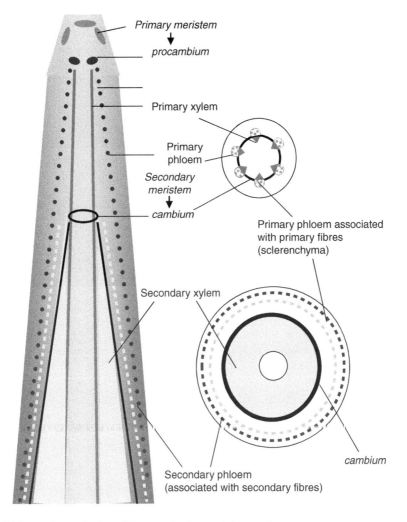

Fig. 3.2. Origins and organization of the vascular tissues in hemp stems.

(secondary extra-xylem fibres are modest in length), a thick cellular wall and a specific chemical composition. The physical, morphological and chemical characteristics of the primary fibres make them highly sought after by industry. The secondary fibres adjacent to the cambium are generally left behind during the process of fibre stripping. In both cases, the non-xylem fibres are closely associated together in bundles orientated parallel to the stem (Fig. 3.3). Once removed, these bundles are commonly called 'technical fibres'. This term contrasts with the unitary fibre that is defined anatomically according to strict morphological criteria (narrow elongated cells with few punctuations). It is a generic term that is used to describe xylem fibres and extra-xylem cells (sclerenchyma).

The cell walls: the major constituent of hemp fibres

The xylem and sclerenchyma tissues are dead and devoid of protoplasm. As cell growth comes to an end, the walls of these cells become thickened by the laying down of the parietal polymers, cellulose, hemicellulose and lignin. This thickening corresponds with the development of rigid secondary walls attached to the existing parietal strata. At maturity, the cell walls are composed of a complex matrix of pectin, lignin and hemicellulose in which is embedded an armoury of cellulose fibrils. The arrangement and orientation of the cellulose microfibrils in relation to the cell's axis allow three closely associated zones to be identified (Roland *et al.*, 1995) (Fig. 3.4). Starting from the outside and moving towards the lumen, the zones are:

- Middle lamella: forms the interface between adjacent cells.
- Primary wall: supple and extendable, this zone allows cells to lengthen when young.

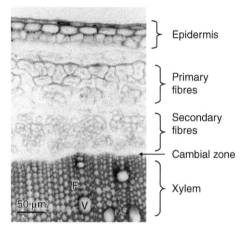

Fig. 3.3. Transverse section of a hemp stem. Lignins are coloured using phloroglucinol and hydrochloric acid. V = vessels; F = xylem fibres.

- Secondary wall: this is put down in the final stages of cell growth and is made up of three layers, referred to as S1, S2 and S3.

The cellular types noted for their fibrous content show a predominance of rigid secondary walls. The width, composition and architecture of the different parietal layers will evolve throughout the maturation of the plant cells and vary according to the cell type, tissue type and organ.

The middle lamella is rich in salts of pectic acid, demonstrates hydrophilic properties in young growing cells and contains little cellulose. This layer glues adjacent cells together, while the deposition of lignin further reinforces this, providing stiffness and rigidity.

The primary wall is composed of cellulose microfibrils embedded in a matrix of pectin and hemicellulose. The cellulose fibres are widely dispersed and aligned at all angles. This parietal layer is generally lignified in fibrous plants.

The secondary wall is the main source of fibres and is characterized by its high cellulose content and the orientation of cellulose microfibres in each of its three layers (S1, S2, S3). In contrast to the primary wall, the microfibres organize themselves in line with the axis of the cell, while showing some variation between layers. In particular, the deposition of cellulose microfibres is virtually completely parallel to the cell axis in the thick

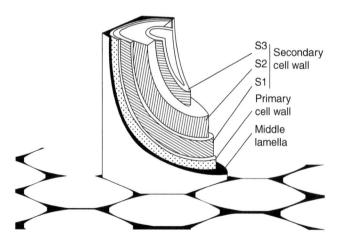

Fig. 3.4. Architecture of the S1, S2 and S3 layers of the tracheid cell wall showing the relative orientation of the cellulose microfibrils in each (from Brett and Waldron, 1996).

S2 layer of hemp. Unlike the plant fibres found in wood, the secondary wall has little hemicellulose and lignin-impregnated matrix. Little information is available about the precise architecture of the secondary walls of hemp. It is known, however, that the extra-xylem fibres of hemp bear a strong resemblance to those of flax, in terms of their composition and histology (Rahman, 1979), the importance of the S2 layer and the presence of a much reduced S1 layer. The secondary wall fibres are characterized by an unusual crystalline form of cellulose that forms microfibrils which lie parallel to the cellular axis (Bonatti *et al.*, 2004).

The formation of the different parietal layers is a sequential dynamic process that accompanies cell growth. The lignification of the walls leads to the parietal structures becoming both impermeable and rigid. The particularity of the non-xylem fibres of hemp lies in their low concentration of lignin. As with flax, lignin is found primarily in the intercellular cement of the middle lamella and primary wall. The deposition of lignin is greatest at the time of flowering and mainly affects the outer layers of the secondary wall.

3.4   Part II: Constituents and Chemical Composition

3.4.1   General points

Plants are highly organized biological systems in which each tissue or organ fulfils a precise function. Generally speaking, it is possible to distinguish two biochemical constituents: (i) those arising from the plant's physical structures and (ii) those dependent on the plant's metabolism.

As in all living organisms, the cell is the basic unit of plants (Fig. 3.5). Plant cells typically possess an envelope or plasma membrane, as well as a fibrous reservoir of microtubules that make up the filamentous cytoskeleton. The cell encloses the cytoplasm and the collection of organelles (mitochondria, ribosomes, reticulum...), of which some are plant specific (chloroplasts, vacuoles), that make up the metabolic machinery of the cell. The organelles are themselves delimited by

membranes. Plant cells are further characterized by an external structure called the cell wall.

The physical structures of the cell are made up of complex assemblies of molecules and/or macromolecules. Almost 100 molecules play a role in the architecture and compartmentalization of plant cells. The life processes of a cell, tissue, organ or, indeed, the whole plant rely on a collection of metabolic pathways working in concert to assure the various functions necessary for a plant's reproduction and survival. These pathways are controlled by the genes and the proteins and enzymes the genes encode. In total, some 50,000 molecules are implicated in the various plant anabolic and catabolic processes (Pichersky and Gang, 2000).

While many of the principal metabolic activities (including respiration, assimilation, transport and differentiation) are common to all cells, there are a number of proteins synthesized by specific cells or tissues. The molecular reservoirs (starch, lipids and proteins) deposited and stored in seeds are of particular importance. Other molecules, known as secondary metabolites, are less important to plant survival. Their presence may, however, have direct or indirect consequences for plants. These include molecules that are attractive or repellent to insects, toxins, allelopathic or antifungal molecules and phytohormones.

Considering the complexity, diversity and multiplicity of biochemical structures implicated in the structure and metabolism of plants, this treatment of *C. sativa* will focus on the principal molecules of an industrial, agro-edible or pharmacological interest. Therefore, the various products will be presented, rather than an exhaustive presentation of the hundreds of molecules identified to date (Turner, 1980).

It should be noted that:

1. With the exception of the cannabinoids, the majority of molecules isolated from hemp are not unique in and of themselves, as they are found in other plants (annuals and perennials).

2. The hemps used for industrial and pharmaceutical purposes differ in their cannabinoid content.

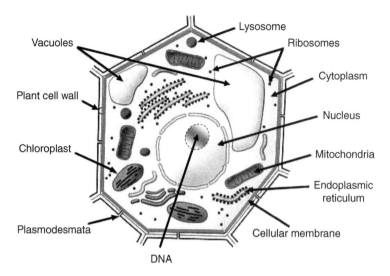

Vacuoles

Lysosome

Ribosomes

Cytoplasm

Plant cell wall

Nucleus

Chloroplast

Mitochondria

Endoplasmic reticulum

Plasmodesmata

Cellular membrane

DNA

Fig. 3.5. The plant cell (from http//www.gnis.fr).

3. New openings for industrial hemp usage will need to take into account the uses that can be made of other hemp products, most notably hemp seeds and chopped hemp.

3.4.2 Chemistry and biochemistry of useful hemp-derived products

Hemp has probably been cultivated in Asia since 5000 BC. It is only much more recently that this tradition arrived in Europe, where hemp is thought to have been introduced and/or domesticated in France and Germany during the Iron Age (700 BC).

Globally, the way in which hemp has been used has changed little. Fibres are used for the production of various materials including rope, textiles, sails and stuffing. Seeds are destined for human and animal consumption but can also be used to produce oil, either for use as an edible oil, as a drying oil in paints or occasionally for fuel. Flowers and leaves are used as psychotropes and for their therapeutic properties (as narcotics, analgesics and antispasmodics).

Modern industrial technology has, of course, allowed humans to exploit the intrinsic properties of the various hemp products better but has had little effect on the end use of these products.

The biochemical properties of hemp include those associated with the cannabinoids, certain of which possess psychotropic properties.

It is this characteristic that has resulted in 'cannabis' becoming the most popular and heavily used illicit product in France, according to the Observatoire Français des Drogues et des Toxicomanies (OFDT; http://www.ofdt.fr/ofdt/fr/pt_canna.htm) (Bello et al., 2003).

Despite the controversy surrounding cannabis usage, it continues to attract attention from researchers seeking to investigate its therapeutic potential (Iversen and Chapman, 2002; Berman et al., 2004). That said, the popular claims that cannabis is an innocuous product should not be muddled with those of the scientific and medical communities, who recognize its therapeutic potential while remaining wary of the dangers associated with its use (http://www.drogues.gouv.fr).

Stem fibres

Industrial hemp is used mainly for its primary fibres and hurds (chopped hemp). Three fibre types can be distinguished histologically: (i) woody fibres from the centre of the stem, derived from cambial activity – these are used to produce hurds; (ii) primary fibres situated outside of the vascular tissues and derived from the primary meristem; and (iii) secondary extra-xylem fibres derived from the cambium (van der Werf et al., 1994).

Fibres are derived from plant cell walls, that is to say the constituents of the envelope that encloses the plant cell (Fig. 3.5, Table 3.1). Cellulose is primarily and structurally the main element of these fibrous walls. The non-cellulose parietal elements include other polysaccharides (pectins and hemicelluloses), lignins as well as proteins and various oligomers of sugar and phenol (Turner, 1980; Jurasek, 1998; Buchanan *et al.*, 2000). The relative proportions of these various constituents depend on the histological origins of the fibres (primary fibre, secondary fibre, woody fibre), their stage of growth and plant maturity, as well as the plant variety. Ecophysiological factors affecting the growth and development of the plant will also cause variation in the constituents of the fibrous cell walls (van der Werf *et al.*, 1994; Cappelletto *et al.*, 2001; Mediavilla *et al.*, 2001).

CELLULOSE. Cellulose is a homopolymer made up of a chain of glucose molecules linked by β1-4 glycosidic linkages. The repeating unit is formed by the dimer of cellobiose (β-D-glucosyl-(1-4)-D-glucose). The linear arrangement of the glucan chains is stabilized by bonds between the hydroxy group at 3 and oxygen (O-5) of the contiguous cycle (Fig. 3.6). The glucan chains then assemble themselves into a crystalline structure by the formation of hydrogen bonds between and within chains. The microfibrils formed are approximately 3–5 nm thick in most plants. They represent an alternating of the crystalline and amorphous zones within each structure. Hemp fibres are particularly rich in crystalline cellulose (60–80%). That said, cellulose crystallinity is less important in the primary wall than in the secondary wall and is related to the arrangement of microfibrils and the composition of the parietal matrix (that is, the other non-cellulose components of the walls).

The organization of the cellulose microfibrils in the various parietal layers has a direct bearing on the macroscopic mechanical properties of the fibres, as is the case for flax and wood (Roland *et al.*, 1995; Girault *et al.*, 1997; Focher *et al.*, 2001; Fratzl, 2003; Bonatti *et al.*, 2004). The properties of cellulose will also be influenced by the mechanical stresses that the plant is subjected to during growth, as well as the methods of extraction used to recover the fibres. Dislocated zones, in which the crystalline cellulose has a different orientation, can be seen as a result.

PECTINS. The pectins are a group of polymers rich in galacturonic acid, rhamnose, arabinose and galactose. These polymers are made up of a primary chain with secondary chains branching off, which may themselves be branching or have substitution (Ridley *et al.*, 2001). Pectins are typically found in the middle lamella and primary walls of dicotyledons (and to a lesser extent in monocotyledons; Robert and Catesson, 2000). The pectin acids are homopolymers of galacturonic acid (polygalacturonan, PGA), which is eventually methylated. The chains are held together by ionic bonds and the resulting structures are stabilized by calcium ions. A rhamnose residue may be introduced into the chain, producing a deviation or 'elbow' among type I rhamnogalacturonans (RGI, Fig. 3.7). More complex rhamnogalacturonans, such as RGII, are recognized by their short branches of xylose and arabinose. Other types of pectin correspond with neutral polysaccharides (β1,4-galactones, α1,3-arabinones, β1,4-galactones with an arabinose substitution at C3) (Chabbert *et al.*, 2006).

Little further information on the structure of Cannabis cell walls can be added to the preceding general overview. A study of isolated fractions of fibre pectins in macerated hemp demonstrated similar arrangements to RGI pectins (Fig. 3.7; Vignon and Garcia-Jaldon, 1996).

Table 3.1. Illustration of the chemical composition of the periphloem fibres in hurds (Vignon *et al.*, 1995).

Weight (%)	Cellulose	Hemicellulose	Pectin	Lignin	Wax	Protein
Periphloem fibres	55	16	18	4	1	2
Hurds	44	18	4	28	1	3

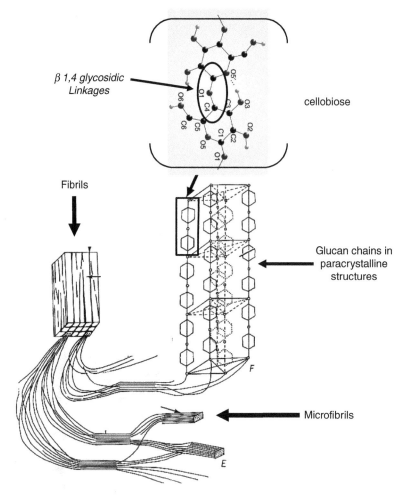

Fig. 3.6. Structure and organization of cellulose fibrils (from Esau, 1977). C^n, O^n = number of carbon/oxygen atoms in glucose units.

HEMICELLULOSES. Hemicelluloses (or glycans) are heterogenous polysaccharides divided into different families according to the wall type and botanical source (Joseleau, 1980). Their structure allows them to interact directly with cellulose microfibres in addition to joining up with other glycans. They can thus form both a matrix and a reservoir of fibres. Among dicotyledons, we find the xyloglucans that are to be found mainly in the primary walls of flax and hemp (Fig. 3.8a). The second large family of hemicelluloses in dicotyledons is made up of arabinoglucuronoxylanes (Fig. 3.8b). The secondary walls of hemp contain primarily glucuronoxylanes, of which a small proportion is arabinoglucuronoxylanes. Finally, other non-cellulose polymers containing mannose are found to varying degrees among all the angiosperms: these are the glucomannans, galactomannans and galactoglucomannans (Buchanan et al., 2000).

LIGNINS. Lignins are complex polymer structures resulting from the polymerization of three phenylpropenoids (lignin-derived phenols): coumarylic, coniferylic and sinapylic alcohols. These precursors give rise to the monomer subunits p-hydroxyphenyl (H), gaiacyl (G) and syringyl (S), respectively, and can be distinguished by the degree of methoxylation of their aromatic rings.

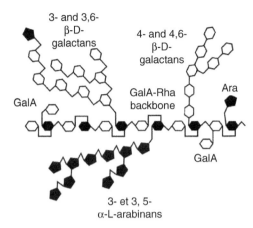

3- and 3,6-β-D-galactans

4- and 4,6-β-D-galactans

GalA

GalA-Rha backbone

Ara

GalA

3- et 3, 5-α-L-arabinans

Fig. 3.7. Schematic structure of a type I rhamnogalacturonan. The primary chain is made up of repeating dimers of galacturonic acid and rhamnose. The branching oligosaccharides of flax are primarily made up of arabinose and galactose (Ara, Gal) (from Ridley, 2001).

The tridimensional polymer can contain up to a dozen different intermonomeric bonds but, in the case of hemp, its structure remains poorly defined (Fig. 3.9). Lignification progresses by a process of encrustation with the deposition of a hydrophobic material within the polysaccharide matrix. The extent of the lignifications varies considerably, however, according to the type of fibre. Primary periphloem fibres are poorly lignified compared to the secondary woody fibres used to make hurds, where nine-tenths of the lignin in hemp stems is found (Han and Rowell, 1997) (Table 3.1).

The lignin of primary fibres encloses type G and type S units, and especially type H material (Cronier *et al.*, 2005). In this sense, the lignins are related more closely to the graminaceae than to leaved trees with an S/G ratio of 0.8 (del Rio *et al.*, 2004).

PROTEINS AND PLANT EXTRACTS. Plant cell walls also contain constituents such as proteins, oligosaccharides and turpenes in relatively small concentrations, although local concentrations may be much more significant. Hemp is no exception to this rule: up to 15% of the dry matter is made up of plant extracts, i.e. soluble molecules that can be extracted from plant material using a suitable solvent (water, ether,

dichloromethane, acetone, etc.; Cappelletto *et al.*, 2001).

All structural proteins capable of associating with pectins and hemicelluloses via their sugar residues contribute to the pool of proteins found and measured in fibres. These include proteins rich in glycine (GRP), proline (PRP), hydroxyproline (HRGP) and arabinogalactan (AGP). Plants respond to an insult by synthesizing proteins that generally will be localized within the injured tissues. That said, the proportion of proteins found in hemp fibres is relatively low and evolves according to the maturity of the plant, attaining a maximum concentration of 1–2% (Vignon *et al.*, 1995; Cronier, 2005).

Among the plant extracts identified in hemp phloem fibres are found a number of lipophilic compounds (extracted using acetone) such as the fatty acids (C4–C30), alkanes (C21–C33), free sterols or sterol esters (sitosterol) and waxes (fatty acids and esterified long chain aliphatic acids) (Gutierrez and del Rio, 2005). Finally, among these extracts, traces of cannabinoids can be detected in industrial hemp fibre (2% of the total extracts) (Kortekaas *et al.*, 1995).

The last important biochemical group of note in fibrous plants are the minerals. These include silica and heavy metals derived from the soil during the plant's growth. Silica is present in hemp stems at low concentrations only (<1%) (Kamat *et al.*, 2002) and is therefore barely present in the extracted fibres. By contrast, heavy metals are taken up by the Cannabis plant and distributed throughout the plant, with residues found in the cortical and xylem fibres. Hurds grown on contaminated ground can therefore have high concentrations of cadmium, lead and nickel (up to 3% by weight) (Linger *et al.*, 2002).

Storage products and secondary metabolites

The full value of hemp can be realized by exploiting a number of co-products other than just hemp fibre. These include seed-derived edible products and pharmacological products. Various substances can be extracted from other plant organs, notably the leaves and flowers of certain varieties of hemp.

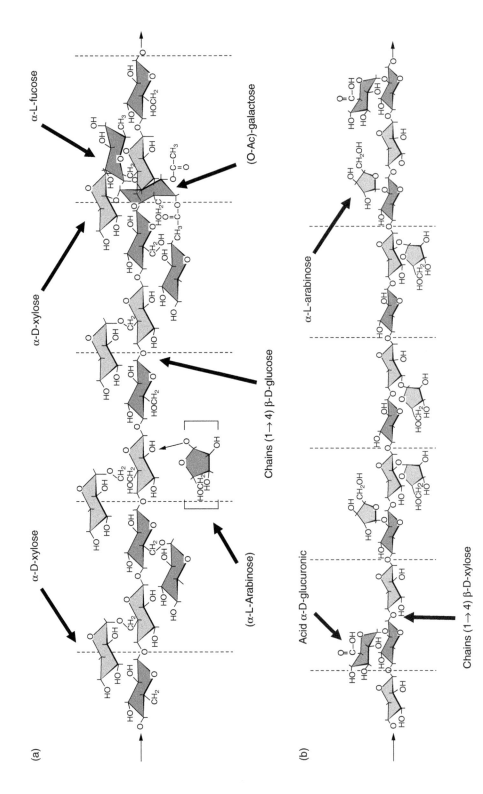

Fig. 3.8. Schematic diagram of the structure of fucogalactoxyloglucan from an angiosperm (a) and an arabinoglucuroxylan (b) from a dicotyledon (from Buchanan et al., 2000).

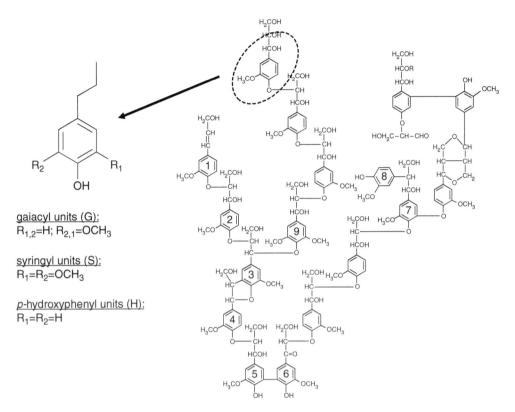

Fig. 3.9. Structure of lignin and its constituent monomers. The different bonds found among lignins are represented by the sets (1,2) and (9,3): α-, β-O-4 benzyl-aryl- and alkyl-aryl-ethers; (3,4) β-5 phenylcoumaran; (5,6): 5-5′ biphenyl; (7,8): β-1 diarylpropane. The (1,2) and (9,3) arrangements are the most common (approximately 40–50% of the bonds) (from Kurek, 1992).

gaiacyl units (G):
$R_{1,2}$=H; $R_{2,1}$=OCH$_3$

syringyl units (S):
R_1=R_2=OCH$_3$

p-hydroxyphenyl units (H):
R_1=R_2=H

Products derived from Cannabis fruit

The fruit of *C. sativa* are known as akenes (or achenes). The embryo is protected by a hard envelope consisting of the testa (embryonic membrane) and the pericarp (fruit membrane) (see the chapter on taxonomy and botanical descriptions). The seed contains variable quantities of polysaccharides, lipids and proteins that will meet the needs of the developing embryo until it is able to photosynthesize.

The achenes are used as animal feed or for the production of oil. A protein-rich flour is produced as a by-product of the oil extraction process and is used as animal feed. The hard fruit envelopes can be used as stuffing materials (Molleken and Theimer, 1997).

Hemp achenes contain 25–35% lipids, 20–25% proteins and 20–30% carbohydrates, of which 10–15% are fibres. In addition to these major constituents, there are a number of minor substances derived from secondary metabolic processes. These include terpenes, sterols, tocopherols and minerals (Kriese *et al.*, 2004).

To the best of our knowledge, there is no information available concerning the polysaccharides found in hemp achenes. In fact, the protective tissues (testa and pericarp) enclose the same parietal polymers found within hemp fibres, together with the constituents of cell membranes (Esau, 1977). The hemp residue recovered following oil extraction contains 15% lignin (gravimetric measurement) and 40% structural polysaccharides (cellulose and hemicellulose) (Mustafa *et al.*, 1999). These approximations reflect the global composition of the seed and its husk. We have seen no report of starch in hemp seed and the nature of the remaining 8% of non-structural carbohydrates is not documented explicitly (Mustafa et al., 1999; Leizer et al., 2000).

The protein deposits found in hemp seed are primarily legumin, with edestin representing 60% of its total quantity (Patel, 1994). Albumin is the second class of protein (30% of the total). From a nutritional point of view, hemp seed represents a complete and balanced source of amino acids as the 20 amino acids, including the 9 essential amino acids, are all present. Certain minor proteins have been isolated and purified, although no functional properties have been attributed (sulfur-rich protein, resembling mabilin IV, found in *Caparis masaikai*) (Odani and Odani, 1998; Hills, 2004).

The lipid fraction of hemp seeds is composed mainly of linoleic, α-linolenic and oleic acids (80% of the lipids extracted in oil) in a 55:20:10 ratio (Fig. 3.10) (Oomah et al., 2002). Palmitic, stearic and γ-linolenic acid make up the remaining (approximately 10%) lipids, together with a number of minor lipids, each making up less than 0.7% of the total (eicosenoic, vaccenic and arachidonic acids, etc.). The distribution of these acids in the tissues of the seed differs from that in the envelope. The envelope contains three times more oleic acid than the seed (Molleken and Theimer, 1997). The excellent nutritional qualities of hemp seed and its oils are due to the linoleic to linolenic ratio of 3:1 (Leizer et al., 2000).

Various other molecules and substances have been described in Cannabis seeds and fruit:

- alkaloids such as the trigonellins, cholines, cannabamines and isoleucin betaines (Bercht et al., 1973)
- cytokinin plant hormones such as zeatin
- tocopherols (including γ-tocopherol), tocotrienols and plastochromanols with antioxidant properties and collectively called vitamin E (Fig. 3.11)
- enzyme inhibitors and antinutrients such as tanins, phytates and sinapins (Mautthaus, 1997).

Finally, it should be noted that the oils extracted from the seeds contain no psychotropic ingredients. Where ingredients such as cannabidiol (CBD) and tetrahydrocannabinol (THC) are identified, they will have originated from the contamination of the pressing process by various other parts of the plant. A thorough cleaning prior to pressing will correct such problems.

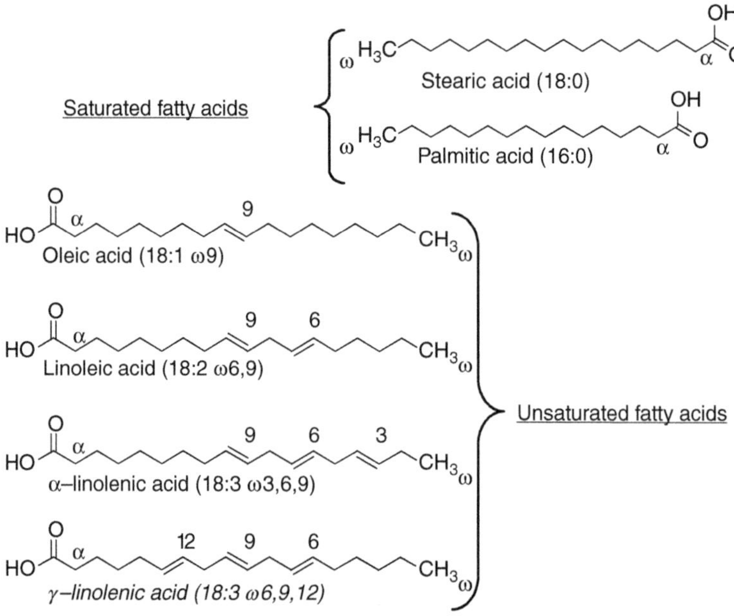

Fig. 3.10. Saturated and unsaturated fatty acids derived from hemp seeds.

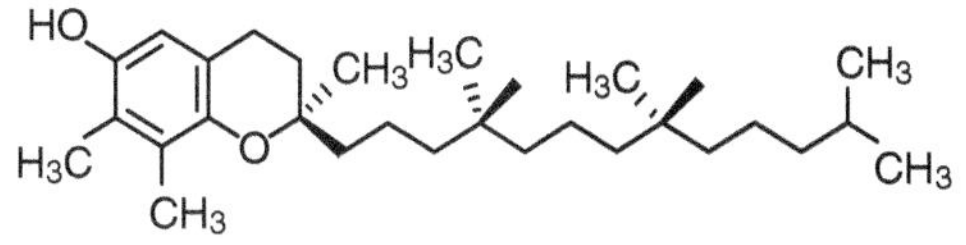

γ-tocopherol

Fig. 3.11. Structure of γ-tocopherol, one of the constituents of vitamin E.

Inflorescence and leaf-derived products

The principal class of products derived from the inflorescences and leaves are the terpenes, of which the cannabinoids are part. These secondary metabolites are synthesized by different types of glandular cells (stalked and sessile) situated on the leaves or on different parts of the male and female flowers (bract, anther, etc.).

A number of flavonoids and alkaloids have also been described in the leaves of *C. sativa*. That said, these products are of relatively little significance, as they have no specific use (Mechoulam and Hanus, 2000).

In terms of their economic value, only the terpenes and cannabinoids (essential oils, fragrances and pharmacologically active products) appear to be of interest.

THE CANNABINOIDS. The cannabinoids are a family of molecules that characterize *C. sativa*. Some cannabinoids possess pharmacological properties and can exert psychotropic effects on humans and animals (Bello *et al.*, 2003).

The cannabinoids are terpenophenolic compounds with a typical 21-Carbon skeleton composed of a phenolic ring (olivetol) with the monoterpene geranyl (Fig. 3.12).

Among the 60 or so known cannabinoids (Turner, 1980; Mechoulam and Hanus, 2000), the most important family members are the cannabidiols (CBD), the cannabigerols (CBG), the tetrahydrocannabinols (THC), the cannabinols (CBN), the cannabichromenes (CBC) and the cannabicyclols (Mechoulam and Hanus, 2000) (Fig. 3.12).

It is thought that only the acid derivatives are synthesized and stored in the plant tissues. The neutral decarboxylated form may be partially formed in the plant or during the extraction process, depending on the protocol used for the leaves and flowers (Fellermeier and Zenk, 1998; Mechoulam and Hanus, 2000; de Meijer *et al.*, 2003). The two chemical compounds with psychotropic and/or pharmacological properties are THC and CBD (Mechoulam and Hanus, 2002). The quantitative ratio of these compounds within the inflorescences allows plants to be grouped into three distinct classes (chemotypes). This classification is used to determine those cultivars destined for fibre production from those destined for drug production (de Meijer *et al.*, 2003).

THCs and CBDs are used and/or tested in their pure form or together when evaluating their clinical usefulness in the treatment of various pathologies. A number of pharmaceutical grade products are available (Berman *et al.*, 2004; Stott and Guy, 2004; Wingerchuk, 2004; Makriyannis *et al.*, 2005).

THE TERPENES. The terpenes are produced by the same glandular structures that produce the cannabinoids. They are polymers of isoprene units (Fig. 3.13). The mono- (10 carbons or 2 units of isoprene) and sesquiterpenes (15 carbons or 3 units of isoprene) are often associated with plant fragrances (Buchanan *et al.*, 2000). Some 60 or so monoterpenes and 40 or so sesquiterpenes have been isolated from Cannabis flowers (Turner, 1980).

In the case of narcotic cannabis (flowers and bracts), two monoterpenes (myrcene and limonene) represent approximately 85% of all the terpenes isolated (Fig. 3.13) (Ross and ElSohly, 1996). Two sesquiterpenes, guaiol and eudesmol, make up approximately 16% (Hillig, 2004). Various other products are found in concentrations below 2%. These include pinene, linalool and caryophylene, together with a multitude of trace terpenic

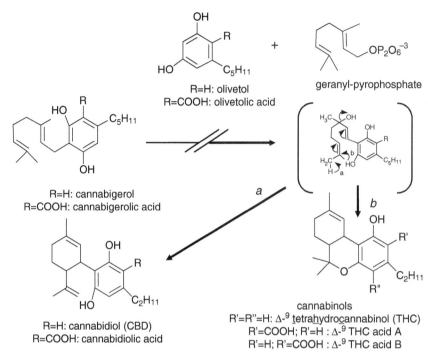

OH

R

HO C₅H₁₁

R=H: olivetol
R=COOH: olivetolic acid

+

geranyl-pyrophosphate

R
HO C₅H₁₁

OH

R=H: cannabigerol
R=COOH: cannabigerolic acid

a

OH
R
HO C₂H₁₁

R=H: cannabidiol (CBD)
R=COOH: cannabidiolic acid

b

OH
R'
C₂H₁₁
R"

cannabinols
R'=R"=H: Δ-⁹ tetrahydrocannabinol (THC)
R'=COOH; R'=H : Δ-⁹ THC acid A
R'=H; R'=COOH : Δ-⁹ THC acid B

Fig. 3.12. Diagrammatic representation of the biosynthesis of tetrahydrocannabinoids (THC) and cannabidiol (CBD) (from Mechoulam and Hanus, 2002).

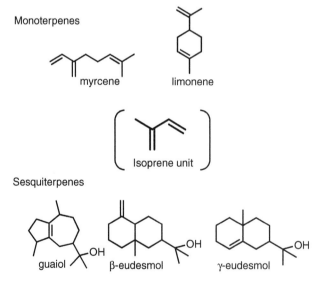

Monoterpenes

myrcene   limonene

Isoprene unit

Sesquiterpenes

guaiol   β-eudesmol   γ-eudesmol

Fig. 3.13. Structure of the principal terpenes of Cannabis.

products. These trace terpenes allow various cultivars to be differentiated by quantitative chemotaxonomic analysis (Novak *et al.*, 2001; Hillig, 2004).

THE FLAVONOIDS. Plant flavonoids are important components, as they are responsible for the pigmentation of both flowers and fruit and are found throughout the plant (Harborne and

Williams, 2001). There are currently in excess of 4500 known flavonoids (Buchanan *et al.*, 2000) in a number of different classes (anthocyanins, flavonols, flavones and isoflavones) (Harborne and Williams, 2001).

Studies conducted on different varieties of *C. sativa*, both those producing fibres and those rich in THC, have identified some 20 glycosylated flavones: orientin, vitexin, luteolin and apigenin glucoronide (Fig. 3.14) (Turner, 1980; Vanhoenacker *et al.*, 2002). Flavonol-type structures have also been isolated and characterized in the pollen of *C. sativa*: kaempferol and quercetin-3-O-sophoroside (Ross *et al.*, 2005). Given the numerous possibilities for substitutions on these molecules, the variety of structures isolated allows them to be used as chemotaxonomic markers in the same way as terpenes (Pichersky and Gang, 2000).

THE ALKALOIDS. The alkaloids are molecules extracted using organic solvents or by distillation following a treatment with alkalis such as soda or potash. These compounds are derived from amino acids (most commonly) or from purine or pyrimidine rings. Alkaloids are often psychoactive and more than 1500 known compounds have been isolated from plants (Buchanan *et al.*, 2000). Alkaloids are, however, present in very low concentrations in the leaves of Cannabis (<0.3%). The compounds isolated can be classified in one of three large groups (Turner, 1980; Mechoulam and Hanus, 2000): spermidin alkaloids (palustrine, cannabisativine), simple amines (hordenine, piperidine) and heterocyclic or non-heterocyclic quaternary ammonium compounds (betain, choline and trigonellin) (Fig. 3.15).

3.4.3 Conclusions

This review of the chemistry of the constituents of hemp is not exhaustive and provides only an overview of the main groups of interesting products derived from Cannabis. Hemp contains fibres of exceptional quality and its

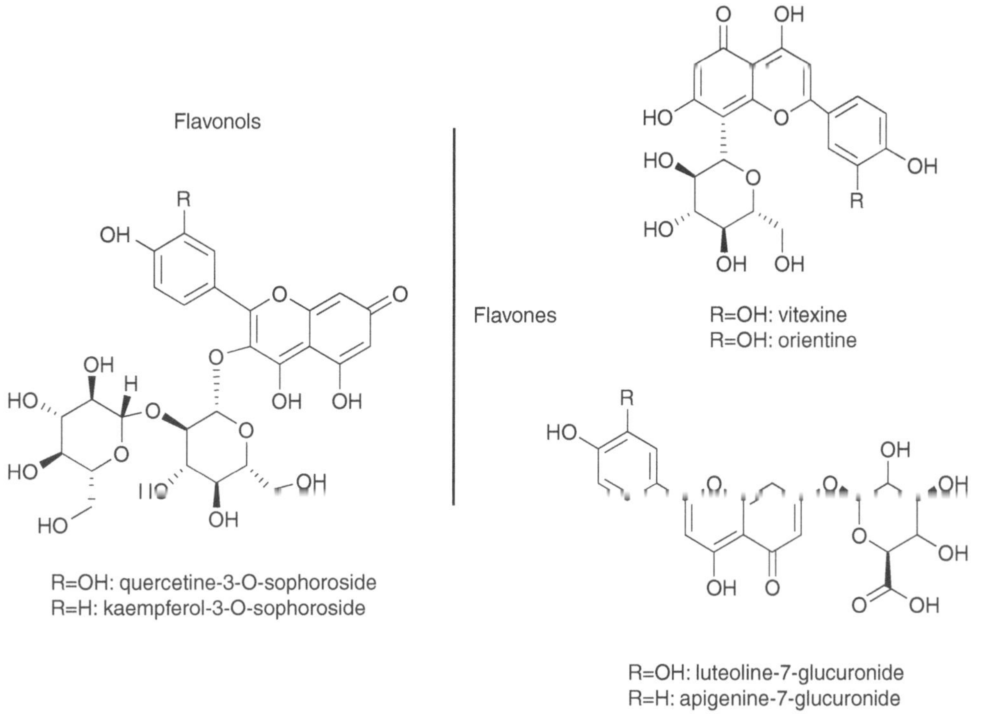

Fig. 3.14. Structure of the principal flavonoids of Cannabis.

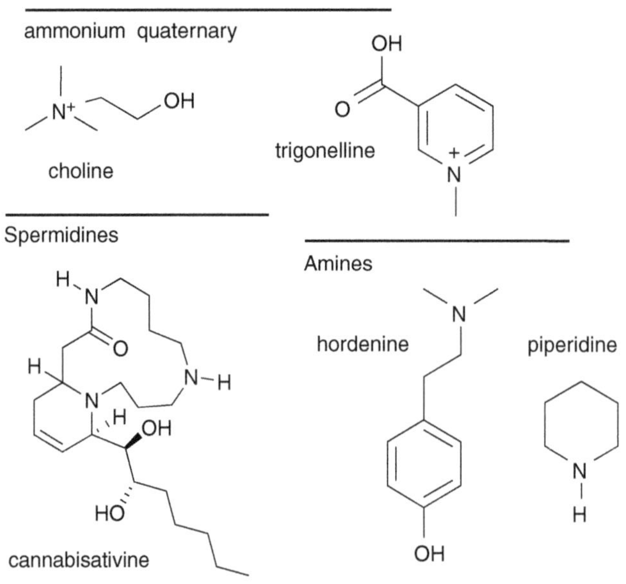

Fig. 3.15. Principal alkaloids derived from Cannabis leaves (from Turner, 1980).

phytochemistry is equally rich, yielding secondary metabolites with uses in both pharmacology and chemistry. With the current focus on the preservation of biodiversity, the opportunities for the controlled and reasoned exploitation of hemp's fibre and molecular potential are clear for all to see.

3.5   Part III: The Vegetative Cycle

3.5.1   Hemp and photoperiodism

Discovery of the phenomenon

As early as the end of the 19th century, a number of authors remarked on the anomalous sex ratio of hemp plants or on the anomalies identified in the organogenesis of hemp plants when vegetating under environmental conditions that differed from those of normal cultivation. The tendency of most authors was to seek the explanation in the ground and in differences in the composition and nutritional properties of the soil. Generally speaking, observations made on a number of species suggested that a 'substance' might be acting to control a number of key vegetative phenomena: including germination, growth

and flowering. These were thought to be influenced by changes in day length.

Starting in 1910, the Frenchman, Julien Tournois, undertook a series of experiments that changed the focus of research in this field. Working initially on the hop (*Humulus*) plant, he discovered that when this plant was grown in the winter, protected by a cold frame, the different developmental phases were shortened and the final flowering phase occurred more rapidly. Extending his study to the hemp (*Cannabis*) plant, he again observed a significant reduction in the length of these developmental phases.

After undertaking various checks, and taking care to compare the cultivars in winter and in summer, with the latter being deprived only of light, Tournois was able to conclude that day length was the only environmental factor responsible for the observed phenomena (Tournois, 1912). He is therefore credited with the discovery of photoperiodism, which is later explained by Garner and Allard (1920), following their work on soybean and tobacco.

Further experimental work by Garner and Allard allowed them to propose a classification system of plants according to their response to day length:

- Long-day plants (hemeroperiodic) – long day, short night produces flowering.

- Short-day plants (nyctoperiodic) – short day, long night produces flowering.
- Day-neutral plants (aperiodic) – indifferent to day length.
- Intermediate plants (amphiperiodic) – requiring a stricter set of conditions. This group was added by Allard (1938).

Hemp, like the majority of tropical plants, belongs to the short-day plant category.

Influence of photoperiod on hemp

The principal effect of photoperiod on hemp is on the flowering of this plant. More specifically, the photoperiod influences the end of the development of the inflorescences, and therefore the opening of the last female flowers. At this point, it can be said that the hemp is fully flowered.

It has been observed experimentally that growth is arrested as soon as the last female flowers are fertilized. This fertilization takes place 1 week after these last female flowers open, and terminates the flowering stage.

The yield in hurds is thus under the strict control of the photoperiod (night length). It is therefore influenced directly by the date of sowing and the latitude, as these two factors influence the photoperiod. The variety of hemp also influences the sensitivity to light, and this factor can therefore also influence the yield.

The start of flowering, corresponding with the appearance of the first female flowers, is also influenced by the photoperiod. This stage is termed the beginning of flowering.

This phenomenon is still poorly understood. It is more complex in nature and is termed photothermoperiodism.

Experimental studies have demonstrated that, as nights lengthen, temperature requirements are reduced:

- Summer nights are short and warm, resulting in flowering occurring more rapidly.
- The nights at the end of spring are long and cooler, resulting in slower flowering.
- The nights at the start of spring are short and cool, resulting in a more unpredictable flowering pattern depending on the climate.

The interaction of these two effects can have specific consequences on hemp cultivation (Table 3.2). No other developmental stage

Table 3.2. Agronomic consequences of photoperiodism.

Timing of sowing	Active growth during days (length)**	Thermal requirements for flowering	Start of flowering	Photoperiodic requirements to bring about the end of flowering	Stage of full flowering
Normal* (15 April–15 May)	Max.imal (end of June)	Maximum	Standard – start of July	Standard (example/ norm)	Standard (start of August)
Late (1 June)	Shortest (start of July)	Weak	Slightly later, although more or less the same	Identical	Identical (to the example/ norm)
Very late (15 June)	Shortest (mid July)	Very weak	Later	Identical (although flowering is poorly developed)	Later
Early (1 April)	Shortest (start of May)	Weak	Early	Standard	Identical (to the example/ norm)
Very early (20 March)	Very short (end of April)	Very weak	Very early	Standard	Identical (to the example/ norm)

* We consider the 'normal' sowing date to require a qualified standard.
** Active growth starts about 4 weeks after sowing.

of hemp is known to possess sensitivity to photoperiod.

3.5.2 Vegetative cycle and growth

The vegetative cycle of hemp is relatively simple to describe, occurring in a few steps, with each step being highly dependent on a number of specific environmental factors.

The vegetative cycle (Fig. 3.16) consists of the following.

Emergence

Emergence can be expected to occur approximately 100°C days, or 4–10 days, after sowing, depending on the soil conditions. This is the most vulnerable phase of cultivation, during which it is possible to lose large numbers of plants, either because they have received too little or, more commonly, too much water.

Implantation (3 weeks after emerging from the ground)

This phase of growth is the slowest part of the cultivation process and corresponds to the extension of ground cover. During this time, the root system is developed and preparations are made to explore the ground further: 3 weeks (25°C days after emergence) are needed for this to take place and for the correct density to be established. By the end of this stage, the plant will have

grown to 30–60 cm high and will have three pairs of leaves.

Active growth

This period stretches from the time of implantation to the start of flowering. It is during this phase of growth that the yield in hurds and fibre is generated and that the water and mineral requirements are greatest and must be met. Providing all these needs are met, growth can attain 1 t of dry matter every 120°C days. This corresponds to a growth rate of 1 t of dry matter/week in an area with a climate similar to that in northern France.

Flowering

The appearance of the first flowers in a hemp crop is dependent on the sum of the temperatures experienced since sowing and by the photoperiodic sensitivity of the variety. During this period, needs are maximal and the seed yield starts to be determined.

Full flowering

This stage corresponds to the opening of the last female flowers and is under the strict influence of the photoperiod. For any given variety grown, cultivated in any given location, with the exception of very late crops, this stage will be attained at a fixed date regardless of the date of sowing and independently of annual differences.

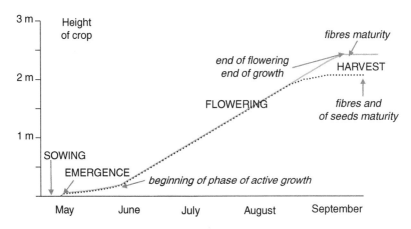

Fig. 3.16. Vegetative cycle of hemp.

End of flowering

The end of flowering is defined by the fertilization of the last female flowers. This stage marks the end of the growth phase and determines the yield in hurds, as well as inducing the subsequent phases of hurd and seed maturation.

The physiological maturity of the seeds is obtained approximately 40 days after full flowering. The timing of the harvest will, however, need to take into account the humidity of the seeds and the capacity of the driers. These factors may result in harvesting being delayed (see Chapter 6, Hemp cultivation in France and Chapter 8, Agricultural Economics). In fact, seed humidity levels are very dependent on the state of cultivation and, in particular, the presence of leaves. This last factor is dependent on the hydration of the crop and its nitrogen fertilization.

Calculating crop yields

The potential hurd yield of a hemp crop can, under normal circumstances, be calculated using the following formula:

$$\text{Potential hurd yield (t MS/ha)} = (\Sigma \text{ temperatures} - 100 - 250)/120$$

The sum total of the temperatures is to be calculated from the day the seeds are sown until the variety has finished flowering. For any given varietal at a known latitude (zone of cultivation), this date will be known and predictable.

NB: The '100' corresponds with the sum of the temperatures required at emergence. During this period, there is no growth of the crop (that is to say – no yield in agricultural terms).

The '250' corresponds with the sum of the temperatures recorded during the implantation phase (growth of the root system and development of the first three pairs of leaves). During this time, the agricultural growth and yield is poor and it is possible to ignore the small quantity of hurd produced.

The active growth starts at the end of the implantation phase and it is only at this point that the sum of the temperatures starts to influence yield.

The agricultural yield must be increased by 15% to take into account the average humidity levels and the payment schemes that reflect this.

Note

The majority of molecules described in the text are indexed, together with their bibliographic links, in the Pub Chem database of the National Center for Biotechnology Information (NCBI, USA; http://pubchem.ncbi.nlm.nih.gov/).

4 Genetics and Selection of Hemp

Janoš Berenji,[1] Vladimir Sikora,[1] Gilbert Fournier[2] and
Olivier Beherec[3]

[1]*Institute of Field and Vegetable Crops, Novi Sad, Serbia;* [2]*Faculty of
Pharmacology, University of Paris-Sud, France;* [3]*FNPC, France*

4.1 Introduction

In order to deal comprehensively with the subject of hemp genetics and selection, Dr Vladimir Sikora and Dr Janoš Berenji, Director of the Institute of Field and Vegetable Crops, Novi Sad, Serbia, present the principles of genetics with a particular emphasis on hemp.

Following this, Professor Fournier, Director of the Laboratory of Pharmacology of the Faculty of Pharmacology at the University of Paris-Sud, together with Olivier Beherec, Technical Director of the FNPC, present in detail the hemp selection methods currently used in France.

4.2 Genetics: A Review[1]

4.2.1 Fundamentals of plant genetics and the varietal selection of hemp

What is genetics?

Founded on a study of heredity and its variations, genetics provides answers to two fundamental questions:

- What are the mechanisms that allow parents to transmit characteristics to their offspring?

- What is the role of genetic material in the creation of a new and unique individual? How does this individual then produce a further new generation?

This chapter provides a recap of the essentials of plant genetics necessary for the reader to develop an understanding of hemp genetics and reproduction.

DNA, genes and chromosomes

During the 19th century, three important developments in the field of biology prepared the way for modern-day genetics.

1. Charles Darwin published *The Origin of the Species*.
2. The internal structure of animal and plant cells started being studied.
3. Mendel published the results of his experiments in plant hybridization.

Mendel interpreted the results of his crossings of sweet peas based on the premise that heredity had a particular base that allowed physical characteristics to be passed, without alteration, from one generation to another. The theory that Mendel developed is today called Mendelism. The intensive study of cells at the start of the last century yielded the discovery of chromosomes; that is to say, the hereditary material found in cell nuclei that is able to self-duplicate. The coming together of cytology and Mendelism in the early 20th century can be viewed as the start of classical genetics. This period saw the theory of chromosomal heredity develop and progress made in the understanding of the mechanisms of heredity of different characteristics in different species.

The discovery of nucleic acids, the constituent ingredients of DNA (deoxyribonucleic acid) and essential components of chromosomes, heralded the era of molecular genetics. DNA is made up of two chains rolled into a double helix, each made up of a series of nucleotides that interlink each chain. The four nucleotides[2] of DNA contain the bases adenine (A), guanine (G), cytosine (C) and thymine (T). According to the rule of matching pairs, bonds are only established between complementary bases: A with T and G with C. In this way, the nucleotide sequence of each strand can be determined from that of its opposite. This is the basis of DNA replication (Fig. 4.1).

A gene is the fundamental functional and physical agent of heredity. It is a sequence of nucleotides, i.e. a segment of DNA situated in a particular location on a chromosome and coding for a specific characteristic. In other words, the gene is the biological unit that transmits hereditary information from parents to their offspring, thereby controlling the appearance of that trait in the new organism.

Qualitative characteristics such as flower colour are coded by only one or several genes. Many other characteristics of agricultural significance, such as yield, precociousness, etc., are quantitative characteristics and are mediated by the simultaneous interaction of a number of genes. The transmission of genetic information appears basically to be the same in all organisms. DNA replication (and therefore chromosome replication) yields two identical copies and allows the transmission of a carbon copy to the next generation during the process of cell division. The separation of the two copies of duplicated chromosome during cell division occurs in such a way that the daughter cells receive an identical copy of each chromosome from the mother cell. This process forms the basis of the development that allows individuals to grow in height and complexity.

At certain stages of the life cycle, individuals reproduce and give rise to new individuals of the same type. *Vegetative*, or asexual, reproduction is based on mitotic cell division. New plants are generated from parts of the reproducing parent plant. The majority of plants reproduce by sexual reproduction, during which new individuals are created from male and female parent plants: by the union of male and female gametes. Gametes are specialized sex cells. The male gamete, or pollen, fuses with the female gamete (ovule) to produce a zygote. Meiosis is the process of cell division in which the number of chromosomes is halved, such that there are half the number of chromosomes in a reproductive cell (gamete) compared to a somatic (non-reproductive) cell. Somatic cells are diploid and contain 2n chromosomes. The number of chromosomes in diploid hemp is 2n = 20. Gametes are haploid cells. Haploid hemp therefore has n = 10 chromosomes. Triploid hemp (3n = 30) and tetraploid hemp (4n = 40) have been created, but does not possess any advantages over the various forms of diploid hemp.

A population is a group of individuals of the same species living in a defined area. The individuals within a population can reproduce sexually by means of the sexual organs of the opposite sex.

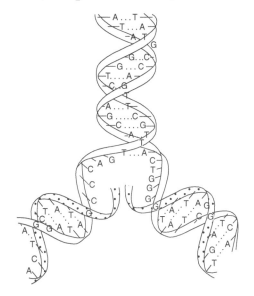

Fig. 4.1. DNA replication. The nucleotides, made up of the bases indicated by the letters A, G, C and T, provide the framework and allow the replication of the double helix.

Genotype and phenotype

The genetic make-up of an individual is termed its genotype. It relates to the sequence of DNA nucleotides carried by the chromosomes. The phenotype describes the physical appearance of the organism and is the result of the combined effect of the genes and the environment in which the organism lives. It is difficult to determine with any precision what are the respective contributions of the genotype (G) and the environment (E) from an examination of the phenotype (P) of the individual. Where qualitative traits are concerned, the contribution of the environment is negligible (P = G). The quantitative characteristics resulting from the interaction of a number of genes can be influenced strongly by environmental factors (P = G + E). The work of plant breeders consists of the development of the genotype of a cultivar.[2] The cultivation techniques used by farmers will need to take into account environmental factors (such as soil, fertilization, light, etc.) on an ongoing basis in order to minimize any divergence between the genetic potential of the plant and the actual performance of the cultivar under production (Fig. 4.2).

Plant genetics and breeding

One of the most important practical applications of genetics is to be found in the selection of plants and animals. Plant selectors use their knowledge of plant genetics and apply this to the development of cultivars in order to meet the needs of farmers, as well as those of the end user of the plant crop. Plant selection is both an art and a science and involves the development of plant genotypes. Humans have, throughout modern history, at first unconsciously and then consciously, domesticated and then selected plants according to their needs. Today, the hemp cultivars are descendants of the hemp that was first cultivated by humans. Their ancient work was conducted as an art, selecting plants empirically with great success. With the discovery of Mendelism, classical genetics was soon applied in plant selection programmes. For the first time, plant selection was conducted scientifically. Recently, molecular genetics has contributed to the development of plants using modern

Fig. 4.2. Plants of the same genotype (Novosadska Konoplja) cultivated at a low sowing density (left) and a high sowing density (right) showing the importance of the environment on plant architecture (phenotype).

biotechnological methods, such as genetic engineering, to produce genetically modified organisms (GMO) or transgenic plants.

Various aspects of industrial hemp breeding are discussed by Hoffmann *et al.* (1970), Berenji and Sikora (1996), Bócsa and Karus (1998), Bócsa (1999), Finta-Korpel'ová and Berenji (2007), etc.

Dempsey (1975) listed 18 institutes involved in industrial hemp research in the USSR (6), Eastern Europe (6), Western Europe (4) and Asia (2).

Some of the frequently mentioned institutions involved in industrial hemp breeding are:

- Fleischmann Rudolf Research Institute of the Károly Róbert College in Kompolt, Hungary (formerly the 'Fleischmann Rudolf' Agricultural Research Institute)

- Hemp Breeding Department at the Institute of Bast Crops, Glukhov, Ukraine
- Fédération Nationale des Producteurs de Chanvre – FNPC, Le Mans, France
- Institute of Field and Vegetable Crops, Novi Sad, Serbia
- Institute of Natural Fibres, Poznań, Poland

4.2.2 The selection of hemp

Reproductive biology of hemp

The selection techniques in use are determined by the reproductive biology of the hemp plant. Hemp typically reproduces by sexual reproduction but, in certain cases, parts of the plant can be used for asexual reproduction. Hemp is naturally dioecious. Dioecious hemp presents as two types of plant, one with female flowers and the other with male flowers (Fig. 4.3). The ratio of sexes (the number of male to female plants) in a dioecious population is close to 1:1. The male and female plants demonstrate

morphological differences and are therefore described as sexually dimorphic.

An important stage in the development of hemp was the discovery of monoecious plants. Monoecious hemp exhibits the same structure of male and female plants as the dioecious plant, except that both male and female flowers are found on the same stem (Fig. 4.4). Monoecious hemp with male and female flowers on the same plant can occur spontaneously in dioecious hemp populations/cultivars in a very low ratio, usually less than 10–20 plants/ha. According to Hoffmann (1938), this frequency is 0.1%. The phenotype of monoecious hemp plants is very similar to that of the dioecious female plants. The monoecious form can be defined as an artificial form of hemp because, even without human interference, it returns progressively to the dioecious state.

Several transitional stages may also be observed between these two basic sexual forms of hemp (dioecious and monoecious).

Dioecious hemp results entirely from cross-pollination: the seeds produced by female plants are the result of pollination of the seed

Fig. 4.3. Sexual dimorphism of dioecious hemp. Male flowers (top left) and female flowers (top right). The appearance of the two plants can be compared side by side (bottom).

Fig. 4.4. Monoecious hemp showing female (a) and male (b) flowers on the same stem.

by pollen from male flowers growing nearby or close enough to allow wind to carry the pollen. Monoecious hemp undergoes 20–25% self-pollination, but is usually treated by selectors as the product of cross-pollination.

The vegetative multiplication of hemp is practised in only a few special cases. A group of plants may be produced from a single plant, giving rise to a number of clones. This procedure is called cloning (Fig. 4.5). Plants issuing from the same clone are genetically identical and will demonstrate the same characteristics as the parent plant. In practice, vegetative multiplication is used only for reproduction in greenhouses, often for drug production or medicinal purposes.

The recently developed cultivation of hemp tissues can also be viewed as a method of vegetative reproduction.

Selection methods for hemp

The methods suitable for cross-pollination have been adapted to hemp selection (Hoffman *et al.*, 1970; Berenji and Sikora, 1996).

MULTIPLICATION. Selection is a process that allows the proportion of desirable genotypes to be increased. Successive generations are subjected to a process of selection that chooses those plants that correspond best to the prescribed objectives. Selection has become the most elaborate form of plant creation. The creator must work with the materials at his or her disposal and will select those characteristics that are best suited to his or her needs.

Cultivated forms selected for production and described as 'indigenous' have evolved from their original wild hemp ancestors, undergoing a series of successive selections.

Mass selection remains one of the most ancient procedures used for the selection of hemp. This method was applied in such a way that all the seed stock derived from selected plants was sown in order to prepare a new generation. This process was repeated over several years. Farmer-selected landraces originally evolved from wild hemp populations by mass selection. At the very beginning of hemp breeding, cultivars were obtained from landraces by selection, resulting in improved landraces (Hoffmann, 1961). In Italy, Carmagnola was improved by mass selection, leading to the development of the cultivars Bolognese, Toscana and Ferrarese (Ranalli, 2004). A similar scenario was described for the breeding of Novosadska konoplja (Berenji, 1992). Mass selection is still used for the maintenance of registered hemp cultivars.

Fig. 4.5. Cloning hemp.

Individual selection is practised by separating the descendants of each plant selected from a given population. The value of a selected plant is evaluated by testing its descendants. Individual selection is basically a selection of genotypes and is capable of delivering significant genetic improvement compared to mass selection, which is essentially a form of phenotypic selection. The efficiency of the selection process is, however, rendered more difficult by the fact that many of the important traits for selection (e.g. hurd yield, fibre percentage, quality, etc.) can only be measured and evaluated after flowering. This signifies that a mixture of pollen from compatible flowers and non-desirable flowers contributes to the pollination of selected plants. The control of cross-pollination has been undertaken using the Bredemann technique (Bredemann, 1942), named after German creator who used it for the selection of hemp. However, undesirable plants will not have been identified and removed and could therefore pollinate the selected plants.

The Bredemann method is based on the determination of the fibre content of the male plant stems, generally measured on a cut branch before the opening of the flowers. Only those male plants with a high fibre content will be allowed to release their pollen and pollinate female plants. The seeds obtained from those females, themselves with a high fibre content, will then be used to prepare the next generation.

Genetic markers identifying specific sections of DNA closely associated with certain traits can help give an idea of the plant's genotype and aid in selection (marker-assisted selection) (Mandolino and Ranalli, 2002). Genetic markers have been identified for female characteristics by Shao and Shong (2003), for male characteristics by Törjék *et al* (2002) and for cannabinoid profiling as well as for other interesting traits.

RECOMBINANT SELECTION. As the variability of indigenous species has been exhausted by selection, new genetic variations for more elaborate selections have been obtained by selected crossings of parent plants carefully. In combination selection, a recombination of parental genes with a unique genotype is undertaken to provide a base for further desirable genotype selection. As the precise performance of each individual and all of its descendants is preserved throughout the whole selection process, combination selection is often called genealogical selection.[3]

SELECTION BY HYBRIDIZATION. the selection of hybrids is founded on the use of the phenomenon known as heterosis, or hybrid vigour. Hybrid vigour describes the increase in size, vigour, productivity and general performance of hybrids relative to the parent lines or race. Bócsa (1954) was one of the first to create populations of hybridized hemp.

Cultivars do not completely transmit all their characteristics to their descendants. Instead, they establish divisions within the next generations. As a result, farmers find themselves having to buy certified hybrid seed stock from breeders every year if they wish to benefit from the advantages of hybrid cultures. Until the use of monoecious hemp,[4] the industrial production of hybridized hemp was unknown. Dr Ivan Bócsa from the Fleischmann Rudolf Agricultural Research Institute in Kompolt, Hungary, was the pioneer in the development and practice of this phenomenon.

Unisex plants have their origin in the crossing of dioecious female plants with monoecious parent plants. The first generation, after this initial crossing, is called unisex and consists of 70–85% female plants, 10–15% monoecious plants and only 1–2% male plants. Manual elimination of all the male plants, originating from dioecious female plants, before pollination allows the chance of crossings to be controlled. The crossing is then undertaken again in order to improve the yield in certified seeds, which can then be used as seed stock for cultivations destined for commercial use. The majority of French cultivars belong to this category.[5]

SELECTION BY MUTATION. Mutations arise suddenly in the genes and chromosomes and heredity changes unpredictably. Depending on the cause, mutations may be spontaneous or may be provoked. The unpredictable nature of engineered mutations makes their use on a large scale difficult as a means of selection.

Spontaneous mutations frequently occur in the somatic tissues of hemp and are situated in identifiable parts of the plant (chimeras; Fig. 4.6).[6] These mutations do not, however, transmit themselves to the next generation.

Where selection is concerned, only the spontaneous mutations arising in the germ cells are used. For example, monoecious hemp is considered to have arisen as a result of a spontaneous mutation in dioecious hemp.[7] Another example is that of the phenomenon described as 'yellow stems' recognized and used by hemp breeders.

TRANSGENIC HEMP. Recent developments in the science of molecular genetics have resulted in increasing sophistication in the techniques proposed for plant selection. Recombinant DNA technology allows a specific gene from an organism's DNA, coding for a desirable characteristic, to be isolated, cloned and inserted into the genetic material of a plant. Feeney and Punja (2003) claim to have inserted the gene coding for *isomerase phosphanamose* successfully into hemp using the bacterium *Agrobacterium tumefaciens* as a vector. This transgenic hemp was of no commercial value, but its creation demonstrated that hemp might be added to the list of plants that could be modified genetically where necessary.

Objectives and results of hemp selection

Hemp is a multi-purpose crop of increasing importance for Europe and other parts of the world (Berenji, 1996, 2004). When hemp is used as a fibre crop, the main objectives of selection are: hurd yield, fibre content and fibre quality. Research has also been conducted to increase oil production with the development of a hemp variety with this characteristic. More recently, hemp selection to meet these dual needs has been prioritized. Specific selection programmes for 'drug varieties of hemp' in order to produce marijuana or hashish as well as medicinal substances have also been undertaken. The production of ornamental hemp is yet another example.

The objectives detailed below concern both monoecious and dioecious hemp. It would appear, however, that dioecious hemp lends

Fig. 4.6. Yellow chimera of hemp leaves, arising from a spontaneous mutation in the somatic cells of the leaf.

itself better to the production of fibres, while monoecious hemp is better suited to the 'dual need' of both seed and fibre. Watson and Clarke (1997) provided an excellent overview of the objectives of hemp breeders, reproduced below.

HURD YIELD. Late-flowering cultivars produce higher stem and fibre yields than early-flowering cultivars, due to a longer growing season (van der Werf *et al.*, 1994). Date of flowering was highly heritable and had a significant relationship with stem yield (Hennink, 1994).

Hybrid cultivars expressing heterosis contributed significantly to the increase of genetic yield potential for stem yield.

FIBRE YIELD. Since the beginning of fibre hemp breeding, bast fibre has more than doubled from about 12–15% to 25–35% (Heimann, 1990, cited in Hennink, 1994). As a result of selection for fibre content, contemporary dioecious hemp cultivars have a genetic potential of 38–40% fibre content in the stem, resulting in 28–30% fibre content in large-scale production (Bócsa and Karus, 1998; Berenji and Sikora, 2000).

Simultaneous selection for bast fibre content and stem yield is the best breeding strategy for improvement of bast fibre yield (Hennik, 1994).

FIBRE QUALITY. The first selection studies modified the ratio of primary fibres to secondary fibres successfully in favour of the latter. Despite the fact that fineness, resistance to traction, etc., are important aspects of fibre quality, it is generally recognized that the cultivation and fibre processing techniques have more potential for improving fibre quality than selection alone (Müssig, 2003). This is an example of the important role played by the environment, compared to the genotypic potential, in the phenotypic expression of fibre quality for commercial hemp production.

CANNABINOIDS. Respecting the increasingly rigorous requirements imposed by legislation (Berenji, 1998), low tetrahydrocannabinol (THC; < 0.2%) and even THC-free hemp cultivars were bred primarily by Ukrainian and French hemp breeders (Beherec, 2000). Illegal, predominantly amateur plant breeders produced marijuana hemp with very high THC levels. There is an increasing demand for adjusting the cannabinoid content and composition, ranging from predomination to complete elimination of certain cannabinoids from the plant, which is especially important in the case of pharmaceutical hemp (Berenji and Sabo, 1998). Pharmaceutical hemp cultivars have been bred in the Netherlands by HortaPharm BV and licensed to GW Pharmaceuticals Ltd for commercial production.

HEMP FOR OIL PRODUCTION. An increase in the yield of oil is of great importance for oil hemp as it must compete with other plants capable of delivering far more oil. It has been observed that 70–90% more oil is produced from monoecious hemp than from dioecious varieties. The use of intravarietal selection for an increase in seed yield, including an increase in oil content, was begun in 1997 by Bócsa et al. (1999).

The Finnish cultivar Finola, originally known as FIN-314, is cultivated specifically for seed production and oil extraction. In its natural environment in Finland, the oil content of this short, early maturing, frost-resistant cultivar is 35% (Callaway, 2002). According to Scheifele (2000), the cold-pressing extractable oil content of this cultivar is 27%. The seed oil of this cultivar contains high levels of the essential fatty acids, gammalinolenic acid (4.4%) and stearidonic acid (1.7%), which are valuable as nutraceuticals and dietary supplements (Laakkonen and Callaway, 1998).

The Hungarian cultivar Fibrol has an extractable seed oil content of over 35% (Bócsa et al., 2005). There are breeding attempts to modify the fatty acid composition of hemp seed oil (Finta-Korpel'ová, 2006).

Hemp cultivars

The majority of hemp cultivars are of open pollination and are both monoecious and dioecious, much as the hybridized cultivars based on the principle of monosexuality. De Meijer (1995, 1999) published an extensive study on commercial industrial hemp cultivars, including French, Hungarian, Polish, Romanian, Ukrainian, Russian, Italian, Serbian, German and Far Eastern cultivars, as well as drug strains.

All new cultivars must be *unique* (clearly distinct from all existing cultivars), *homogenous* and *stable* in their essential characteristics. DHS[8] tests are designed to ensure that these requirements are met. A compulsory prerequisite for all new cultivars is that they demonstrate improved performance relative to existing varieties for the traits of interest. In order to recognize the agronomic value of a new cultivar, its traits, and therefore its genetics, must be preserved post-selection (Fig. 4.7). With hemp, the operation to preserve these traits during the selection process requires as much effort as the operation devoted to the creation of a new product. The most commonly employed method for preserving the genetic identity of a cultivar is mass selection.

Production of certified seed stock

The commercial seed industry consists of all the people, businesses, seed companies and commercial enterprises involved in the production and commercialization of certified seed for farmers. During the production of seeds, the identity and purity of cultivars are evaluated in the field. The male/female ratio is monitored closely, as is the THC concentration, aspects that are uniquely characteristic of the hemp

Fig. 4.7. During the operation to conserve hemp cultivars, the isolation of specific genotypes from other genotypes is often undertaken.

seed industry. The control of seed properties (humidity, purity, germination, sanitation) both before and after the multiplication of the seed stock and before its conditioning allows the necessary information to be gathered for certification. Post-control tests involve the sowing of a sample from each batch in order to validate the controls undertaken in the previous season. From this point of view, the control of the sex of the plant is particularly important.

Over the past few years, the production of certified seed of organic hemp has become a challenge. A lot of attention is directed towards the delivery of certified seed stock that not only is of constant quantity but also of consistent quality and price.

Genetic resources used in the selection of hemp plants

Germplasm – the genetic resources a breeder employs to create new varieties – may be drawn from indigenous seed stocks, from new or ancient cultivars, or from populations created by breeders, or from special genetic stocks such as those carrying mutations. Overuse of only a few varieties, together with natural attrition, can lead to the loss of germplasm. It is vitally important that plant genetic resources be preserved assiduously so breeders can continue to develop improved cultivars into the future. This is true of all plants, not just hemp (Berenji et al., 1997). The effort to collect, classify and conserve hemp germplasm from all over the world led to the creation by N.I. Vavilov of the enormous collection held at the N.I. Vavilov Institute of Plant Industry in St Petersburg, Russia, consisting of 397 entries covering a tremendous genetic range (Kutuzova et al., 1997).

4.3 The Selection of Hemp

The creation of varieties and the production of seed (both for cultivation and industry) are complementary activities that should never lose sight of their common objectives. They both aim to satisfy the demands of the market, in terms of both the quality and quantity of seed produced. Both strands of production must therefore work closely together.

Depending on the species, these two production activities may be assured by one enterprise or by two distinct operations. The hemp industry is relatively small and its constituent parts are integrated and assured by a small number of operators (Fig. 4.8).

4.3.1 Objectives in the creation of varieties

The situation that hemp cultivation in Europe currently finds itself in is a direct result of the events of the early 1960s, at which point hemp almost disappeared from the countryside and was saved only by a vigorous genetic (as well as an agricultural and mechanical) response.

At this time, the rural exodus deprived the land of the workforce that previously would have assured the traditional harvesting of hemp. The harvesting process, including cutting, drying and pressing, had to become mechanized.

The traditional dioecious varieties were unable to cope with this change and a considerable amount of work was conducted in order to create monoecious varieties that would be earlier maturing and better suited to mechanized cutting.

The same rural exodus and the arrival of alternative exotic natural, as well as synthetic, fibres led to the disappearance of many traditional textiles. The paper industry, which had been using hemp (in the form of worn-out fabrics) as a raw material, developed its own cultivation programme in order to meet its own needs in cellulose fibres.

The declared needs resulted in an increase in cellulose fibre yields per hectare. These demands therefore translated into an optimum hurd yield and a maximal total fibre concentration.

The existence of a market for hemp seeds has also resulted in adaptations to the varieties of fibre-producing plants so that they have become earlier maturing, allowing the seeds to mature simultaneously. This reflects the challenges of producing a crop with two different products (seeds and fibre).

The absolute prohibition of the cultivation of hemp, for public health reasons, in a number of countries is due to the presence of the psychotropic chemical delta-9-THC. In countries where cultivation is permitted, the content of this chemical is strictly controlled and has led to specific selection programmes that have sought to reduce or eliminate this chemical from cultivated varieties.

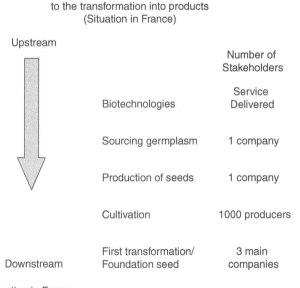

From the creation of hemp varietals
to the transformation into products
(Situation in France)

		Number of Stakeholders
Biotechnologies		Service Delivered
Sourcing germplasm		1 company
Production of seeds		1 company
Cultivation		1000 producers
First transformation/ Foundation seed		3 main companies

Upstream
Downstream

Fig. 4.8. Hemp production in France.

Finally, the difficulties subsequently encountered in the development of varieties led to a refocusing of work, in particular towards monoecious varieties. More specifically, work concentrated on the monoecious trait in order to try and eliminate masculinized monoecious forms.

and female plants differ both morphologically and physiologically. The vegetative phase of the male plant is shorter by approximately 6 weeks than that of the female plant. This sexual dimorphism results in a crop that is not uniform, demonstrating marked differences in plant maturity (Fig. 4.2).

4.3.2 Biotechnology

Given the disproportionately heavy investment required for biotechnological work to be conducted in the hemp industry, relative to the size of the industry, it is not surprising that little such work has been undertaken. The results of this work will be presented in Chapters 5 and 6 of this book.

There have been no studies to date seeking to produce genetically modified hemp, though it is technically possible.[9] There are therefore no GMO varieties of hemp.

Table 4.1. Comparison between monoecious and dioecious varieties – D19 is dioecious; Fedora 19 is monoecious. (i) Identical maturation. Superior yield seen in monoecious varieties. (ii) Other advantages: mechanical harvesting facilitated; homogenous crop obtained. Data from the FNPC.

	PSP	EC (%)	PSE	PG
D19	7.77	38.42	2.99	6.32
Fedora 19	9.18	37.29	3.42	9.73

Note: PSP = yield in dry stems (t/ha); EC (%) = percentage of bark on the stems; PSE = yield in bark (t/ha); PG = yield in grain/seeds (100 kg (ql)/ha).

4.3.3 Selection criteria definitions

The monoecious varieties

RECAP OF THE BACKGROUND TO THIS SUBJECT. The disappearance of the labour force necessary for the harvesting of hemp led to its mechanization. Both the production of agricultural seed and the industrial cultivation of hemp fibre were affected by this change.

- Seed production: saw a 35–40% reduction in seed production as there was no equipment available for the harvesting of seeds.
- Industrial hemp production: inferior yields (Table 4.1) seen due to the heterogeneity of the harvested materials and the absence of harvesting equipment. This led to the harvesting of plants during the flowering period, before the male plants aged and dried and well before the maximum yield for hurd and fibre was attained.

SEX OF THE PLANT. Hemp is naturally dioecious. That is, the male and female flowers are not found on the same plant (Fig. 4.9). The male

Fig. 4.9. Monoecious hemp.

Within dioecious varieties, a small proportion of naturally occurring plants will be found bearing flowers of both sexes. The incidence will vary between one plant in a million and one in a thousand. These plants are termed monoecious.

MONOECITY. The first monoecious plants to be recognized within a dioecious line were identified by the German researcher Bredeman, between World War I and World War II, and subsequently by Von Sengbusch and his collaborators. These plants do not show heterogeneity when the sexes occur separately and yield a greater quantity of seeds, as all the plants are fruit bearing.

The expression of monoecity varies between individuals in relation to the number of male and female flowers borne by each plant. A classification system was proposed by Neuer and Von Sengbusch in 1943 to describe these types:

1. From 1+: one cross, monoecious plant with a majority of male flowers, and therefore heavily masculinized.
2. To 5+: five cross, monoecious plant with a majority of female flowers, and therefore heavily feminized.

The complexity of the genetics behind these characteristics has not, however, allowed a homogenous sexual phenotype to be conserved across the generations.

DIFFICULTIES ARISING FROM THE CREATION AND PRESERVATION OF MONOECIOUS TYPES. The first problem to be solved was that of eliminating the male plants at the time of seed increase in order to avoid a rapid return to a dioecious state, or one with a high proportion of male plants. To this day, the problem persists, although these male plants can be managed and are being kept in check.

The second problem to be addressed was that of creating and maintaining monoecious varieties with a high expression of female flowers. This is important because masculinized varieties produce less fruit and are more prone to fungal parasitism with *Botrytis*.

That said, studies have shown that the monoecious state is highly variable and not very stable. It is very sensitive to pedoclimatic factors. Photoperiod has a feminizing effect on plants when days are short, and nitrogenous fertilization has been shown to have a masculinizing effect.

Different strategies have been introduced by breeders to obtain and maintain this characteristic:

1. Varieties have been created in which 100% of the population is monoecious. All the monoecious types are represented, including the masculinized lines. The sexual expression of these plants varies according to pedoclimatic factors. Developed by all researchers working on the monoecity of hemp, these varieties are still used today in Poland and Ukraine.
2. Varieties have been created in which the population is mixed, consisting of a mixture of female and monoecious plants. The production of seed and seed stock is essentially assured by the female plants and, to some degree, by the feminized monoecious plants, while pollination is assured by the monoecious plants. This variety allows the impact of pedoclimatic factors on sexual expression to be managed and the proportion of masculinized monoecious plants to be kept small. This practice was developed in France in the 1960s, in parallel with the original practice, namely that of the dioecious–monoecious hybrids.

Today all varieties developed and cultivated in France are mixed, consisting of a mixture of female plants and monoecious plants.

THE DIOECIOUS–MONOECIOUS HYBRID OPTION. Faced with the challenge of producing and maintaining monoecious varieties, research focused on an original solution. This work was conducted in collaboration with Professor Bócsa, of Hungary, and sought to hybridize dioecious female plants with monoecious plants. This hybridization produced a novel F1 population made up of a very high proportion (>95%) of female plants.

The difficulties encountered in trying to produce seed stock of these varieties, and, in particular, the need to eliminate the entire population of dioecious male plants manually, meant that this variety could not be distributed widely.

A number of techniques were found that allowed the peculiarities of these variants to be used inexpensively.

The dioecious varieties

Today, across Europe, few breeders work on dioecious varieties. We know only of the Hungarian and Serbian programmes that work exclusively with these varieties, while the Italian programmes make use of a mixture of dioecious and monoecious strains.

Maturity and its relation to environmental conditions

Stem productivity is the essential criterion in hemp breeding programmes. It varies with the duration of crop growth and the variety grown.

The end of the growth period corresponds with the end of the flowering period (see Chapter 3), which itself depends on the plant's photoperiodic response. With days shortening during the summer months (and nights lengthening), early-maturing plants require shorter nights than late-flowering plants.

The following equations describe the relationship:

Early plant = needs short night = short growth phase = reduced stem growth.

Late plant = needs long night = long growth phase = tall stem produced.

The breeder must work to ensure that the photoperiodic requirements within a variety of plants are homogenous. Similar requirements will ensure that there is no spread of maturity among plants and that yields are optimized.

The breeder needs to know where to set the limits according to the desired maturation. If there are too many early flowering plants, the hurd (and therefore fibre) yield will suffer. If there are too many late flowering plants, they may not be harvestable or be harvested too early.

The mechanical harvesting techniques in use require that the crop be dried after cutting. This takes place in the fields, leaving crops vulnerable to bad weather. In France, they must therefore be harvested no later than the month of September, when temperatures fall and rainfall increases.

An interesting aspect of hemp's sensitivity to short days is that maturation is adapted to the latitude at which they are cultivated. In effect, a variety responding to a night length of 8 h, for example, will finish its growth earlier in Italy than in France or in Sweden.

Certain breeders have therefore suggested moving the production of seed southwards, away from the zones of cultivation, in order to produce late-flowering varieties that will have a higher hurd and fibre yield. As indicated previously, the lateness of flowering can still pose problems with regard to harvestability, and this must be limited.

Today, those varieties developed and cultivated in France reach full flowering (at the latitude of Le Mans) between 31 July and 15 August. Seed of the earliest varieties will reach physiological maturity between 15 and 25 September.

Increasing fibre concentration

While the economic value of the hemp crop is dependent on all the products derived from the crop (fibre, seed and hurds), hemp is cultivated primarily for its cellulose-rich cortical fibres, and in particular for the cellulose in its fibres (Fig. 4.10).

These fibres are located in the cortical parenchyma and are termed primary fibres.

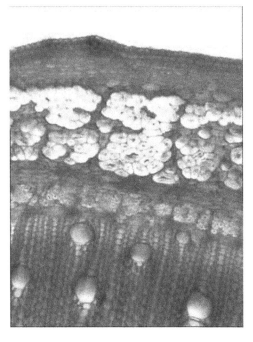

Fig. 4.10. Histological section of hemp.

Those in the phloem or secondary phloem are termed secondary fibres. They differ from each other and from the fibre of wood by their diameter, length and the degree of lignifications (Fig. 4.11).

The proportion of primary and secondary fibres within the bark of the stem of the hemp plant is dependent on the agronomic conditions of the crop and does not appear to be of importance with regard to the use of these fibres in the paper industry.

An increase in the concentration of total cortical fibres of hemp varieties remains an essential selection criterion.

This trait is highly heritable and this heritability has simplified selection. From a practical point of view and taking into account how difficult fibre is to extract from the rest of the cortical part of the stem, measuring techniques were simplified as early as the 1960s by evaluating the proportion of the stem made up of bark, a factor that correlates closely with fibre concentration.

Today, the varieties developed and cultivated in France show proportions of bark in the stem between 30 and 35% (depending on agronomic conditions) compared with values of 20–25% before these selection programmes were initiated. Two new varieties of hemp, called Santhica 27 and Santhica 70, show a proportion of bark as high as 35–40%.

Selection of varieties with low concentrations of delta-9-THC

Both EU and French legislation impose a maximum level for the constituent psychotrope, delta-9-tetrahydrocannabinol (Δ9 THC). Consequently, as early as the 1970s, a selection programme was instituted in order to make available to producers varieties that conformed to the legislation.

The maximum levels were:

* 0.30% in an average mixed sample made up of the upper third of 200 plants (female and monoecious) collected in the 10 days following the end of flowering (up until 2000).
* 0.20% in an average mixed sample made up of the top 30 cm of 50 plants (female and monoecious) collected during the period extending from 20 days after flowering up until 10 days after the end of flowering (from 2001).

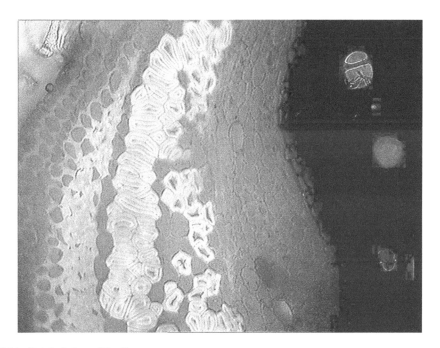

Fig. 4.11. Detailed view of the fibres.

The original varieties cultivated in the 1960s and 1970s, in France, contained between 0.10% and 0.30% delta-9-THC.

Within a short space of time, only those varieties with a THC content less than 0.15% were being preserved. This eliminated the natural variation seen in response to pedoclimatic conditions and guaranteed that levels would not exceed the norms (Fig. 4.12).[10]

During the course of the testing programme put in place, and following the initial sifting period, a further series of steps have allowed us to:

- Arrive at a point where all plants have an average value close to 0.05%;
- All the varieties cultivated in France today (Férimon, Fédora 17, Félina 32, Epsilon 68 and Futura 75) have been derived from these plants;
- These varieties all have cannabidiol (CBD) as their main cannabidiol. This precursor of delta-9-THC is devoid of psychotropic properties.

The discovery, analysis and development of particular mutations then followed. Plants were identified whose biosynthetic pathways went no further than the production of cannabigerol (CBG). The varieties that were developed from these plants had no delta-9-THC content; that is, the levels fell below those that currently could be measured. This was true for the entire cultivation cycle. The varieties Santhica 23, Santhica 27 and Santhica 70 arose from this programme of research.

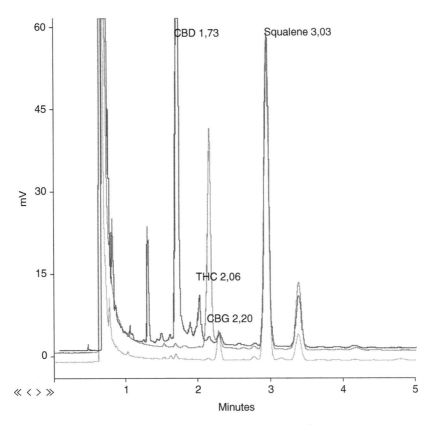

Fig. 4.12. Graph showing the elimination of natural variation due to pedoclimatic conditions. Chromatographic analysis of three kinds of hemp varieties. *x*-axis = time of retention in minutes; *y*-axis = electrical signal triggered by the molecule during its passage in the detector in mini-volts.
Key:
Dark grey: variety with THC content around 0.05% (Fédora 17).
Medium grey: variety without THC, but containing cannabinoids (Santhica 27).
Light grey: variety without THC or cannabinoids (no registered varieties)

Those plants in which the biosynthetic pathways stop at the CBG precursor have also been shown to produce olivetolic acid. For these plants, the delta-9-THC concentration is zero according to EU methods of analysis. Further information is required about this particular chemical and further checks are required to establish that delta-9-THC levels do not change from zero during the crop cycle. Cultivated varieties have yet to be produced from this programme.

These various studies have allowed a number of varieties that meet legislative requirements with regard to psychotropic substances to be placed on the market, thus ensuring the enduring viability of this crop.

Promoting stem and seed productivity

Stem productivity remains the most important selection criteria for hemp. It has, up until now, been improved through work conducted on the plant's response to photoperiodism.

Within hemp populations, there is a significant natural variability in growth characteristics. Current work is focusing on the selection of plants demonstrating strong growth within lines that are of known and dependable maturation.

The current aim of the programme requires us to pass from the first to the second situation.

- Situation 1 = fixed maturity = normal growth = normal stem productivity.
- Situation 2 = fixed maturity = strong growth = improved stem productivity.

These avenues of investigation were ignored for many years because breeders had limited means and their time was taken up with the more urgent or easier challenges and problems, as described above. Furthermore, the polygenic nature of these characteristics necessitated the use of more sensitive selection techniques than those used previously to develop and select for traits.

Seed productivity is also an important factor in agricultural revenues, both for those mixed maturity crops and for the production of stock seed. The development of modern programmes to improve stem productivity has had to take seed productivity into account. That said, this aspect of selection remains difficult to realize, given the difficulties experienced in determining heredity. Furthermore, this trait is also the most sensitive to the conditions of cultivation and experimental results can be very imprecise. This makes the selection procedure for the best parent plants very intricate.

Additional selection characteristics

The hemp industry has additional needs and seeks additional characteristics, which have been taken into account, researched and evaluated by breeders over the past few years.

YELLOW HURDS. The predicted development of markets for hemp fibre into new (non-paper) areas has created a demand for cleaner fibres (with a lower residual hurd content).

The production of cleaner fibres is dependent on a number of mechanical, agronomic and genetic factors. It is also very much dependent on the material used by mills and the rate of throughput/processing. Studies have demonstrated that well managed nitrogen fertilization, attention to hurd humidity and retting all play a role and facilitate the work of mills.

A review of existing hemp types and varieties has allowed a Hungarian line to be discovered (or rediscovered). This line originates from the Lebanon (Kompolti sargazu) and produces yellow hurds, for it is less able to assimilate nitrogen than traditional varieties. The poorer assimilation of nitrogen allows the fibre to detach more easily during decortication. This results in a cleaner decorticated fibre.

Originally poorly adapted to the cultivation requirements and conditions now present in France (e.g. dioecious, late flowering and with an excessively high THC concentration), this yellow-stemmed variety is now being subjected to a selection programme that seeks to capitalize on this trait. A parallel evaluation programme of the market potential for this kind of material is also being undertaken prior to commercial development.

EARLY VARIETIES. The current tendency is to use increasingly early varieties in order to integrate hemp production better into modern farming practices. The aim is to harvest as early as possible in order not to delay the sowing of cereal crops during the same season.

The range of available early-maturing varieties is increasing and there is a movement in favour of the development of very early varieties that will produce very little hurds and free up the ground earlier.

These varieties potentially may be best adapted to the production of hemp seed. Their use will require the validation of an economic model for a hemp crop that does not produce fibre.

HEMP OIL. Hemp is known to bear an oil-rich seed.[11] Considerable natural variability in this oil is seen between plants. During the 1990s, the potential for a selection programme was evaluated. This programme focused on the oil content (%), and specifically the oleic oil content. In particular, attempts were made to increase the content in γ-linolenic acid.

Initial results were very encouraging. Lines were identified with seeds containing up to 45% oil compared to the more usual 30–35%, while lines containing >5% γ-linolenic acid (2–3% found in classic lines) were also identified. As of the 1990s, this material had not been brought to commercial use because the market was not developed and its size remained uncertain.[12]

The selection of industrial hemp must take into account a number of important criteria. These must respond and adapt to the needs of the two lines of cultivation, while producing end products that can satisfy several industrial uses.

This represents a difficult challenge and requires that reasonable compromises be struck in order to satisfy the different end users while still maintaining economic viability.

4.3.4 Selection techniques

Mass selection method

Given the lack of means available and the availability and usefulness of simple techniques, mass selection techniques have long been used to breed hemp. Two such techniques are still used today.

When selecting for maturity and monoecity, selection is effected by recurrent selection of the individual phenotype (Fig. 4.13). The phenotype is therefore chosen based on the visual recognition of genetic characteristics.

The chosen plants reproduce by cross-pollination.

This technique presents the advantage of being cheap and efficient, as it works through a large number of individuals and has a short reproductive cycle (in terms of generations).

Selection for THC concentrations and fibre concentration using recurrent phenotypic techniques produces good dependable results (Fig. 4.14).

In practice, this technique requires mother plants to be selected according to phenotype. The outcome can then be tested and studied in a specially adapted plot. Testing of descendent plants allows a definitive choice to be made and the selected families are retained and used as stock for future generations.

This technique is cheap and rapid (although slower than the technique described previously). Unfortunately, it is less easy to manipulate and must be reserved for the functions to which it is best suited.

In both cases, these techniques do not allow a significant homogenization of genetic material, the resulting varieties remain improved populations (i.e. genetically heterogeneous).

Creation of composite varieties

A composite variety (also known as synthetic) is an artificial population resulting from controlled reproduction across several generations to produce a limited number of lines that generally are not pure lines and are chosen for their own desirable agronomic properties, as well as those they offer to cross mixes.

The creation of composite varieties (Fig. 4.15) requires:

• The selection of a starting group of phenotypes through a process of mass selection.
• The creation of partially homozygous lines through self-fertilization across several generations. The number of generations will vary in relation to the economic importance of the species and the hope of a return on the investment. This process allows favourable plant characteristics to be preserved and undesirable characteristics eliminated, leading to a homogenization of the plants. The best parents are selected simultaneously.

DIAGRAM OF SELECTION OF MONOECIOUS HEMP
Recurrent Phenotypic Selection - Individual

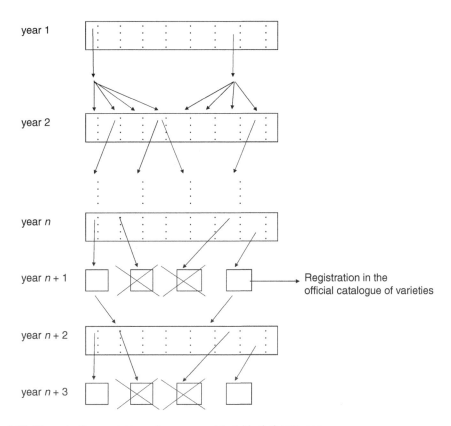

year 1

year 2

year *n*

year *n* + 1   → Registration in the official catalogue of varieties

year *n* + 2

year *n* + 3

Fig. 4.13. The selection model based on recurrent individual phenotypes.

- The best lines are crossed in order to arrive at a hybridized result. The choice of the number of parents is influenced by various factors, but generally lies between 4 and 10.

Subsequently, the seed is increased to produce the seed stock required for industrial production.

This technique is best suited to allogamous species of little economic importance. Such species are unlikely to be the subject of programmes seeking to create controlled hybrid varieties (in the case of hemp, true hybrids pose technical and economic difficulties).

The use of this technique to improve both hurds and grain is a relatively recent development in France.

Selection using molecular markers

This technique (or techniques) relies on a molecular study of the genome and seeks specifically to identify particular genes within certain plant varieties. The genome of the plant species must therefore be known, together with the identity of any desirable genes. This technique may therefore complement larger programmes working on the polygenic factors that affect yield.

These techniques are now widely practised and have been used on almost all species, including, of course, hemp.

Certain research projects of varying ambition have been undertaken on hemp:

- Molecular markers for male plants have been identified. They have been used in the production of seed stock but have not

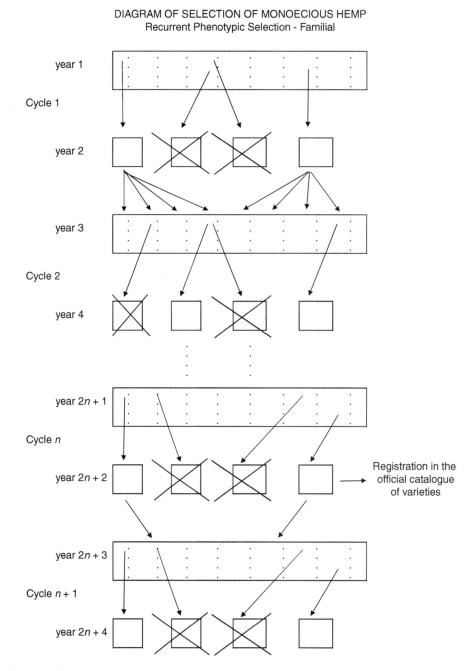

DIAGRAM OF SELECTION OF MONOECIOUS HEMP
Recurrent Phenotypic Selection - Familial

Fig. 4.14. Recurrent phenotypic selection model for families.

been used in the creation of varieties, where their use remains somewhat controversial.

- Work has been undertaken in Italy to identify markers for THC. Published results are encouraging.

- Research on the biosynthesis of fibres during the growth cycle has been initiated in the Netherlands.[13] The aim of this work is to determine the genes involved, to study their variability and any potential for the creation of new varieties.

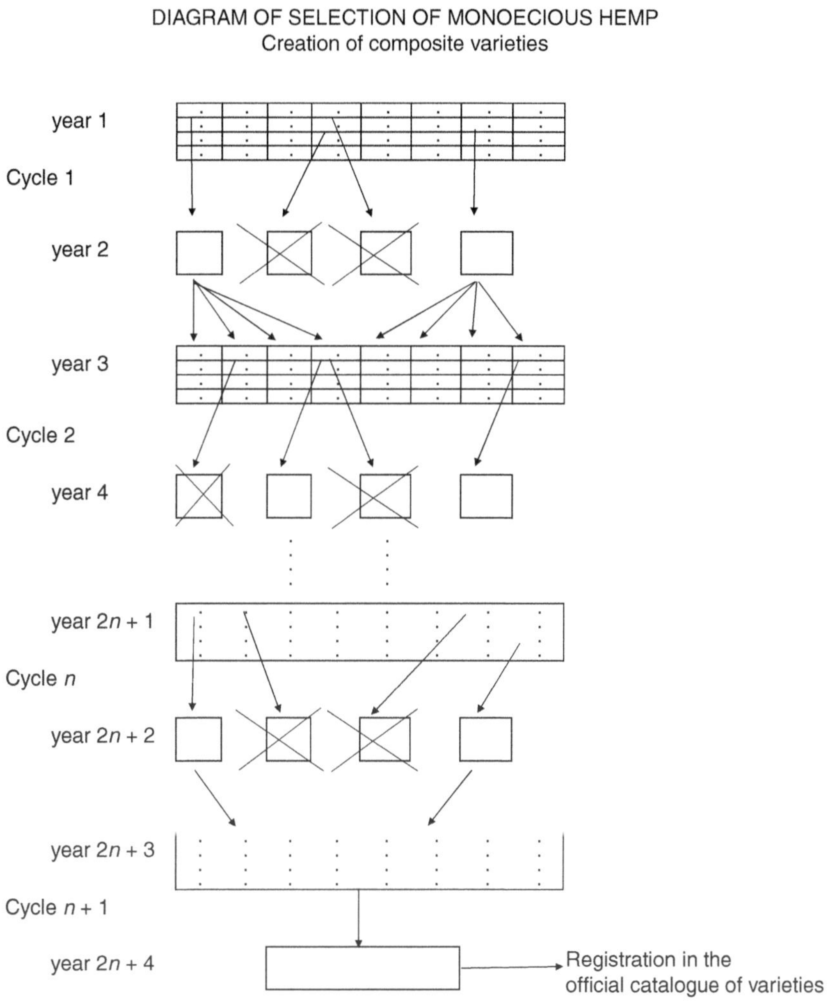

DIAGRAM OF SELECTION OF MONOECIOUS HEMP
Creation of composite varieties

Fig. 4.15. Creation of composite varieties.

There has been little work in hemp that applies the recent advances in biotechnology. There has been no attempt to create transgenic hemp (GMO, 'genetically modified organism').

4.3.5   Production of stock seed

The creation of a crop variety is only one of a number of stages permitting farmers to improve the genetics of their crop. Once realized, it is necessary to institute a programme that will preserve the variety, including the production of stock seed that remains identical to the original seed used. A production programme for commercial seed stock must be developed that is capable of supporting the large-scale production of a given variety.

All breeders will be conscious of the problems posed to growers by each variety. In particular, they will be aware of the need to reproduce genetic material without it engendering a higher economic cost compared with other varieties.

The grower will need to consider how to:

• maintain the varietal characteristics
• permit the variety to express itself through a high quality of stock seed

- ensure a constant supply, maintaining both quantity and quality.

Different production schemes exist for hemp. They are adapted to the variety used and the qualitative objectives being sought.

Multiplication of varieties

The multiplication of certified monoecious hemp seed (Fig. 4.16) has a distinct advantage: the multiplication factor improves with each generation and allows the industry to be supplied by a small number of seed-producing crop cycles.

Fields sown for hemp stock seed are sown at a very low density, of the order of 1.2 kg of seed/ha. This yields an average harvest of 1.25 t seed/ha (an increase of ×1000). In addition to this multiplication factor, the low sowing density also allows the fields to be inspected and the crop further refined manually by culling in order to obtain optimal quality.

Refinement consists of the:

- removal of plants that do not conform to the variety at the vegetation stage
- elimination of male plants (residual dioecious) at the flowering stage to remove their gametes, which would effect a rapid return to the dioecious state

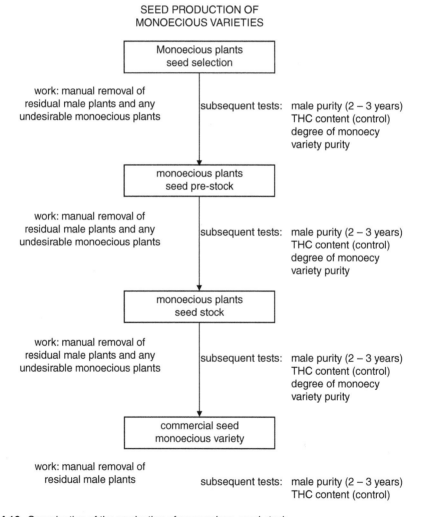

Fig. 4.16. Organization of the production of monoecious seed stock.

- elimination of undesirable monoecious masculinized plants during the flowering stage, leading to foundation increases. This further reduces the presence of undesirable monoecious masculinized plants in the final commercial seed.

Between each cycle it is essential that strict tests be conducted to identify and select the right groups of plants for multiplication:

- Varietal conformity must be checked. Generally speaking, this is a formality, but must be conducted in order to ensure the absence of undesirable outcrossing and/or mixing after harvesting.
- The THC content of batches must also be checked. Here again this is a formality, as the THC content remains stable between generations. These checks do, however, provide an additional check against undesirable pollination.
- Male purity must also be checked. It is, in effect, impossible to eliminate male plants completely before the flowers open. There is, therefore, always a small amount of pollination by dioecious plant pollen. This must be controlled and may require a reselecting of plants at the next cycle. The efficiency of this selection process is only to be relied upon if the proportion of male plants to be eliminated is small.

These checks necessitate 2–3 years of controls between each generation. One check in particular is that of the male purity of batches (while ideally the batch would be free of male plants, the fact is that some always remain – seed producers must have an idea of the level of remaining plants in order to ascertain if the batch is commercially viable). This increases greatly the time taken to place a new variety on the market.

Multiplication of monoecious–dioecious hybrid varieties

We have already seen how the complexities inherent in the formation of *monoecious–dioecious hybrids* alter the seed production scheme.

In the first step, monoecious plants are used as male (pollinators) and female plants of a dioecious variety as female (seed bearers) (Fig. 4.17). Unfortunately, female plants do not exist alone in a dioecious stand and it is therefore necessary to 'rogue' (cull) the male plants (amounting to some 50% of the stand). This work is onerous and labour-intensive; it can therefore only be realized on a small scale. If done well, the result of this cross is a generation of predominantly (>95%) female plants.

In the following generation, this F1 is crossed with a monoecious variety that provides the pollen necessary for fertilization. The next generation, corresponding to the commercial seeds, is produced by open pollination, as described in the previous chapter.

The quality and control restriction are as described in the previous description and will cause similar delays to the marketing of these varieties.

Multiplication of monoecious varieties

On today's market, there are no seeds of dioecious varieties in widespread use. For these varieties, the constraints are very different:

1. No a posteriori controls on male purity are required in these batches. Controls are less onerous and varieties easier to get to market.
2. There is no need to eliminate male plants and production is less labour-intensive, and therefore cheaper. The elimination of plants that do not meet requirements is still necessary, but this problem is true for all species.
3. The seed yields are a lot lower (approximately 35% lower), which places the benefits detailed in 1 and 2 above into perspective. In addition, seed increase is not faster or cheaper.

Finally, it should be noted that the biggest difficulty with monoecious varieties lies in the absence of a mechanical harvesting system that is appropriate for crops of this type. In effect, by the time the seeds are mature and ready for harvest, the male plants will have been dead for about a month and will be completely retted. Classical harvesting systems, needed for the cutting of fruit-bearing female plants, are unable or inefficient at harvesting male plants. The latter will be dry at the time of harvest and will often roll around the machine axle, thus jamming the operation.

In France, this problem has been solved experimentally (and on a small scale) by eliminating 90% of the male plants manually, as early as the end of the fertilization period.

SCHEME FOR SELECTION OF MONOECIOUS
HEMP 'HYBRID' METHOD

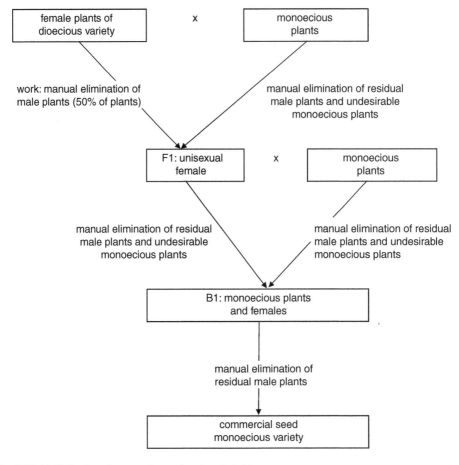

Fig. 4.17. Multiplication of monoecious–dioecious hybrids.

This technique can be extended and used in the production of foundation seed. But it has yet to impact industrial seed production.

In countries where dioecious varieties are in use, the male plants are pulled manually before the harvesting machines pass through (J. Berenji, personal communication).

4.3.6 Conclusion

Despite its small size and the difficulties it has to contend with, the hemp industry has been able to develop a system of producing plant varieties capable of meeting the technical and legislative challenges that have presented themselves over the past 40 years.

The scientific tools available today will enable the development of improved varieties with enhanced agronomic performance, while addressing new demands from industry and emerging markets.

There is little doubt that the coming years will see a surge in the creation of new hemp varieties to support the development of the hemp industry in France and elsewhere in Europe.

Notes

[1] These footnotes have been added by P. Bouloc. Definitions are taken from the *Petit Larousse*.

[2] A cultivar is a plant variety resulting from a selection, mutation or hybridization (whether occurring naturally or provoked) and cultivated for its agronomic qualities.

[3] This method is similar to the creation of a population with a broad genetic base, allowing work to be conducted without the risk of impoverishing the variability of the population necessary for constant progress.

[4] Which still does not exist, as no one knows how to produce it at a reasonable cost.

[5] This is no longer the case today. This route is too complicated and without any real advantage, as several generations are required in order to produce industrial quantities of seed.

[6] Chimera: organism consisting of two (or, rarely, several) cell types with different genetic origins.

[7] Hemp is naturally a dioecious plant. The monoecious variety is a mutation that has taken on a significant selective advantage.

[8] The French catalogue uses the term DHS, whereas the English use the term DUS (distinctness, uniformity and stability).

[9] See later in the chapter for further discussion of this point.

[10] The figures presented are copies of gas phase chromatographs (GPCs). The GPC allows the constituent parts of a sample to be separated and their concentrations calculated. In this case, THC is being tested. In this example, the different cannabinoids found in hemp are being separated. The equipment is regulated so as to separate out these constituents, while excluding as many non-cannabinoids as possible.

For any given setting of the machine, each component will have a set retention time. This results in it being detected at precise times (as shown on the graph). The abscissa represents time in minutes and allows the identity of each component to be determined.

The purpose of the analysis is to control the concentration of each constituent. The ordinate shows the intensity of the signal in mini-volts. The regular use of a control allows the intensity and weight of the constituent to be checked and correlated.

By dividing the recorded intensity of the chosen constituent by the intensity of the reading recorded for the same constituent when injected at a precise dose, it is possible to determine precisely the concentration of the constituent in any given sample.

Hemp is classified easily according to its psychotropic potential (see text). This classification is based on the relative concentrations of two or three cannabinoids (CBD, THC and CBG). The graphs produced show several different types of hemp: one variety with as much THC as CBD (intermediary drug), one sample with high concentrations of CBD and little THC (fibre type), a third sample with no THC but with CBG and a fourth sample of hemp with none of these cannabinoids.

[11] For more on this subject please refer to Chapter 3, 11 and 16 of this book.

[12] Refers to the French market only.

[13] Work conducted at the offices of Plant Research International (PRI) in Wageningen, the Netherlands.

5 Factors Affecting the Yield of Industrial Hemp – Experimental Results from France

Sandrine Legros, Sébastien Picault and Nicolas Cerruti
Institut Technique du Chanvre (ITC), France

5.1 Introduction

This chapter summarizes the results of performance trials undertaken by the following institutions: La Fédération Départementale des Groupements d'Etude et de Développement Agricole de l'Aube (FDGEDA), La Fédération Nationale des Producteurs de Chanvre (FNPC), Institut Technique du Chanvre (ITC) and the cooperative La Chanvrière de l'Aube (LCDA) and Le Syndicat des Producteurs de Chanvre (SPC).

Cultivated by humans for several thousand years, hemp has long played an important role in the European agricultural landscape. In France, in the 19th century, up to 200,000 ha was devoted to its production. As demand for its end products was gradually extinguished by competition from exotic or artificial fibres, it almost disappeared from the scene.

Despite these obstacles, work on hemp selection and agronomic research never ceased completely. Over the past 40 years, a considerable number of field trials and experiments have been conducted within organizations. The objective was to understand the plant and its agronomic performance better. These efforts focused on both improving yields and meeting legislative rules. The paper industry that between the 1940s and the 1990s was the only industrial outlet for hemp was able to continue sourcing its raw material. Today, new opportunities are opening up as markets in

building materials and plastics grow and develop. It has now become necessary to update, refocus and coordinate hemp research in order to meet the needs and expectations of these markets, which demand improved yield and quality of the raw material.

This chapter will provide an account of the work conducted on hemp in France between 1970 and the present day. Which research areas have received attention? What significant results have been obtained in this time? What can we consider to be the current state of our understanding? What are the latest research issues and where are they leading? These are the questions that we will address. This chapter does not purport to be a source of scientific references, but rather seeks to provide an overview of the experimental work conducted by the FDGEDA de l'Aube and the Chambre d'Agriculture de l'Eure, the ITC and the FNPC. This treatment endeavours to blend our current knowledge with that we have inherited from the past but which has often been scattered and unformalized. It is hoped this will be of use to those working with this crop in the future.

5.1.1 Technical sources

This chapter has drawn on the following sources:

* FDGEDA de l'Aube: field trials undertaken between 1980 and 1999.

- Chambre d'Agriculture de l'Eure: field trials undertaken in the 1990s.
- FNPC: replicated trials undertaken between 1970 and 2002.
- ITC: replicated trials undertaken by the FNPC in 2003 and field trials undertaken by the SPC Seine-Saône in 2003.[1]

Experimental protocols differ from one year to the next and between organizations. This is true for most subjects studied and makes the comparison of results problematic and the calculation of averages across large numbers of experiments difficult.

For this reason, we have chosen to construct certain graphs using relative averages:

Relative average yield = [(yield × 100)/ maximum yield]/*n*

Relative average fibre content = [(fibre content × 100)/maximum fibre content]/*n*

where *n* is the number of trials.

5.2 Physiology of Hemp

5.2.1 Growth

Several systems are in existence today allowing the stage of growth of the hemp plant to be determined. The two most popular systems are those of Médiavilla (Table 5.1) and the FNPC (Fig. 5.1). At present, the former is suited best to producers, for it allows a computer record of the observations made during the growth of the plant.

We divide the growth of hemp into four phases (Fig. 5.2; FNPC, 1999, 2000):

1. Establishment (3 weeks).
2. Active growth.
3. Slowing in growth at the start of flowering.
4. Growth arrest at the start of full flowering.

The only characteristic that distinguishes varieties is the *earliness* of the crop, i.e. the time between emergence and flowering for female plants.

Whether we study early or late maturing varieties, the growth curves before flowering are strictly identical. A difference is seen only at the start of flowering: the early varieties flower and stop growing earlier than the late maturing varieties. The growth curves are again parallel after the end of flowering.

The daily growth rate is dependent on the stage of growth and follows a parabolic curve. Growth is slow during the establishment phase and then increases regularly (although slower and slower) up to an average of 6 cm/day, before then falling off (FNPC, 2000); see Fig. 5.3. The early and late developing varieties both behave in the same manner. The earliness of a variety is judged by the date on which the daily growth rate is maximum. The curve for the late maturing variety shows an inflection later, this point appearing to correspond with the start of flowering for this variety.

The fastest relative speeds (per cent of daily growth relative to the size of the plant) are attained during the establishment phase (1.0–15% growth/day). This is maintained during the active growth phase before falling off rapidly to zero at the end of flowering.

5.2.2 Establishing yield parameters

Dry weight of hurds

During the active growth phase, the yield of fibre, hurds and seeds develops as a linear function of growing degree days (GDD) (Fig. 5.4). This stabilizes from the start of flowering and, in the 4–5 days following this date, the final hurd yield is definitively established.

NB: The temperature requirements to produce hurd dry matter vary depending on the amount of nitrogen applied. The more nitrogen, the lower the GDD requirement (less than 120 GDD for nitrogen doses of >150 units/ha). The lower the amount of nitrogen supplied, the greater the GDD requirements: up to 300 GDD where no nitrogen fertilizer is supplied and there is an absence of nitrogen in the soil (FNPC, 2000, 2002).

Fibre levels

This corresponds to hurd yield.

Seed dry weight

The seed dry weight increases regularly from the point of full flowering to the point of grain

Table 5.1. The different stages of development of *Cannabis sativa* L. according to the Médiavilla method.

Code	Definition	Remarks
Germination and emergence		
0000	Dry seed	
0001	Appearance of radicle	
0002	Appearance of hypocotyl	
0003	Opening of cotyledons	
Vegetative stages		
1002	First set of leaves	1 node
1004	Second set of leaves	3 nodes
1005	Third set of leaves	5 nodes
10xx	nth set of leaves	
Flowering and seed formation		
2000	Changing point	Change of phyllotaxis on the stem
2001	Flower buds	
Dioecious plants		
Male		
2100	Flower formation	First closed staminate flowers
2101	Beginning of flowering	First opened staminate flowers
2102	Flowering	50% opened staminate flowers
2103	End of flowering	95% of staminate flowers open or withered
Female		
2200	Flower formation	First pistillate flowers; bract with no styles
2201	Beginning of flowering	Styles of first female flowers
2202	Flowering	50% of bracts formed
2203	Beginning of seed maturity	First seeds hard
2204	Seed maturity	50% of seeds hard
2205	End of seed maturity	95% of seeds hard or shattered
Monoecious plants		
2300	Female flower formation	First pistillate flowers; perigonal bract with no pistils
2301	Beginning of female flowering	First pistils visible
2302	Female flowering	50% of bracts formed
2303	Male flower formation	First closed staminate flowers
2304	Male flowering	Most staminate flowers open
2305	Beginning of seed maturity	First seeds hard
2306	Seed maturity	50% of seeds hard
2307	End of seed maturity	95% of seeds hard or shattered
Senescence		
3001	Leaf desiccation	Leaves dry
3002	Stem desiccation	Leaves dropped
3003	Stem decomposition	Bast fibres free

Source: From Médiavilla *et al.* (1998) in *International Hemp Association* 5(2), 65–72.

maturity. The curves for different varieties are offset in time but parallel; this is explained by differences in maturity.

Nitrogen fertilization is critical for seed yield: in the same way as for hurds, the more nitrogen, the higher the seed yield.

Leaf dry weight

Regardless of the variety, the leaf dry weight develops vigorously at the start of the growth cycle and attains 1.5 t/ha on average by 1 July. This increase continues during the flowering period and reaches an average of 2.5 t/ha by the end of August. Leaf

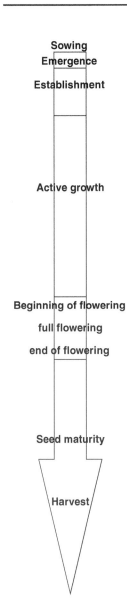

Emergence (0 to 4–9 days)

Occurs after approximately 80 GDD (growing degree days), which is usually 4–9 days after planting depending on soil conditions. This is the most critical phase of cultivation.

Establishment (4–9 days to 3 weeks)

Slow growth phase of the crop during which the canopy is established. It can take up to 3 weeks to establish the complete canopy. At the end of this stage, the crop will measure 25–30 cm in height and each plant will have three leaves.

Active growth phase (3 weeks to 2.5 months)

Period falling between the end of establishment and initiation of flowering. Fibre production takes place during this phase and mineral and water demand is greatest and must be met. It is during this period that the plant is most sensitive to lodging.

Flowering

The appearance of the first flowers in a hemp crop is related to the total temperature during emergence and to the photosensitivity of the variety. Growth slows during this period.

Full flowering

This stage corresponds to the opening of the last female flowers and is strictly dependent on the photo period. A particular variety grown in a particular zone will reach this point at a specific date, irrespective of the date of sowing or of any variation in annual conditions.

End of flowering

This is defined as the point at which the last flowers are pollinated. This point marks the end of growth and determines the final yield of fibre; it initiates the maturation of the fibre and seeds.

Seed maturity

This is obtained 40 days after full flowering. The harvesting will be timed, however, to take into account the moisture level of the seeds and the capacity of the driers (an essential stage) and this can lead to considerable delays between this stage being reached and actual harvest.

Fig. 5.1. Growth stages of *Cannabis sativa* L. according to the FNPC method.

loss accounts for the reduction in dry matter seen at the end of the growth cycle.

Number of nodes

The number of nodes increases regularly from the start of the growth cycle through flowering. Varieties start to show differences at the start of flowering. Late maturing varieties show the largest number of nodes. An increase in the number of nodes is highly dependent on accumulated heat units (GDD); no effect from other influences (e.g. nitrogen) has ever been shown.

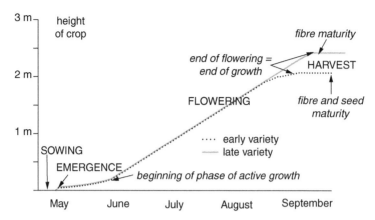

Fig. 5.2. Growth of hemp. Example of a crop sown in early May showing growth of an early and a late developing variety.

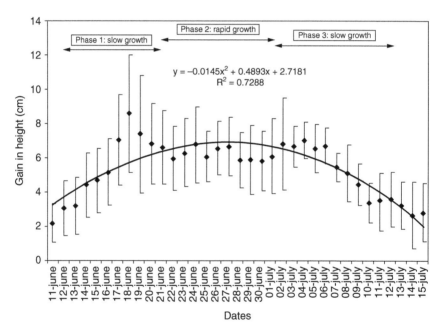

Fig. 5.3. Daily growth rates for hemp. ADG = average daily gain in height (FDGEDA de l'Aube, 1999).

5.2.3 Absorption of organic elements and minerals by hemp

The breakdown of mature hemp plants into constituent organic compounds and minerals is presented in Table 5.2.

Nitrogen

All plants contain a certain amount of nitrogen. It is an essential constituent of the cytoplasm of cells and plays an important role in plant metabolism. This element is obtained, in its mineral form, from the soil.

Nitrogen has a significant influence on plant growth. It is found in particular in young plant tissues and in certain storage tissues. It promotes the development of the stem and leaves, which become dark green as a result of their high chlorophyll content (a molecule high in nitrogen). The green parts of the plant allow carbon to be assimilated into the plant

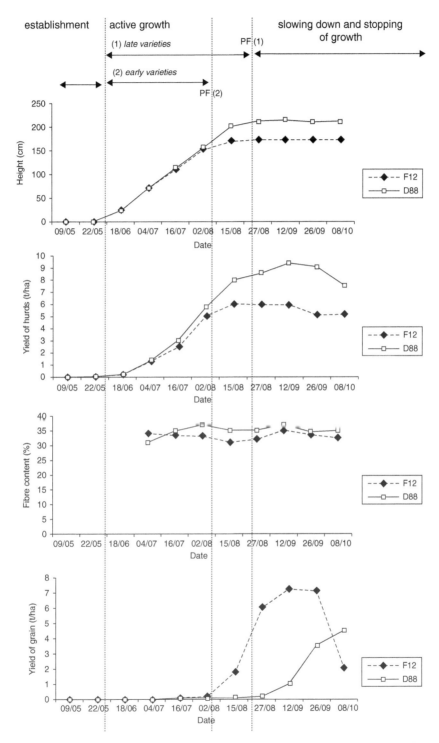

Fig. 5.4. Development of hurds, fibre and grain yield during the growth cycle of the hemp crop. This example shows a crop sown on 9 May, consisting of an early variety (F12) and a late variety (D88). F = full flowering. (FNPC, 2000).

Table 5.2. Quantity of organic compounds and minerals taken up and captured by hemp at maturity. (Average of eleven tests: Chambre d'Agriculture de l'Eure, 1977; FNPC, 1978–1986; FDGEDA de l'Aube, 1993).

Organic elements and minerals	Exports kg/t dry matter (DM)
N	8.43
P_2O_5	3.73
K_2O	17.56
CaO	13.03
MgO	2.13
Na	0.167
Mn	0.515
Cu	0.005
B	0.011
Zn	0.015

and organic compounds synthesized. The green tissues of the plants are the producers.

Where nitrogen is lacking, leaves become yellow and plants are stunted; nitrogen is a key determining factor in plant yields.

Nitrogen is first absorbed very rapidly by the plant during the early growth phase and then more slowly until the end of the cycle (FNPC, 1986, 2000, 2002). Nitrogen levels in the leaves, fibre and hurds are very high early in the cycle (6%, 1% and 1%, respectively) and then decrease progressively with time. At the end of flowering, consumption increases and most of the absorbed nitrogen is used to fill the seeds. At maturity, more than 50% of the nitrogen is found in the seeds.

Phosphorus (P_2O_5)

Phosphorus is a constituent of all plants. It plays an important role in the transformation of chlorophyll, glucose metabolism and the metabolism of lipids and proteins. It is a growth factor and is found abundantly in growing tissues. While its effects may be less obvious than those of nitrogen, it is equally indispensable for plant development. Its presence allows plants to absorb and use nitrogen more readily. It has a positive influence on fertilization, fructification and seed formation. It also increases the rigidity of cereal stems and helps resist any tendency to lodge, as well as improving the lignification of the stems.

Phosphorus consumption increases slowly from emergence to the end of flowering. During this period, it is found mainly in the leaves. From the end of flowering, the amounts absorbed are much less important and are destined (>70%) primarily for the seeds.

Potassium or potash (K_2O)

Often mistakenly called potash, potassium is involved in the formation of all plant tissues. It is found in significant quantities in their ash.

Potassium is critical to the process of photosynthesis, contributing to the synthesis of sugars, and it facilitates the migration of products formed in the leaves to the storage organs: the tubercles, fruit, seeds and roots, etc., thus improving plant growth and quality.

Nitrogen, potassium and phosphorus (NPK) are the 'big three' constituents of commercial fertilizers, the 'macronutrients'. Potassium contributes to resistance to the 'hazards of production' (accidents and disease) and is indispensable to plant growth, enabling cells to use more nitrogen and phosphorus and thus improve crop yields.

Potassium is absorbed rapidly at the start of the growth cycle, peaking at the start of flowering. Found predominantly in the leaves, it is shed with the leaves as senescence sets in. Its absorption is linked strongly with that of nitrogen: if nitrogen fertilization is poor, the plant produces few leaves and therefore uses little potassium.

Calcium (CaO – calcium oxide)

The role played by calcium in soil properties is well known. It is essential to plant nutrition, though not to the extant that NPK is.

Like nitrogen, calcium absorption increases throughout the growth cycle. It is especially rich in the leaves, fibre and hurds of hemp, but there is little in the seeds.

Copper, zinc, boron (Cu, Zn, B): the trace elements

Trace elements, as the name implies, are only found in small concentrations in the soil and, while essential, are only required in small amounts by plants. The principal trace elements are iron, boron, manganese, copper, zinc, molybdenum and chlorine.

Where the uptake of trace elements is inadequate, deficiency will result in disease. This can manifest in different ways according to the element concerned. In general, though, foliage is affected and may appear deformed, discoloured, necrotic or weak. Mineral deficiencies may be due to a deficiency in the soil (true deficiency) or to a block of the element in question by an antagonistic factor. Such factors typically interfere with the absorption of the element by the plant.

The effect of deficiencies is often hard to detect as it is easily mistaken for a lack of nitrogen: producing a yellowing of the leaves and a shortening of the internode distance.

We have no experimental data on the precise trace element requirements of hemp. We know only that these three minerals – Cu, Zn and B – are represented equally throughout the plant and that their concentrations vary little over time.

Manganese (MgO)

Manganese levels increase during the growth cycle in all parts of the plant. Its absence can give rise to chlorosis and deformity in the leaves.

5.2.4 Physical and chemical characteristics of hemp and hemp seed

An analysis of the chemical composition of each of the component parts of hemp is provided by B. Kurek, of INRA, in Part II of Chapter 3 of this book.

5.3 Varieties

The varieties of hemp tested in France by the FNPC and the FDGEDA of l'Aube, Saône and Loiret are presented in Tables 5.3, 5.4 and 5.5.

The best hurd yields are obtained from the late developing varieties, the poorest from the early varieties. This is due to the growth cycle of hemp and the fact that vegetative growth stops at the point of full flowering. The opposite is true where seed yields are concerned. The fibre content varies slightly according to the variety but does not appear to be related to maturity, even if fibre content is higher in the late developing varieties.

Among the varieties tested over a 12-year period:

Table 5.3. Average yields over a 12-year period for the hemp varieties tested by the FNPC between 1990 and 2001 (non-exhaustive list).

		FNCP 2002		
	Maturity	Hurd yield (t/ha)	Fibre content (%)	Seed yield (ql/ha)
F12	Early maturing	9.06	43.18	12.79
F17	Early maturing	9.06	45.27	12.77
Béniko	Early maturing	9.26	46.93	9.90
S27	Average	9.32	52.60	6.075
F34	Average	9.62	51.88	11.55
Bialobzrsekie	Early maturing	9.81	60.09	8.50
Kompolti	Late maturing	8.94	30.73	
Cannakomp	Late maturing	9.51	44.23	
E68	Late maturing	9.74	34.25	
D405	Late maturing	10.28	35.30	
Ilosa	Late maturing	10.37	37.35	
F75	Late maturing	10.45	42.61	
Carmagnola	Very late maturing	10.87	35.76	
Lipko		11.38	49.35	
Fibranova		11.38	45.67	
D88	Very late maturing	12.29	46.23	

Table 5.4. Characteristics of the hemp varieties tested by the FDGEDA de l'Aube in 1999.

		ITC 2003 (FNPC)		
	Maturity	Hurd yield (t/ha)	Fibre content (%)	Seed yield (ql/ha)
Finola		1.60	18.60	20.33
S302		8.21	39.30	59.00
S301		8.24	37.00	58.00
Glukho 46		8.27	38.10	66.67
Zolo 11		8.61	38.60	41.67
USO 31	Early maturing	8.75	44.50	64.33
F17	Early maturing	9.75	33.70	7.30
Glukho33		9.79	36.80	82.67
F12	Early maturing	9.80	39.20	64.33

Table 5.5. Results of the tests on the most recent varieties (FNPC, 2002; ITC, 2003).

		ITC 2003 (SPC)		
	Maturity	Hurd yield (t/ha)	Fibre content (%)	Seed yield (ql/ha)
F12	Early maturing	5.80	31.90	10.43
F34	Average	5.83	30.60	9.20
F17	Early maturing	6.07	27.87	9.27
S27	Average	6.50	34.60	7.73

- the hurd yield is highest for E68, F74, F75 and D88 (dioecious variety)
- the highest fibre levels are obtained from F12, S204, S201 and D88
- the best seed yield is obtained from F19, F17 and F12.

Among those varieties tested most recently (2004–2005), F75 and E68 offer very good hurd yields. Similar results for hurd yield are obtained from the newly tested varieties, Lipko, USO and Fibranova (dioecious varieties), with the best fibre yields to date. At the time of writing, the varieties F32, F12, F34 and Santhica 27 showed the highest fibre yields, while the Kompolti and Carmagnola varieties performed worst in these trials.

The varieties with the best seed yield are E68, F75, delta 405 and Kompolti. All such encouraging results are being subjected to further tests.

A comparison of the diameter of the stems, taken at the base, as a function of the variety was examined by the FDGEDA de l'Aube in 1999 (Fig. 5.5).

The protocol used by Frédéric Michel (European FAIR programme, 1999) used four replications of 20 varieties sown to produce a final stand of 150 plants/m². The stem diameters were measured on the third day of harvesting and taken at the base of the plant, 5 cm from the soil line. Twenty-five stems were measured on two replicates and averages were computed.

The average diameter varied between 6 and 9 mm, with a standard deviation sometimes close to zero (i.e. plants were very uniform). It appeared that D88, F75, E68 and F17 were very uniform according to this criterion. By contrast, significant difference was observed between the slender varieties (Fasamo, Bialobaszie and USO31) and the stockier varieties. However, these results must be viewed with caution as they are based on only two replications. This trait is of great importance to industry during fibre extraction and has also been studied in L'Eure and by the FNPC.

Generally speaking, it has been observed that the stem diameter is strongly dependent on stand density (plants/m²), which is a function of seeding rate and germination. This can be summarized as follows:

Many stems = fine diameter

Few stems = large diameter

The height of the plant is secondary.

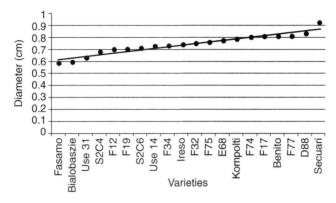

Fig. 5.5. Stem diameter as a function of plant variety (FDGEDA de l'Aube, 1999).

5.4 Agronomics

5.4.1 Date of sowing

Emergence takes place at 85 GDD[2] after sowing (the temperature is recorded from the ground at the level of the bed where the seed is sown). When rainfall is abundant after planting, emergence may be delayed and poor. Even if plants emerge precisely 90 GDD after sowing, the density at emergence declines when soil temperature is low (FNPC, 2000).

Whatever the date of sowing, flowering will only begin once 800–900 GDD have accumulated. The start of flowering varies from May to August, depending on the date of sowing. This stage is primarily thermophotoperiodic (FNPC, 2000, 2002).

For both early planted (February and March) and late planted (end May, June and July) seeds, the total accumulated GDD requirement is close to 900. For planting in April and May, the requirement approaches 1100 GDD.

Regardless of the date of sowing, full flowering and termination of flowering will occur on an identical date for a given latitude and variety, with the following exception: planting after mid-June may not achieve the temperature accumulation required for flowering. In this case, flowering is determined solely by photoperiod. The duration of flowering is thus largely dependent on sowing date, that is, flowering will be longer with early sown crops and more restricted for late sown crops.

The effect of an extended flowering period is non-uniformity of plants in the stand at harvest and this can affect the quality of the seed crop adversely.

Effect of sowing date on yields

Hurd yield is determined by the duration of the vegetative growth phase, which ceases at flowering; hence, yield is diminished the later the planting date (Fig. 5.6). However, planting date does not appear to affect the *percentage* of fibre in the hurds (assuming the stand is well established). The best grain yields are obtained from crops sown in April, when the climatic conditions are the most favourable. If sown too early, flowering will be too spread out and will result in a large proportion of unripened seeds at harvest.

5.4.2 Density of sowing

Density of plants at emergence and density at harvest

A positive correlation exists between the density of seeds sown and the population at harvest, as well as between the density sown and the density at emergence (Fig. 5.7).

In all the trials, loss of population density in the stand was recorded between sowing and emergence and averaged around 10%. Higher losses were found where the seeding rate was elevated, reflecting the fact that at low density there is less competition between plants.

Losses in density between emergence and harvest were more important (up to 50% in

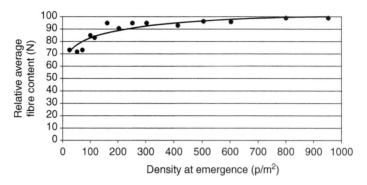

Fig. 5.6. Relative fibre content as a function of the seeding rate expressed in plants/m². A synthesis of the trials undertaken by ITC (2003) (SPC, FNPC) and FNPC (1975, 1976, 1977, 2000, 2001, 2002) and FDGEDA de l'Aube (1982, 1989).

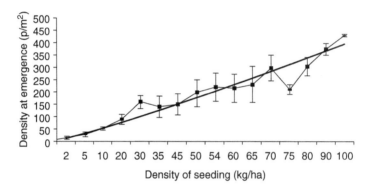

Fig. 5.7. Density at emergence (plants/m²) as a function of the seeding rate (kg/ha).

certain tests) and were highest for the high seeding rate, as described above.

The height of the crop correlates weakly with population density. High density produces shorter plants, due to competition for water and nitrogen. Where the seeding rate was less than 50 kg/ha, the height of the crop at harvest declined with increasing density (with 20 cm difference between the shortest and tallest plants). Where the seeding rate was higher (>50 kg/ha), further increases in seeding rate did not produce a reduction in plant height at harvest. This result is not, however, reflected in the height gained during canopy establishment and the early stages of the growth cycle. During this period, a low seeding rate resulted in the fewest plants, due to competition for light.

In the early stages of growth, this elongation at high seeding rate resulted in thinner stalks and a concomitant elevation of relative fibre content at harvest, independent of any losses.

Influence of seeding rate on hurd yields

Hurd yields rose rapidly with increased seeding rate when the latter was less than 20 kg/ha. At this rate, hemp was not able to establish complete canopy and smother weeds. At a seeding rate of 20 kg/ha, the hurd yield was 90% of its potential (Fig. 5.8). At higher densities, the yield continued to grow with increasing density, but much more slowly, to the point that the seeding rate became immaterial. At a seeding rate of 35 kg/ha, the hurd yield was 95% of its potential, and 97% at 50 kg/ha.

Influence of seeding rate on fibre content

Fibre content increased rapidly with increasing density at emergence, as long as this was less than 200 plants/m² (corresponding to a seeding rate of 45–50 kg/ha, Fig. 5.9). At this density, fibre production achieved 90% of potential (fibre content approximately 35%). Above a density of 200 seedlings/m², fibre content continued to rise with increasing density but very slowly (1–5% increase). However, since fibre content of the stalk varies very little from one variety to another, the increase of a few percentage points is not insignificant; fibre percentage is more a function of seeding rate than of variety.

Influence of seeding rate on hemp seed yields

The best hemp seed yields are obtained at a low seeding rate. When the seeding rate is very low, however (i.e. <20 kg/ha), hemp is unable to smother weeds and yields tend to stagnate.

Given that hurd yields are essentially independent of a seeding rate above 20 kg/ha, farmers may favour a reduction in the seeding rate.

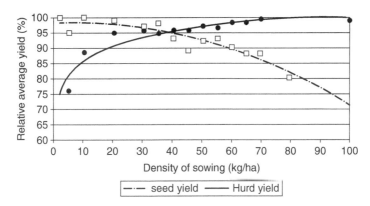

Fig. 5.8. The relative hurd and seed yields as a function of the density of sowing (expressed as kg/ha). This is a synthesis of the tests undertaken by ITC (2003) (SPC, FNPC) and FNPC (1975, 1976, 1977, 2000, 2001, 2002) and FDGEDA de l'Aube (1982, 1989).

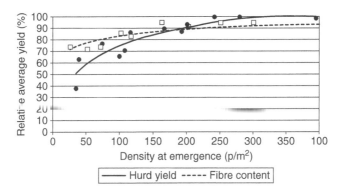

Fig. 5.9. Hurd yields and relative fibre content as a function of seeding rate expressed as plants at emergence/m². The hurd yield results are a synthesis of the tests undertaken by ITC (2003) (SPC, FNPC) and FNPC (1975, 1976, 1977, 2000, 2001, 2002) and FDGEDA de l'Aube (1982, 1989). Relative fibre content results are a synthesis of the tests undertaken by the FDGEDA de l'Aube (1989).

Where the seeding rate is low, however, stalks are thicker and taller, and therefore harder to harvest and process.

In the case where seed is to be harvested with a combine harvester, the experimental results indicate the optimal seeding rate at approximately 20 kg/ha. Unfortunately, at that density, canopy is insufficient to smother weeds and infestation is a hazard of production.

Finally, where the crop is to be harvested for both seeds and fibre, a compromise seeding rate is required in order to optimize both the seed yield and the fibre content. For farmers wishing to produce fibre (for crops where seeds are not harvested first), it is possible, therefore, to advise them on the density of sowing that will afford them an easy harvest and processing. Recent trials (which would benefit from additional replication) suggest a seeding rate of 50–60 kg/ha.

5.4.3 Nitrogen fertilization

The absorption of nitrogen occurs primarily between the three-leaf stage (50 cm) and the end of flowering (FNPC, 2001).

Hemp's nitrogen requirements have been estimated as between 14 and 15 u/t DM when water resources are not limited (FNPC, 1999, 2000; Chambre d'Agriculture de l'Eure, 2001). Nitrogen assaying enables the producer to adjust the fertilization rates for his or her crop, depending on the residual nitrogen levels in the fields.

When applied in excess, nitrogen is used for leaf production rather than fibre production. The plant then stays green longer. This can lead to difficulties in cutting the crop, increases the drying time and renders fibre processing more difficult.

As a rule of thumb, the more nitrogen applied, the greener the stems; and the more difficult will be the processing of the fibre.

Influence of nitrogen on density, height and colour of the crop

Increasing applied nitrogen results in a reduction in the density of the harvested crop due to increased competition between plants during the vegetative period. Nitrogen applied at planting allows plants to gain a little in height at harvest.

It was noted in these trials that heavy application of nitrogen caused plants to stay green longer even when colour variation was minimal.

Influence of nitrogen on yields

Hurd and seed yields increase with the dose of nitrogen and it would appear that excess nitrogen in the soil promotes seed production (FNPC, 1999, 2000, 2001). Hurd yield stabilizes at a certain nitrogen titre (100–120 units), regardless of the concentration in the soil (FDGEDA de l'Aube, 1982; FNPC, 1984, 1986; SPC, 2003; Fig. 5.10).

These figures are to be considered with the amount of dry matter (DM) exported:

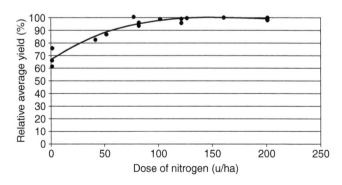

Fig. 5.10. Relative hurd yields as a function of nitrogen fertilization of the sown crop (soil nitrogen content unknown). Synthesis of the tests undertaken by ITC (2003) (SPC, FNPC) and FDGEDA de l'Aube (1982).

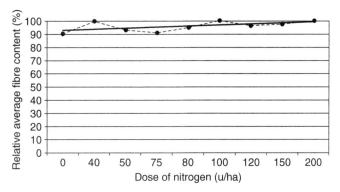

Fig. 5.11. Relative fibre content as a function of nitrogen fertilization of the sown crop (soil nitrogen content unknown). Synthesis of the tests undertaken by ITC (2003) (SPC, FNPC) and FDGEDA de l'Aube (1982).

8.43 kg/t DM. The fibre content appears to be relatively unaffected by nitrogen levels (Fig. 5.11).

Influence of the date of fertilization and splitting of fertilization on yields

Nitrogen fertilization at planting increases hurd yield compared to fertilization at the 50 cm stage (Figs 5.12 and 5.14). By contrast, the date of fertilization does not appear to affect fibre content, even when an increase in content is observed following a late fertilization (Figs 5.13 and 5.14). A late fertilization improves seed yields much as is seen for cereal crops.

A divided fertilization (X units at planting and X units at the 50 cm stage) appears to have no effect on hurd and seed yields when compared to a dose of 2X units at planting. It also appears to have no effect on fibre content, even when an increase in content is observed in individual trials where a split fertilization is used. Recall that late fertilization can make fibre processing more difficult, as the hurds stay green longer.

5.4.4 Sulfur and magnesium fertilization

Sulfur is necessary for the formation of many plant proteins. It is also a catalytic element that participates in essential metabolic reactions including chlorophyll formation.

Sulfur is found in mineral form in the ground as well as in organic materials. When mineralized, it forms sulfates. Magnesium is a constituent of chlorophyll. A magnesium deficiency will result in colour loss in the leaves and the appearance of necrotic lesions.

For the same nitrogen dose and date of fertilization, hurd and seed yield – and, to a lesser extent, fibre yield – appear to benefit from magnesium sulfate fertilization (Fig. 5.15). It promotes hurd and seed yield when applied at planting (compared to the 50 cm stage). However, the opposite is true where fibre content is concerned.

Sulfur does not appear to affect hurd or fibre yield and increases seed yield only if added at planting. These conclusions all require confirmation by additional research, as this synopsis is based on only 3 years of tests, but the deviations observed are relatively small.

5.4.5 Potassium fertilization

Potassium fertilization was investigated between 1977 and 1989 at Le Chambre d'Agriculture de l'Eure. The results indicate:

- uptake of potassium by hemp in the order of 200 kg/ha for a yield of hurds of 10–11 t/ha;
- excess potassium is not absorbed;
- hemp is able to draw on potassium held deep in the ground when availability in the tilled layer is inadequate;
- potassium fertilization only affects hurd yield in particularly difficult years. Under such conditions, potassium supplementation can improve growth (Fig. 5.16).

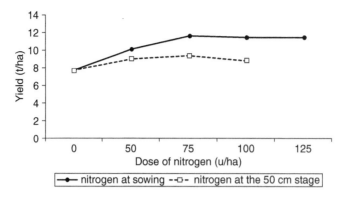

Fig. 5.12. Hurd yield as a function of nitrogen dose (added in addition to existing soil nitrogen content) and the date of fertilization (at sowing and at 50 cm stage) (ITC, 2003 (FNPC)).

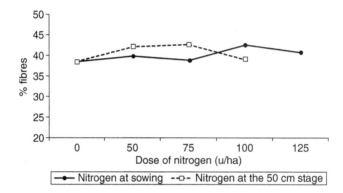

Fig. 5.13. Fibre content as a function of applied nitrogen (in addition to nitrogen content of soil) and date of fertilization (at planting and at 50 cm stage) (ITC, 2003 (FNPC)).

This series of trials reveals a progressive drop in yield when hemp is grown as a monoculture (cultivated on the same fields for 12 years). This is a real-life example.

5.4.6   Harvesting date

Hurd yield is higher if harvest is very early (c.20 August; See Fig. 5.17). At maturity, hurd weight declines slightly but the weight of the fibre component is relatively stable; thus, the percentage of fibre content is slightly higher for later harvest dates (September; See Fig. 5.18).

For seed, the later the harvest, the greater the yield; more so for late flowering varieties (Fig. 5.19). Seed harvest is not started until greenlighted by a laboratory inspection of samples to determine seed maturity so that the timing is optimized. Harvest must be accomplished within a narrow window of only 4–5 days around that date.

5.4.7   Retting

When the cut crop is retted on the ground ('dew retting'), hurd weight remains constant over the 2–3 weeks following cutting, irrespective of the harvest date (FNPC, 2001, 2002). It then declines rapidly, with a loss in dry matter of 30–50% in 1.2–2 months. The hemp hurd decomposes and residue remains on the land after bailing.

Change in the fibre concentration during the retting process appears to depend on the date of harvest. Later harvest favours a constancy in the fibre content, with levels remaining high

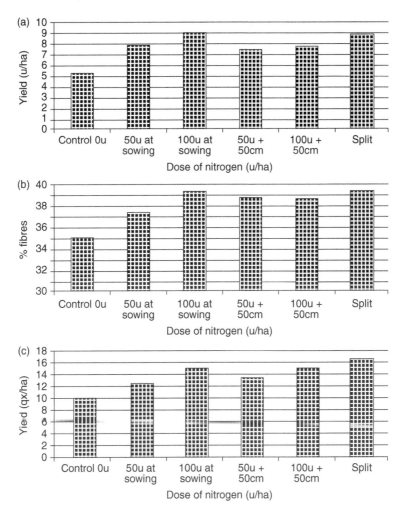

Fig. 5.14. Comparison of the effects of different nitrogen fertilization regimens on hurd yield (a), fibre content (b) and seed yield (c) (FNPC, 2001, 2002; ITC, 2003 (FNPC)).

for longer before falling. Earlier harvests have a much shorter period during which fibre content remains constant. Harvests as early as the end of July have next to no such period. These results suggest that lignification of the stems renders them more resistant to breakdown, although this remains to be confirmed experimentally.

5.4.8 Crop cultivation factors and their influence on fibre processing

Between 1999 and 2002, the FNPC and Le Chambre d'Agriculture de l'Eure investigated the variables that affected processing of fibre from hemp stalks.

Different processing regimes were scored according to the cleanliness of the fibres, as indicated by the number of passes through the rollers required to obtain the lowest possible residual hurd content. The quicker this level was reached (corresponding to the lowest number of passes through the milling/rolling process), the easier the processing.

The variables explored so far are: (i) length of retting; (ii) nitrogen fertilization; (iii) seeding rate; (iv) date of harvest; and (v) variety.

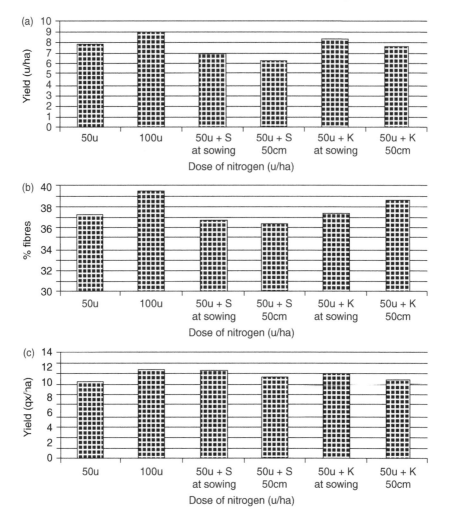

Fig. 5.15. Comparison of the effects of adding sulfur or magnesium sulfate (MgSO$_4$) at the time of nitrogen fertilization on hurd yield (a), fibre content (b) and seed yield (c) (FNPC, 2001, 2002; ITC, 2003 (FNPC)). K = MgSO$_4$; S = sulfur.

Influence of retting

Increased retting times will facilitate fibre processing and yield smoother fibres. In all tests, hemp from fields retted from 5 and 7 weeks was faster to separate and process. Fields that have been retted for only 1–2 weeks are always harder to process.

After full flowering, the timing of harvest does not seem to have a strong influence on the efficiency of the retting process (as September is usually dry) or the ease of processing.

Influence of applied nitrogen

Heavy doses of nitrogen result in stems remaining green longer. This makes fibre separation more difficult.

Influence of seeding rate

The seeding rate is, together with the length of retting and the nitrogen dose, one of the most critical factors affecting the ease of fibre separation. At higher densities, stems are thinner and fibre separation is easy, and vice versa.

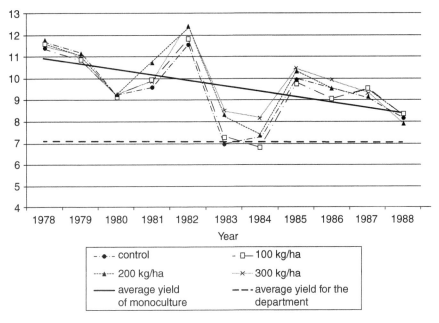

Fig. 5.16. Hurd yield as a function of potassium fertilization over an 11-year trial period conducted in L'Eure. The effect on yield of monoculture is compared to the average yield for the department (Chambre d'Agriculture de l'Eure, 1989).

These results indicate that fibre processing is easiest with a seeding rate of 10 kg/ha.

Influence of the harvest date

Early harvest makes for more difficult fibre separation. This is especially true if the plant is not retted. When harvest is undertaken late, differences in the ease of fibre separation are less marked.

Influence of the variety

The influence of hemp variety on fibre separation is relatively small, although further in-depth studies on this subject need to be undertaken. Some advantage does appear to have been demonstrated for the F17 variety and a slight disadvantage for S27.

5.4.9 Influence of crop cultivation factors on the morphology and biochemistry of fibres

A study of the influence of harvest date on the anatomical and biochemical properties of hemp fibre is currently being undertaken by Brigitte Chabbert's research team at INRA in Reims. The preliminary results of this study are presented in Chapter 3 of this book.

5.4.10 Soil type and its effect on yield

The theoretical potential of the crop is important and is of the order of 14 t of dry matter (DM)/ha. In some situations, the yield has been as much as 20 t/ha. Table 5.6 presents yield against soil type observed in trials in de l'Aube. The nitrogen application corresponds with the recommended levels for the area and is based on empirical practice rather than the results of a scientific study.

5.5 Crop Protection

5.5.1 Treatment of the seed stock

Until 2002, a single seed treatment had been used by the FNPC. In this experiment, however, ten preparations were tested (Table 5.7).

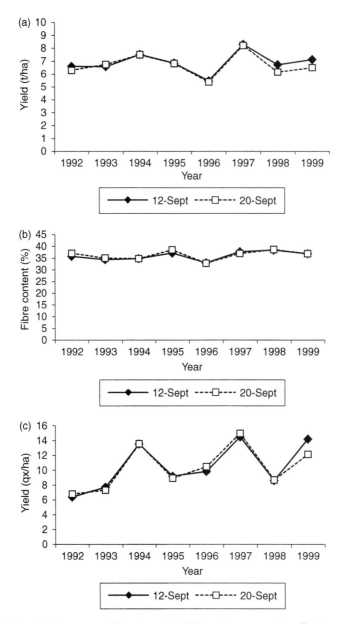

Fig. 5.17. Hurd yield (a), fibre content (b) and seed yield (c) of the early variety, F19, harvested on two different dates: 12 and 20 September (FNPC results, 1992–1999).

The Jumper treatment showed itself to be relatively phytotoxic for hemp (loss of population at emergence). The best stands were seen for the product called IST Enrobé.

With regard to emergence kinetics, the Combicoat 1, Captane and Thiram products produced better results than the untreated control. Both ISTs, together with the two Téprosyn products, gave excellent results, although these still need to be verified. By contrast, Combicoats 2 and 3, together with Jumper, gave poor results when compared to the untreated control.

This trial showed that the products IST and Téprosyn appeared to be very interesting. Further

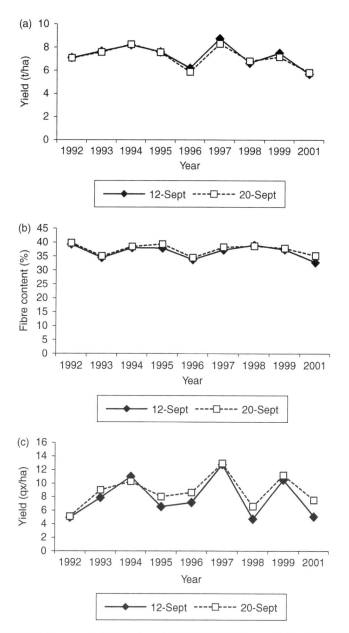

Fig. 5.18. Hurd yield (a), fibre content (b) and seed yield (c) of the intermediate variety F34, harvested on two different dates: 12 and 20 September (FNPC results, 1992–2001).

trials will allow these observations to be confirmed. It was also possible to observe that seed treatments did not improve hurd and seed yields directly but permitted an improvement in germinative vigour, and therefore stand establishment.

5.5.2 Weeding

The main weeds identified in hemp crops are sinapis, chenopodiums, mercurials, linarias, fallopias and poaceaes.

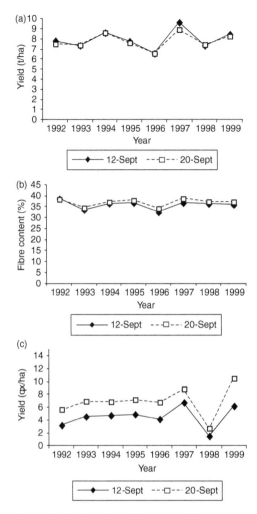

Fig. 5.19. Hurd yield (a), fibre content (b) and seed yield (c) of the late variety, F74, harvested on two different dates: 12 and 20 September (FNPC results, 1992–1999).

No herbicide is currently approved for use on hemp. A few herbicides that could be used are listed in Table 5.8; these have been tested by the FDGEDA de l'Aube and by the FNPC.

Different trials have shown that hemp crops are sensitive to herbicides, especially when they are applied early. All the products tested until now, with the exception of Lontrel, induce plant losses if applied early. With this in mind, it is clear that few products discriminate in favour of hemp, that is to say, leave it unharmed while dealing with weeds. That said,

these products have shown an acceptable degree of efficiency and selectivity on industrial crops producing higher densities. They have potential, which should not be discounted.

The last selectivity trials were undertaken on the following herbicides in 2002: Etnos Lenazar, Bastille WG, Primus, Nikos, Dactal W75, First, Nirvana, Boréal and Calliste. Only Lenazar managed not to damage the hemp, but it also showed no efficiency against weeds. All the others caused hemp growth to stop.

5.5.3   Fight against hemp broomrape (*Orobanche ramosa*)

Hemp broomrape (*Orobanche ramosa*) is a parasitic weed that infests many crops, including tomato, rape, tobacco, melon, potato and hemp, as well as several weedy species including black nightshade (*Solanum nigrum*), fool's parsley (*Aethusa cynapium*), yarrow (*Achillea millefolium*), pale persicaria (*Persicaria lapathifolia*) and the chenopodiums. It would appear that its propagation and development are favoured by monocultures. In l'Aube, the spread of this parasite has been monitored closely by Le Chanvrière de l'Aube since 2002. There are a number of ways of fighting broomrape, based on its biology, but they are of limited use.

Physical methods of control

- Burying: at a depth of 50 cm the germinative power of the *orobanche* seeds is diminished and the roots of the host plant will only reach these depths relatively late in the season.
- Heating: the seeds of *orobanche* are killed at high temperatures.
- Manual removal and disposal.
- Sow early to reduce infestation.
- Crop rotation.
- Fertilization: an increase in the dose of nitrogen reduces the amount of infestation.
- Use of false hosts on which *orobanche* can germinate but not establish itself.
- Plant traps (to be destroyed once attacked by the parasite).

Table 5.6. Hurd and seed yields according to soil type (SPC Seine-Saône, 1997).

Type of soil	Potential hurd and seed yields	Type of crop[a]	Seeding rate	Nitrogen
Barrois				
Topsoil on a limestone substrate	6 t/10 ql	1	50	110
Topsoil on a fissured limestone substrate	6.5 t/10 ql	1	45	120
Moderately deep chalk soil	7.5 t/10 ql	1	45	135
Deep chalk soil	9 t/11 ql	1 or 2	45 (1); 55 (2)	150
Deep silt	10 t	2	55	160
Alluvium soil	12 t	2	55	140
Chalky area of Champagne				
Chalk gravel	7.5 t/10 ql	1 or 2	45 (1); 55 (2)	130
Chalk on surface	8 t/10 ql	1 or 2	45 (1); 55 (2)	145
Without surface chalk or chalk grey or coloured	12 t/11 ql	1 or 2	45 (1); 55 (2)	160
Humid valleys in the Champagne region				
Clay + 45%	8 t/10 ql	1 or 2	45 (1); 55 (2)	140
Limestone clay on a bank	6 t/10 ql	1 or 2	55	160
25% clay, silt	>10 t	2	55	160
Sand	6 t/10 ql	1 or 2	55	160

Note: [a]1 = grain harvested; 2 = hurds harvested.

Biological methods of control

- Use of mycoherbicides (*Fusarium oxysporum* is effective against *orobanche* on tomatoes).

Chemical methods of control

- Injection of gaseous ethylene into the ground to provoke the degeneration and death of *orobanche* seeds.
- Use of glyphosphate at low doses.

The most efficient way of controlling *orobanche* in hemp is a system of integrated pest management (IPM) that entails a combination of chemical, biological and physical strategies together with resistant or highly tolerant cultivars and/or the creation of associations.

The FNPC began investigating *orobanche* control in hemp in 1998 and continues to do so to this day. Three techniques have been evaluated: solarization, glyphosate and false seeding. Unfortunately, the results obtained are incomplete and do not allow conclusions to be drawn about the efficacy of these

Table 5.7. Treatment of seed stock.

Name of product	Fungicide	Insecticide	Growth starter
Combicoat 1	*		
Combicoat 2	*	*	
Combicoat 3	*	*	
IST enrobé	*	*	
IST pelliculé	*	*	
Téprosyn 2498	*		*
Téprosyn 3090	*		*
Captane	*		
Thirame	*		
Jumper	*	*	
Untreated	*		

techniques. These trials do, however, allow us to conclude that:

- Solarization – using a plastic sheet spread over the ground for 3–4 weeks – is very difficult to accomplish on a wide scale. It also poses a problem for hemp producers

Table 5.8. Herbicides tested by the FDGEDA de l'Aube and the FNPC since 1989.

Product tested	Year of test	Observations
Challenge	2003[b] 1999[a] 1998[a] 1990[a] 1989[a]	Very selective in the pre-emergence phase, becoming phytotoxic after emergence and can destroy the crop completely when added to Tramat. Selective results and efficacious at a dose of 3l/ha.
Lasso	1990[a] 1989[a]	Acceptable selectivity but weakly efficacious on hemp weeds.
Tramat EC	1990[a] 1989[a]	Highly phytotoxic to hemp.
Extoll	1998[a]	Highly phytotoxic to hemp.
Defi	1998[a]	Highly phytotoxic to hemp.
Basagran	1998[a]	Highly phytotoxic to hemp.
Gratil	1998[a]	Highly phytotoxic to hemp.
Racer	1989[a]	Highly phytotoxic to hemp.
Assert	1999[a]	Efficient against field mustard and good selectivity.
Milagro	1999[a]	Highly efficient against field mustard and good selectivity, especially for the young stages.
Lontrel	2002[b] 2001[b] 2000[b] 1999[b]	Highly selective for hemp. Limited efficacy and risky.
Vip	2000[b]	Discriminates in favour of hemp. Limited efficacy.
Kerb Flo	2000[b]	Discriminates in favour of hemp.
Cent 7	2002[h] 2001[b] 2000[b]	Discriminates In favour of hemp. Not efficacious regardless of the stage of application.
Emblem	2002[b] 2001[b] 1998[a]	Very efficient (start of soiling one month after treatment, i.e. at the 5 cm stage). Good compromise between selectivity and efficiency.
Harness	2001[b]	Discriminates in favour of hemp. Average efficiency.
Pyron DE	2002[b]	Not efficacious regardless of the stage of application.
Andiamo	2002[b]	Efficient (start of contamination by weed infestation 1 month after treatment, i.e. at the 5 cm stage).

Note: [a]Product tested by the FDGEDA de l'Aube; [b]product tested by the FNPC.

as the industrial processing of the plant requires that it be free of contamination from plastic.

- The use of glyphosphate at weak concentrations (20–80 g/ha) can provide a solution eventually but its toxicity to hemp at these levels requires further investigation.
- An evaluation of the efficiency of the technique of false seeding necessitates that a system be developed for checking that the soil has been decontaminated.

This will involve estimating the number of *orobanche* seeds in the soil.[3]

5.6   Other Areas of Research

5.6.1   Irrigation

Hemp's water requirement was evaluated by the FNPC in 2001 and 2002 and found to be 30–50 mm/t of dry matter.[4]

5.6.2 Stem shorteners

Trials were undertaken by the FNPC and the Chambre d'Agriculture de l'Eure at the end of the 1990s. Their aim essentially was to research a method for producing textiles made of hemp. Using an idea now routine in the cereal world, they attempted to slow down the growth of the plant, reducing the production of hurds and increasing the production of grain. In the case of hemp, this was intended to produce shorter plants and thus make the crop easier to work. They failed to produce any conclusive results and needed further research. The FNPC also conducted a number of tests, but without any results to date.

5.7 Provisional Conclusions: Looking to the Future

At the end of this long period of development and research into the key factors on which hemp production depends, it would appear that there is now a good understanding of how to manage its cultivation; the majority of elements required to make the necessary technical choices for production are freely available. That said, there is much work to be done to further our knowledge and understanding of the mechanisms that affect and account for the variability in the yield of this crop.

Today, work focuses on the furthering of our knowledge of the physiology of this plant and our understanding of population characteristics (height, stem diameter and population density), factors which become important in determining yields. This can be achieved through a better understanding of the mechanisms controlling flowering (a key factor in the development of yields), as well as through an in-depth study of the water-related behaviour of this plant. There also remains much to learn about crop fertilization – and, in particular, how nitrogen can best be used – in terms of establishing optimum times for fertilization and formulations. Lack of uniformity in the stand causes variability in the yield. It is also important that a better understanding of the losses occurring between planting and harvest be developed, whether these losses arise at the time of emergence or during the course of cultivation. These losses are related closely to the dynamics of crop establishment and have a strong bearing on final yields. The major objective in terms of developing our knowledge of hemp cultivation further must be to develop a model that can predict yields better.

We should not forget the importance of a good understanding of the plant–soil relationship. In all cases, the development of yields is a product of the plant's potential and how this interacts with farming practices and soil conditions.

Notes

[1] Additional sources are available from other organizations, including the Chambre d'Agriculture du Loiret, as well as from other French and foreign research centres, but have not been exploited for this account due to time constraints.
[2] Growing degree days: the accumulated degrees according to a formula.
[3] For further information, see the developments made by François Desanlis in Part I of Chapter 6.
[4] It is interesting to note the findings of our Australian co-authors reported in Chapter 6, which show how watering can increase yields considerably. The decision to water is an economic and ecological one.

Appendix 1

Table A1. Vegetative stages of hemp.

Code	Definition	Remarks on leaves and flowers	Size	Duration	Total temperature	Cycle
Germination and emergence						Sowing and lifting
0000				0 d = seedling		
0001	Emergence of the radicle				85° Cj	
0002	Emergence of the hypoacetabular cup					
0003	Opening of the cotyledons			+/− 4–10 d	100° Cj	
Vegetative stages						Planting
1002	1st pair of flowers	1 node/1 pair of flowers				
1004	2nd pair of flowers	3 nodes/2 pairs of flowers				
1006	3rd pair of flowers	5 nodes/3 pairs of flowers	30–60 cm	+/− 25–31 d		
10	pairs of flowers	nodes/pairs of flowers				
						Active growth (from end of planting to early flowering)
		1 T of MS all of 120° Cj either 1T/ week		From 25–31 d to 49–65 d		
Flowering and formation of the seed						Flowering
2000	Point of change	Alternating leaves				
2001	Flower buds					
2300	Formation of female flowers	1st pistillate flower		From 25–31 d to 49–65 d	800° Cj (early or late sowing) to 1100° Cj (medium seedlings)	First flowering
2301	First female flower	1st styles visible				

Continued

Table A1. Continued.

Code	Definition	Remarks on leaves and flowers	Size	Duration	Total temperature	Cycle
2302	Female flowering	50% of bracts formed				Full flowering (judge the development of the inflorescences)
2303	Flowering of male flowers	1st flowers with closed stamens				
2304	Male flowering	Most flowers with open stamens				Finish flowering
2305	First mature seeds	1st hard seeds		From 90 to 105 d		
2306	Seed maturity	50% hard seeds				
2307	End of seed maturing	95% of seeds hard and opening		From 105 to 130 d		
Senescence						
3001	Leaves drying out	Dry leaves				
3002	Stalks drying out	Stalks fallen to the ground				
3003	Stalks decomposing					

6 Hemp Agronomics and Cultivation

François Desanlis,[1] Nicolas Cerruti[2] and Philip Warner[3]
[1]*Consultant, France and South Africa;* [2]*ITC, France;* [3]*Managing Director, Ecofibre Industries Operations Pty Ltd, Australia*

6.1 Introduction

In this chapter, François Desanlis, an agronomic engineer and hemp producer from the Champagne Ardenne area of France, provides an original account of the diverse subject of the agronomics and cultivation of hemp. He describes hemp cultivation in the most important hemp producing area of Europe since the rebirth of this industry in France in the 1960s. This account draws on his long experience in the industry and blends this with a thoughtful and well-reasoned approach to cultivation.

Next, the machinery, and in particular the harvesting procedure, is presented by Nicolas Cerrutti, an expert from the Institut Technique du Chanvre in France, thus updating the concepts presented in the French edition of this text. For this, he is thanked warmly.

Then follows a contribution from Australia, which only started producing hemp again in 2001. Mr Philip Warner, Managing Director of the company Ecofibre Industries Operations Pty Ltd, explains how they approach this venture and deal with the many problems that invariably present themselves when there is no experience to draw on (as is the case after a gap of 50 years).

Finally, Nicolas Cerrutti and François Desanlis present the harvesting process developed by the *Chanvriers de l'Est* in France in 2005.

6.2 Hemp Cultivation in France

6.2.1 Methods of cultivation

Hemp is an annual plant, sown in the spring and harvested at the end of the summer or the start of the autumn. Under the conditions usually encountered in Western Europe, the vegetative period lasts from 80 days for the earliest varieties (FIN134) to over 150 days for late varieties (such as Dioica 88, Novosadska and Kompolty).

The progression of the vegetative cycle, and therefore of the dates of sowing and harvesting, are, however, affected by the sum of temperatures (or 'growing degree days': the accumulated degrees according to a formula) where the induction of flowering is concerned and by reductions in day length where full flowering is concerned (Beherec, 2009).

Place in crop rotation

Hemp can find its place in a crop rotation system. There are no absolute rules as there are multiple possibilities, depending on the region and the needs of the producers.

Hemp monoculture was tested over a 12-year period (Chambre d'Agriculture de l'Eure), and on the flood plains of certain valleys in the Champagne region (Seine and Aube), France, the yields declined progressively. This was attributed to the uncontrollable development of self-propagating weeds, especially *Chenopodium album*, *Sinapis arvensis* (wild mustard), *Orobanche ramose* (hemp broomrape) and increasing plant death due to disease.

As with all spring crops, hemp allows a break in the autumn rotation of crops and can therefore break the reproductive cycles of certain difficult-to-control weeds, including *Bromus strilis* (bromegrass) and *Alopecurus myosuroides* (black grass) in cereals and *Calepina irregularis* (calepina) or *Capsella bursa-pastoris* (shepherd's purse) in rape. For farmers of sugarbeet, hemp can reduce the challenge from nematodes (*Heterodera schachtii*) in the ground. Finally, the absence of pesticide use in the years when hemp is grown allows the microflora to develop and become functional. This promotes the development of soil microfauna, which in turn improves the fertility and function of the soil.

According to producers, hemp cultivation is recommended after sugarbeet and maize, both before and after wheat and before spring barley.

Organic farmers appreciate hemp for its ability to leave the ground free of weeds in the autumn and for improving the soil's structure. They plant hemp after a leguminous crop and before wheat. In effect, there are many combinations and hemp can find its place easily in a crop rotation system.

Needs

Hemp can grow under all conditions and copes well with difficult situations. It is a rustic plant that has preserved many of its wild characteristics. In particular, the farmer will sow populations rather than a pure line, as is done for most cereals.

It is clear that results are better where the climate is favourable and the soil provided with the necessary factors for growth, water and fertilizing elements. Historically, our ancestors reserved manure for hemp crops because the hemp fields were closest to the farmhouse and therefore easiest to fertilize with manure and keep an eye on.

A HEALTHY SOIL. Wet soils are to be avoided, as these do not allow hemp to grow normally.

The cultivation of hemp on the soils found in the *Barrois champenois* region of France, and particularly those soils derived from certain geological ages (such as the Kimmeridgian or Hauterivian from the Mesozoic era), demonstrates this well. The plot presents a heterogeneous character as a function of the sequence of more or less wet soils and the annual pluviometry (Fig. 6.1). The height of hemp varies from close to zero to a normal height over a distance of a few metres. The same phenomenon can be seen in valleys, close to watercourses.

A NEUTRAL PH. The pH of the soil should fall between 6 and 8 for optimal results. Liming of the soil will be required if the pH is too low.

EFFECTS OF ALTITUDE. Tests undertaken in 2003 and 2004 in the Trièves (a region south of Grenoble, France) demonstrated that water provision was a more important limiting factor than altitude (trials limited to between 600 and 1000 m) and that mountain microclimates could provide perfectly adequate locations for production, providing the soil and climate were favourable.

THE TEMPERATURE. Hemp's sensitivity to temperature is known to a certain extent. The zero vegetation point is situated around 1–2°C; hemp is sensitive to frost (up to five pairs of leaves) at temperatures below –5°C, while the optimum temperature for vegetation is situated between 19 and 25°C. The time required for plants to grow to 90 cm in height is 30 days at a temperature of 19°C, but increases to 90 days at a temperature of 10°C (van der Werf, 1994).

Based on the observations made by the FNPC, emergence requires an accumulation of between 80 and 100°C days; the implantation phase (up to three pairs of leaves) requires 250°C days. The active growth phase that finishes at the end of flowering allows the production of 1 t of hurds for 120°C days. Seed maturity requires 40 days from the end of the flowering stage. Thus, for a variety of average maturity, the entire vegetative cycle requires between 2500 and 3000°C days in order to attain complete maturity.

Fig. 6.1. Hemp does not appreciate wet soils.

Excellent results are observed in South Africa and Australia, where the climate is much hotter and closer to the equator, providing the crop is sown 3–4 weeks before the summer solstice, thereby ensuring that flowering goes well.

WATER REQUIREMENTS. During the course of its vegetative cycle, 1 kg of dry matter (DM) mobilizes between 300 and 500 l of water; this corresponds to a pluviometry of 30–50 mm/t DM produced. The evaluation of the water reserves in the ground at the start of the vegetative cycle allows the pluviometry required during the course of vegetation to be evaluated. While this may be somewhat subjective, the efficacy of the pluviometry needs to be taken into account, for rain that intervenes after a period of drought will have a significant positive effect and will make up for any delays due to the water deficit. A water deficit that lasts right through until the end of the growth period will penalize hurd production heavily, although it will not affect the potential hemp hurd yield. In the same way, hydric stress at the time of flowering can interrupt flowering and consequently will advance the harvesting dates for the hemp seed.

FERTILIZER APPLICATION. The figures provided below summarize the results of the experiments conducted by different organizations between 1977 and 2004: the FDGEDA Aube, APVA Haute Marne, SCPA, Chambre d'Agriculture de l'Eure and INRA (life cycle analysis, 2005).

For 1 t of DM, the vegetation process mobilizes:

- 18–24 kg of nitrogen
- 5–10 kg of P_2O_5
- 20–40 kg of K_2O
- 30–40 kg of CaO
- 8–10 kg of MgO

Less fertilizer is lost than the plant uses to grow, especially as the leaves and roots remain behind in the field. The expenditure for 1 t of DM produced corresponds to:

- 9–12 kg of nitrogen
- 6–8 kg of P_2O_5
- 12–19 kg of K_2O

- 20–25 kg of CaO
- 2–4 kg of MgO

No data currently exist on hemp's requirement for sulfur and other micro- or oligo-elements. It would appear that where sulfur is deficient, it is the production of hemp seed that suffers.

The period of contribution and formulation of the fertilizer must be adapted to the soil and climate. The objective is to provide the plant with as much nutrition as is required to meet its growing needs during the most active phase of growth between the middle of May and the end of June.

The use of organic manure is possible within reason. A late mineralization of the organic matter at the end of the vegetation period can lead, in some years, to difficulties in achieving the maturity needed for hemp seed to be harvested in good condition (trials conducted by Trieves, 2003 and 2004). Hemp plants are able to continue growing for several weeks if the conditions are favourable; nitrogen supplementation plays an essential role in this phenomenon. The management of organic manure brings with it concerns over how to choose and time nutritional contributions so that the plant is able to meet its needs at the right time and in the right quantity.

Preparation of the soil

As with all spring crops, the preparation for sowing needs to ensure that the soil structure is not compacted, while still being sufficiently smooth. The taproot, while powerful, is stopped easily in its development by obstacles such as compaction layers, or 'pans'. In such case, the taproot takes on an L-shape, which reduces its efficacy in times of drought.

Thus, ground that contains a large proportion of clay should be ploughed or harrowed at the end of autumn or the start of winter. This will allow the ground to winter and will optimize its structure. Where the soil is silty, ploughing can wait until spring. Ploughing can be restarted once the ground is dry. A false sowing is advised in order to manage weeds and favour a warming up of the seedbed.

Simplified cultivation techniques produce identical results, providing all other conditions remain unchanged. A superficial treatment at the start of the spring, on ground that has been dried, will ensure that weeds germinate. These can then be destroyed easily, either mechanically or by means of a non-selective weedkilling treatment. Where the ground has been poorly prepared and has compacted areas, it is advisable to break up these zones in order to allow plants to root correctly. For the driest areas, the stubble should be ploughed up in the autumn in order to help preserve water.

SOWING. Producers claiming subsidies are obliged to use certified seed stock only. The use of farm-produced seed, even if it originates from authorized varieties, does not qualify for CAP subsidies.

Choice of variety. The choice of variety from within the list of authorized varieties published annually in the journal of the European Union is based on the markets and chosen objectives.

The main supplier is the Central Cooperative of Hemp Seed (Coopérative Centrale des Producteurs de Semences de Chanvre) for French varieties (produced by the FNPC) and USO31 (a Ukrainian variety).

Other listed varieties can also be sown, providing they are multiplied and available from producers and/or importers.

The quality of the seed falls if it is poorly stored. This does not pose a problem for seeds obtained that same year. Where unused seed is stored, however, it must be placed in a cold room. Ideally, germination rates should be ascertained for seed stored on the farm and storage for more than 1 year should be avoided.

Date of sowing. The sowing date is dictated on the one hand by the temperature of the ground and on the other by the need for the plants to develop sufficiently to achieve full flowering. The soil temperature must be sufficiently elevated (8–10°C) to ensure that vegetative growth gets under way quickly and any competition from weeds is limited. In practice, hemp is sown from mid-March in early zones up until the last week of May. Attention must be given to the quality of seedbed preparation because seeding in wet soil produces poor stands. It is better to plant a little later, as long as there is season enough.

Where the southern hemisphere is concerned, trials conducted in Bathurst, South Africa, in 2003, produced similar conclusions, but with at an interval of 6 months.

Depth of sowing. The planting depth for hemp is similar to that practised for other crops: the depth should be eight times the length of the seed. When conditions are particularly dry, a depth of 2–3 cm places the seed in soil that is still humid. At a later time, when conditions are more humid, shallower sowing is recommended. The germination rate declines rapidly as soon as the sowing depth increases beyond 3 cm, particularly in silty soil that has been compacted by rain.

Sowing type. The means of spreading hemp seed, either in line or randomly, does not affect the behaviour of the crop. The degree to which the hemp crop smothers weeds does not appear to differ between either seeding method. It is therefore not necessary, as it is for flax, to use special equipment on the seeding machinery.

Quantity of seed sown. The quantity of seed, and therefore the density of the crop, must be adapted to the objectives set. It is common to use a sowing density of between 35 and 80 kg/ha.

Trials undertaken in the Champagne area (AVPA Haute Marne 1985 and confirmed by FDGEDA Aube and ITC) demonstrate that the amount of dry matter produced per hectare does not differ for densities as far apart as 30–300 plants/m². For the lowest densities, there is a tendency to favour the production of hemp seed. The morphology of the plants is also very different, in particular at the level of the stems: these can become as wide as 3–4 cm and develop branching that is not favourable to fibre production. Finally, a low density favours greater growth, producing taller plants that may develop problems during the vegetative phase. In particular, losses may be seen following large storms and during harvesting where the hemp seed is collected using a combine harvester according to the 'French method'.

The weight of 1000 seeds is typically somewhere in the region of 15–17 g. There is no relation between the size of the seeds,

their germination and the vigour of the young plants.

ROLLING. Rolling can be necessary in order to smooth the soil after sowing and to promote germination under dry conditions. It is recommended, however, that prudence be exercised, in particular on silty soil subject to rain battering or rolled and then heavily rained on, as these will penalize emergence. Rolling can also be of value where the soil has a calcareous-clay character, as this procedure will push stones into the soil. This makes cutting the crop easier at harvest and reduces the number of stones picked up with the hurds and incorporated into the bales. In these cases, rolling can be undertaken later, when the hemp has already emerged, up until the 10 cm stage.

Fight against weeds, diseases and other pests

WEEDS. Hemp's ability to smother out weeds is weak during the early stages of growth but rapidly becomes significant until the point where it can suffocate all weeds. Until that point is reached, when the crop covers the whole ground, there is a real risk that weeds can gain the upper hand. This is particularly true of *brassicaceae* and *chenopodiaceae,* whose speed of growth is greater when conditions do not favour growth. It should be noted that late developing varieties, and in particular dioecious varieties, are most vigorous at the start of vegetation and cover the ground more rapidly (LCDA/SPC 1999 tests). No herbicide is approved for use on hemp crops. Even where trials undertaken by the FDGEDA de l'Aube and the FNPC have found evidence to support the use of a solution to these early problems, it is recommended that the risk be minimized by practising a false sowing and sowing in soil that has been sufficiently warmed up in order to allow the hemp to overtake the weeds.

The preceding chapter provides details of the herbicides tested by the FDGEDA de l'Aube and the FNPC since 1989 (Table 5.8).

Infestations of broomrape (*Orobanche ramosa*) are a recurring problem in most production areas (the Loire valley, Champagne Crayeuse, Haute Saône). No solutions currently exist to deal with this chlorophyll-free dicotyledenous plant that also attacks other species. It appears that this hemp parasite is the same species that parasitizes tomato and melon, as well as rape and sunflower. A number of other non-cultivated plants also host broomrape, although few further details are available. Ongoing work at the Universities of Nantes and Jussieu hopes to provide further information over the coming years.

DISEASES. Breeding over the centuries has never led to the commercialization of a pure line variety. The varieties supplied to farmers are heterogeneous populations in terms of their resistance to pathogens. Consequently, only a small fraction of plants are actually affected by a health threat. Thus, each year, a variable percentage of plants die during the vegetation phase, forming 'dead roots'. Roots suffering from a range of diseases can be seen, including *Botrytis cinerea*, *Sclerotinia sclerotinium* and *Rhynchosporium secalis*. Other pathogenic agents are probably also present, although their effect on yield is negligible.

Although a number of plant diseases have been reported on hemp, major disease outbreaks are uncommon. The most important disease of hemp is grey mould, caused by the fungus *B. cinerea*. This disease attacks hemp stems under conditions of cool to moderate temperature and high humidity. In severe cases, *B. cinerea* can destroy a *Cannabis* crop completely within a week. Hemp canker, a fungal disease caused by *S. sclerotiorum* (Lib.) Mass., attacks hemp stems, resulting in wilt and stem breakage. Under cool, wet growing conditions in the Netherlands, severe hemp yield reductions have been reported recently due to *B. cinerea* and *S. sclerotiorum*. Fungicide applications every 10–14 days were required to keep the crop disease free. Seed and soilborne diseases such as *Pythium* and *Fusarium* are commonly controlled by seed treatment with fungicides prior to planting (Dempsey, 1975). Many other minor diseases have also been reported on hemp.

Further information can be found on the Oregon State University website (http://extension.oregonstate.edu/catalog/html/sb/sb681/#Yield).

OTHER PESTS. Hemp is usually little affected by insects. Various studies have shown the effect of cannabinoids as insect repellents. Some attacks by larvae of the insect genus *Tipula* are seen occasionally. These larvae destroy plants by attacking the base of the stem. In 1993, France witnessed a spectacular invasion of larvae of the cabbage moth (*Mamestra brassicae*), with almost three-quarters of fields requiring insecticide treatment. While the apparent damage appeared impressive at the end of May and the beginning of June, there was no significant difference in the yield between those fields that were and were not treated. Some attacks on the cotyledons by insects of the genus *Psylliodes* (flea beetles) have also been reported. Finally, the Eurasian corn borer (*Ostrinia nubialisa*) also finds refuge among hemp without provoking any significant damage.

Game can sometimes cause damage, as is the case for large populations of rabbits living close to hemp fields and using these as a promising food source. Wild boar can also cause certain problems for cultivators at the start of the vegetation phase or just before the hurd bundles are collected, when their foraging for earthworms and insect larvae can produce holes that interfere with the harvesting process.

Seeds can be treated with bird repellant to help reduce the attractiveness of hemp seed. The seeds can prove just as attractive at harvesting time, when clouds of starlings (*Sturnus vulgaris*) can cause heavy losses (as in the Rhine valley in 2003) through the consumption and shelling of hemp seed a few days before it is due to be harvested.

Slugs can provoke some losses. While it may be worth being particularly vigilant on the edge of woodland and beside paths, it is rarely necessary to intervene, as the loss of a few plants is unlikely to impact on the final yield. The management of risks is similar to that required for other crops.

Climatic accidents

HAIL. A distinction must be made between losses during the vegetative phase (before flowering) and those sustained after flowering.

Spectacular as they may be, losses sustained during the growth period will impact negatively on hurd production, whereas seed production is usually unaffected. Stems sectioned by hailstones continue growing by developing several branches at the point of sectioning. The quality of the hurds is altered and the fibres are often of poor quality.

The same is not true of hail that strikes after the end of the vegetative phase. Excepting shelling losses, the damage caused to the stems, particularly if weather conditions remain humid, can serve as entry points for disease that are detrimental the hurds. In such cases, the crop is virtually lost.

LODGING.   Wind associated with rain can flatten plants during their vegetative phase. In the majority of cases, plants can right themselves and recover. If the lower part of the stem remains on the ground, however, this can make reaping difficult.

Lodging during early stem elongation has a relatively small effect on yield, as the plant can right itself and reform the canopy.

Lodging is aggravated by good growth conditions, particularly where stem elongation is unusually fast. This can lead to temporary stem weakness. This problem can be provoked by excessive nitrogen fertilization.

SEED LOSS.   Hemp seed is very sensitive to seed loss (or 'shelling') by the wind. As soon as the plant reaches maturity, the seeds dehisce easily. The morphology of certain closely grouped fertile heads of certain varieties (e.g. Finola, USO) can limit this phenomenon.

The decision to harvest must take this risk into consideration. Thus, in practice, harvesting starts before all the seeds attain maturity. This minimizes the risk of seed loss.

SELF-THINNING, OR DEAD STALKS.   A distinction needs to be made between losses sustained at the time of emergence, where seeds will not give rise to an emerging plant, and losses sustained during the growth phase.

It is normal to see a halt in the growth of some growing points, a decline or even a premature end to the growth cycle. There can be many causes of this phenomenon. Generally speaking, these plants have suffered an insult that affects their growth. Neighbouring plants gain the upper hand and overwhelm them. It is equally possible that the ground structure does not allow the root system to develop sufficiently. The decline in growth may be seen early in the active growth period and the first dead stalks at the start of flowering. An attack by one or several pathogens can generally be observed on these stalks, although it is usually impossible to tell whether these have caused the problem or whether they contaminated the root after it went into decline.

The significance varies, but a number of consistent elements can be identified. Thus, the proportion affected is low where plant densities are low, increasing from a density of 50 plants/m^2, at which point 50% of the plant's stalks may be found to be atrophied or dead.

Climatic accidents have a tendency to amplify the phenomenon, either through excess water or drought.

Cultivars are observed (LCDA, FNPC 1999 trials) to differ in their propensity to self-thin.

6.2.2 The crop's itineraries

The crop's itineraries are determined by the end purpose, or use, and therefore by the markets that stipulate the technical requirements to be met. Fibre quality varies significantly as a result of the options taken. By contrast, the quality of the hemp hurds remains unchanged.

- Thread production is the oldest market. In Europe, the Danube valley has preserved its knowhow. Chinese production is oriented primarily towards this market.
- Paper manufacturing developed as a market following the invention of printing. This represents the main destination for hemp fibres produced in France, with a focus on the production of specialist papers: Bible paper, cigarette paper and paper for bank notes.
- Plastic manufacturing is a developing market with mineral and/or synthetic fibres being replaced by plant fibres. This is the main outlet for German and English producers.

The specifications can be summarized as follows (for more information, the reader is referred to Chapter 9):

- The desirable qualities for thread production are essentially its fineness, suppleness and solidity.
- The paper market is very concerned with contamination-related problems, in particular, those posed by plastic materials and weeds. The presence of chlorophyll pigments in fibres can also pose a problem: the pulped paper requires bleaching and this results in the production of chlorine-based waste that must then be disposed of.
- Fibres destined for the plastics industry must, depending on the industry, be virtually free of hurds for temperature-hardened plastics. For thermoplastics, fibres must be graded by length according to their use and according to the technology.

The end uses of the fibres influence the production techniques.

These techniques meet the need to produce first-class fibres for the companies that are adapted to the use the extracted fibres will be put to.

For hemp seed, the quality required imposes precise specifications that take the seeds' fragility into account.

Production of textile fibres

The standards for hemp cultivation for the textile industry essentially come from the Danube valley, which is the last area in Europe where this form of production is still in existence.

In order to obtain fine and supple fibres, it is necessary to cultivate dioecious varieties to benefit from fibres derived from the male plant. These fibres are fine, comparing well with flax fibres, whereas fibres from the female plants (or from monoecious plants that present the same characteristics) are eight to ten times wider. Although the ratio of male to female flowers is close to one, the total weight of female fibres is double that obtained from male plants. The male fibres, however, are fine and can produce good-quality thread.

The sowing density aims to produce a high density of plants and thereby limit stem diameter. It should be noted, however, that the diameter of the stems is independent of the fineness of the fibres. That said, a close correlation exists between the size of the fibre bundles and the size of the stem. Knowing that the mechanical preparation of textile fibres (essentially carding operations) will break up the bundles without going as far as to isolate individual fibres, it is recommended that bundles of small diameter are produced. It is also true that retting is more efficient if stems are thin. In practice, it is very difficult to exceed 300 plants/m², which would allow stems of 5 mm diameter to be produced. Producers of hemp for the textile industry sow at around 80 kg of hemp seed/ha.

Harvesting is undertaken before maturity, at the end of flowering. Dioecious varieties are late and the optimal stage is reached during the second fortnight in August. Taking into account the time of reaping, it is not possible to harvest the hemp seed. The stems are bundled into sheafs and tied with stems. These sheafs are loaded into the trailer and taken to retting baths. Two techniques are used for retting: either the trailers are submerged in the retting baths or they are emptied into the baths. Retting lasts for 8–10 days for a water temperature of 20°C. At the end of this period of immersion, the sheafs are dried and then stored under shelter or in stacks.

An alternative to the traditional method is being developed in Italy, where hemp production seeks to produce plant stands of a similar height to those of flax. This allows the hemp to be harvested and fibres extracted using the same techniques as are used for flax. A shortener (a mix of chlorocholine chloride and ethephon) is sprayed over the crop at the start of the rapid growth phase, thus limiting growth. A total weedkiller (glyphosphate or sulfosphate) is then used to stop the vegetative phase at the very start of flowering. The harvesting, retting, baling and fibre extraction are all then conducted in a similar way to that practised for flax.

'Bench top' or in laboratory retting methods (using enzymatic or chemical processes, or a combination of the two) were tested in France approximately 15 years ago following fibre extraction to try to find an alternative to the retting bath. No industrial applications resulted.

Traditionally, up until the middle of the 19th century, male stalks were torn out first as early as the end of July. These were submerged immediately in retting pools (known in French as *roises* or *rôtière*) in order to start the retting process. Female plants were then torn up, dried and deseeded before also undergoing retting. Fibre from the male stalks was sold and supplied small farmers with an important part of the revenue generated by a hemp crop.

Fibre production for the paper industry

Processes originate essentially from the cultivation methods used in France:

This type of production requires that the harvesting of hemp seed be combined with the harvesting of the stems (combined).

Hemp cultivation competes with other spring crops, including peas, sunflower, maize and potatoes, for its place in crop rotation systems. It must yield similar revenue if it is to attract the interest of farmers. In this respect, hemp seed is an important part of the revenue from this crop.

Another argument militating in favour of the twin harvesting of hemp seed and hurds is that it is easier to obtain high-quality hurds (yellow to grey white) when a mature crop is harvested than from plants that are still green and have a higher water content. Statistically, there is more than one chance in two of obtaining poor-quality hurds when the crop is not combined: the hurds will be either green, as weather conditions have not allowed the minimum amount of retting, or black, because reaping has been followed by a lot of rain, resulting in excessive retting. In both cases, quality suffers. An epiphenomenon is also observed with hurd conservation problems, as the seed attracts rodents. Finally, the fibre extraction of non-combined hurds will introduce small quantities of seed into the hemp hurds. This can cause problems where the hurds are used as animal bedding (three-quarters of the hemp hurd market) for the hemp seed renders it more appetizing and encourages animals to eat their bedding. Where the hurds are used as mulch, the appearance of hemp plants among flowerbeds is little appreciated.

Setting aside fibre colour, quality hemp hurds will be clear in colour and will sell well in the market place, particularly as bedding material.

Precocity (earliness) is an interesting factor. It allows harvesting work to be undertaken at the earliest opportunity, taking advantage of favourable climatic conditions and freeing up the fields early. In most cases, hemp is used in a rotation system before autumn wheat.

In this context, monoecious varieties are preferred. The sowing density is a compromise between the price of the seed and the technical constraints of obtaining a crop that is sufficiently able to smother out the weeds and the plants that fail to develop well, thereby allowing the combine harvester to pass easily during the harvesting of hemp seed. The amount generally used is approximately 50 kg/ha.

Hurd harvesting requires the use of plant string (in order to avoid the risk of synthetic fibre contamination). This currently makes the use of high-density parallelipipedic bales (termed square bales) impossible at the present time, as these must be bound using synthetic string.

The production of technical fibres

These sources are recent and from Germany.

There are many similarities between the production of fibres for the paper industry and the production of technical fibres. The constraints are somewhat less demanding where plastic contamination is concerned. This makes the use of high-density parallelipipedic bales possible, and therefore storage and transportation easier.

After fibre extraction, the fibres destined for use in the production of temperature-hardening plastics (a mat, or pre-form, of mixed hemp fibres and plastic fibres such as polypropylene, which can then be temperature fused and moulded) must have as low a load of hurds as possible (<1%). By contrast, those fibres destined for use in thermoplastics (compounds made of very short fibres mixed with polypropylene and thermo-injected) can have a similar composition to those used in paper manufacturing.

6.2.3 Harvesting techniques

The harvesting of hemp seed

Hemp seed is very fragile. Despite the shell that protects the cotyledons, the shock required to crack the seed is small. This then opens the way for the oxidation of lipid stores and, consequently, deterioration in the quality of the product, as the triglycerides are altered. Whether the seed is destined to be sown or sold as a foodstuff, the quality demands are similar. The quality control standards require close control of all stages of harvesting, drying, storage and conditioning. Evaluation of the quality of the hemp seed can be established simply by measuring germination following storage for a few months at ambient temperature.

A more objective criterion that is used as a basis for commercial transactions is the measurement of the acidity index (oleic acidity, OA) and/or the oil peroxide index of the sample to be evaluated. The maximum oleic acidity permitted for alimentary oils is 2% (measured on the final product). The minimal OA of hemp seed that has been optimally collected is already 0.3%. Drying will then increase the index by 0.1% and storage will increase it by 0.05% per month. The objective is to commercialize a hemp seed with an OA below 1%. In commerce, this is the value commonly accepted as determining that hemp seed is food grade (Box 6.1).

Hemp, though it has been cultivated for thousands of years, has preserved the characteristics of wild plants: not all the seeds reach maturity at the same time. We will not revisit the traditional methods of harvesting seeds by hand, though these are still used in certain areas to collect seedstock for resowing.

An antiquated harvest method that would fall between harvesting manually and modern practices employed two stages. First, hemp was cut and placed in sheafs according to a characteristic pattern that regrouped and crossed the heads of the plants in the centre of the sheaf. Reaping was undertaken just as the first seeds arrived at maturity. A period of 1 week to 10 days was then required before the seeds could be collected. In this way, the hemp seed continued to mature while the hurds were drying. The major inconvenience of this technique lies in its vulnerability to climatic conditions (in 1984, for example, a lot of hemp seed actually germinated within the sheafs).

The development of combine harvesters for hemp dates back to the 1980s and the arrival of axial flow combine harvesters that used the principle of centrifugal seed collection. Today, the majority of combine harvesters can be adapted for the harvesting of hemp (Fig. 6.2). Given the amount of material that must pass through the machine, only the heads are cut and combined, with the cutting undertaken as high as possible on the stalk. This capability represents significant progress, as the farmer is now much less dependent on weather conditions. The farmer must, however, determine the optimal time for harvest. This is a compromise between the yield of mature hemp seed and the weather conditions, which, were they to deteriorate, could result in shelling at best and germination

---

Box 6.1. Sampling method for the validation of the quality of oleic acid in hemp seed.

Each batch received at the silos is sampled in order to determine the acidity index and oleic index. The acidity index reflects the extent of fatty acid hydrolysis. These characteristics are useful indicators of the quality of raw (or virgin) oil and allow the refining process and its efficacy to be monitored carefully (measure of residual acidity).

Two methods are used to verify these indices: the titration method and the potentiometer method. These measurements concern all the fatty substances, whether they are of plant or animal origin, as determined by the European standard NF EN ISO 660.199 (approved in France in July 1999).

The sampling and tests conducted at the silos allow the peroxide index to be verified. This index reflects the formation of primary compounds of oxidation. This test is easy to conduct and is therefore widely used in units that do not possess sophisticated laboratory equipment.

Finally, it should be noted that in order to comply with alimentary standards, the oleic acidity level must be below 2% where the product is intended for human consumption. In concrete terms, taking into account the manipulations that the product undergoes after harvesting and in the factory, the processor will ensure that this level stays below 1%.

In this way, the end product can be offered for human consumption.

in situ at worst. The optimal time is very much dependent on the weather conditions during any given year (excepting a period of drought, which can stop vegetation during full flowering, as was the case in 2003). The determinant factor is the variety's precocity (earliness), defined as the date of its full flowering. Approximately 40 days can then be added to this in order to determine the optimal time to harvest the seed.

Thus, for early French varieties and under the conditions typical of north-eastern France, harvesting starts around 12 September, with a peak around 19/20 of that month. At this time, four-fifths of the seeds are mature. In order to ripen the fraction of seeds that are still green, a further 10 days would be required, together with the accompanying risks outlined above. USO31, which is the earliest by 10 days, arrives at an optimal point around 10 September (Fig. 6.3).

The setting up and adjustment of the combine harvester can also impact on the quality of the hemp seed. It is necessary to turn the arms at a slow speed (approximately 300 turns/min) in order not to hit the grains too hard. The counter beater must not be too tight (although it must not be too loose either, as there must be room for manoeuvre where blockages occur) and cleaning must be sufficiently energetic to eliminate the maximum amount of impurities (empty seeds, green seeds, leaf fragments and hemp hurds). When the hopper is emptied, the emptying screw must not turn too fast in order to avoid any further unnecessary damage.

As a general rule, the moisture of the hemp seed is above 20%. Producers from the LCDA are required to deliver a product with a moisture level of less than 27.5% and/or an impurity content below 25%. The quality of the hemp seed can deteriorate very rapidly and

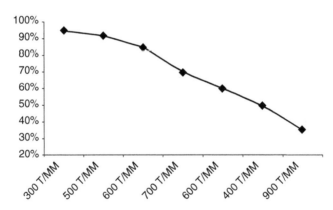

Fig. 6.2. Control of the combine harvester. The germination rate is correlated with the speed of the beater.

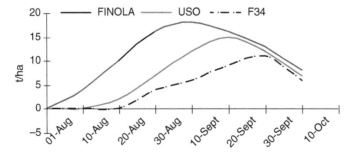

Fig. 6.3. Evolution of the yield as a function of the harvesting date and earliness (modified from Densanlis, 2001).

it must be dried and cleaned as quickly as possible. Ideally, drying should be started within a few hours of harvesting. Where producers are a long distance from the drying silos, the containers can be ventilated via a perforated base, which can help prevent overheating. The driers are modified, again with the objective of avoiding unnecessary shocks, by placing rubber bibs for the hemp seed to fall on. The drying temperature is extremely important; the seeds must not be allowed to reach a temperature above 40°C.

Particular mention must be made of those farmers undertaking drying on the farm. If an air-heating device is used together with an aerating/ventilating device, it is essential that the gas produced by the hot air generator does not contact the drying seed. Hemp seed is easily impregnated with combustion smells and this is likely to render it unsaleable.

Under these conditions, hemp seed can preserve its qualities and will make a first rate raw material for oil extraction or dehulling for incorporation into energy bars, chocolate or cakes.

Hemp seed can be stored when its humidity is at 8%. Commercial standards stipulate 9%, but experience suggests that there is still a risk of deterioration at this level.

The method of storage, especially if this is undertaken at the farm, must pay due attention to the attraction hemp seed represents to birds and rodents. Conditioning in big bags on palette boards and protection from rodents by rat bait may be a suitable answer.

Hurd harvesting

MOWING. As the crop matures, the lignin content of the fibres increases and the stems become increasingly difficult to cut. In the past, the stems were pulled up with their roots, a harvesting method that is still used for flax.

Today, the majority of reaping machines placed on the market by manufacturers are disc mowers. These perform well when harvesting fodder as their forward speed is not limited in the way that it is for section mowers. That said, the impact cutting technology is not adapted to hemp harvesting. The discs are made for cutting fine fluid material such as hay and straw, which must not clog up and block their rotation. Furthermore, this sort of cut generates projectile material and tarpaulins must be used around the blades in order to protect the users. These tarpaulins, while necessary for health and safety reasons, can get in the way of cutting hemp due to the unusual height of this crop.

Knife mower or section mowers. Consequently, cutting blades are the best adapted for mowing hemp. They have the advantage of producing a clean cut, release the material quickly and achieve a high reaping speed (between 15 and 20 km/h) with a working width of 2.4 m.

The system of blades most commonly used in France is the double knife Busatis-bidus system (Fig. 6.4). This allows reaping to be conducted within a few centimetres of the ground, leaving the earth clean while harvesting the maximum amount of hurds.

Unfortunately, the weak point of this system is the need to sharpen the blades frequently in order to maintain optimal performance. The time between sharpening depends on the height of cutting and the presence of dead stalks or stones in the field. Depending on the conditions of work, a newly sharpened set of blades can cut between 1.5 and 15 ha efficiently.

In order to satisfy the demands of producers working a non-combined system (no hemp seed harvested) with plants growing to 4 m in height (late flowering varieties with no cutting of the tops by a combine harvester), some manufacturers market mowers with two, three or even four cutting heights (Figs 6.5 and 6.6).

Fig. 6.4. Double knife-cutting system (Busatis, ITC©).

Fig. 6.7. Automatic 'swather cutter', initially used for harvesting lucerne (ITC ©).

Figs 6.5 and 6.6. Mowers cutting at two and three levels with a working width of 4 m. These are manufactured by Tebeco (Tebeco ©, 2009).

Fig. 6.8. Conditioning mower (ITC ©).

MOTORIZED SWATHER CUTTING AND MOWER CONDI-TIONERS (MOCO). This equipment is often old and dilapidated and is used mainly in the oldest areas of hemp production (e.g. Interval/Eurochanvre) where the equipment has been in use since the time that hemp production started. These machines must be bought second-hand and finding spare parts for them can be difficult. Adapted and transformed using dividers, protective metal sheets or deflectors for use when harvesting hemp, they do not represent the future, particularly where the new production areas are concerned. Furthermore, their work capacity is very limited where yields are high (Figs 6.7 and 6.8).

Nevertheless, this machinery does have one advantage: it is able to cut a swath and ultimately condition the hurds in one single passage. This reduces the amount of stones incorporated into the swaths, especially on fields that have a lot of stones.

MOTORIZED SILO FILLER WITH MODIFIED ROTORS. The first company to use a silo filler for hemp harvesting was the Dutch company, HempFlax, with the same objective of cutting the hurds into pieces as the multi-level mowers. In order to reap hemp, a classic silo filler has to be equipped with a specific rotor block. The rotors currently available on the market allow hurds to be cut into 20–80 cm pieces. The speed of the rotor's chainbelt is very low (around 300 turns/min) and the number of blades reduced to one or two. Equipped with a Kemper beak, these machines allow high-yielding hemp to be harvested. In 2009, these silo fillers were used in the south-west (Euralis/Agrofibre) and in the Vendée (CAVAC), France (Fig. 6.9).

MODIFIED COMBINE HARVESTERS. Combining equipment has been developed by the German fibre

Fig. 6.9. Harvesting in the region of Toulouse by the company, Chanvre de Garonne (Sébastien Garcia ©, Trait d'Union Paysan, 2008).

Fig. 6.11. Swathing the hurds with a giro swather (ITC ©).

Fig. 6.10. Combine harvester BAFA (ITC ©).

extraction company, BAFA (Fig. 6.10). It allows hemp seed to be harvested, the hurds to be cut into pieces of equal length (approximately 50 cm) and its organization into swaths in one passage. This means of harvesting allows the hurds to be pressed into bales and has a processing rate of 1.3 ha/h.

Given that the material transits through the machinery, problems can occur in fields where the yield approaches 10 t of DM/ha. In such cases, the processing rate is lower.

SWATHING. This procedure is performed almost exclusively by mono-rotary swathers (Fig. 6.11). In addition to collecting the hurds before compressing them, swathing softens the hurds, thereby facilitating drying and retting. According to the weather conditions and the crop yield, two or three passages may be realized.

The introduction of turning equipment and twin rotor swathers that deposit the hurds centrally is being studied by the ITC in order to improve yields and harvesting performance.

PRESSING/BALING. The choice of pressing into round or square bales depends on the market to be targeted (technical fibre or paper fibre) and the design of the fibre extraction process. In order to be transformed into paper pulp, hemp hurds must be free of all plastic contamination. When square bales are tied, however, small pieces of twine are cut and mixed with the hurds. Natural string made from sisal is not elastic enough and breaks when the bale exits from the compression canal. All the hemp produced by the LCDA and Interval/Eurochanvre are pressed into round bales using equipment with differing chambers (Fig. 6.12).

More recent productions, such as that of Euralis/Agrofibre in the region of Toulouse and that of the CAVAC in the Vendée, have chosen to press hemp into bales for transport and storage reasons. For this sort of storage, the cutting of hurds into short pieces is a prerequisite.

6.2.4 Yields

Depending on the objectives specified, yields will vary. A number of constants exist, however.

As mentioned earlier, yield is virtually independent of the sowing density across a range starting at 40 plants/m², with an increase

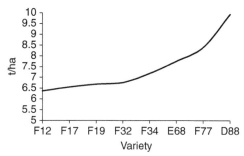

Fig. 6.13. Hurd yield for French varieties according to their precocity. (ITC summary, 2004).

Fig. 6.12. Pressing the hurds with a round baler in the Aube (ITC ©).

in seed yield where sowing density is low (as is the case where seed is sown in lines).

The commonest yield-limiting factor is water availability, which particularly disadvantages small stony fields. There are no references on the subject of hemp irrigation. The critical period for hurd yield is the rapid growth phase between the end of May and the end of July. The hemp seed yield is affected by climatic conditions at the end of the vegetative period. Any significant stress at this time can shorten the flowering time and therefore the yield in hemp seeds.

The hemp seed yield for early French varieties cultivated in France and Germany is very regular, between 900 and 1100 kg/ha, and no marked distinction can be made between shallow and deep soil. The yield is noticeably higher by 150–250 kg/ha for USO31, which is earlier by 10 or so days. The yield in hemp seed increases with advancing maturity, attaining its optimum when approximately 80–90% of the seeds have taken on a brown colour. At this stage, the risk of shelling must be taken into consideration and, in most cases, it is not worth waiting to harvest, as losses will exceed any benefit.

Hurd yield is much more variable (Fig. 6.13). The duration of the growth phase plays a role in the development of the yield. Thus, where an early variety is sown (under good conditions), hurd production will be favoured. Late varieties exhibit a longer growth phase and, in this respect, perform better than early varieties. The latest varieties are almost all dioecious, but this does not allow us to conclude that hurd yield is lower for dioecious varieties than for monoecious varieties, as male plants are much lighter than female plants.

In areas of eastern France where the potential is poor, the hurd yield is generally situated between 4 and 7 t/ha. In areas more favoured by nature (deep soil with good water reserves), the yield varies between 8 and 13 t/ha, whereas wheat yields vary between 8 and 11 t/ha.

The hurd yield for dioecious varieties used to produce fibre for the thread industry falls between 13 and 15 t/ha.

6.2.5 Retting

According to the dictionary definition, retting is an operation that facilitates the separation of textile fibres by destroying the material that binds these fibres together. This can be done through macerating in water, exposing to dew or to humid heat.

Whereas retting in water is still practised and used in the Danube valley and in China for the production of textile fibres, it has been abandoned completely by producers in Western Europe.

Field retting is recommended by the majority of fibre extraction workshops. Producers will, however, prefer to take advantage of favourable weather conditions and bring in dry hurds; retting therefore occurs as a result of climatic circumstance rather than through a deliberate act. A significant financial incentive would be needed to advance this practice.

Retting facilitates fibre extraction. While the quality of the fibre can remain stable for several weeks, this is not true of the hemp hurds, which deteriorate rapidly and change from a white colour to a more or less dark grey, resulting in the loss of some of their market value

6.2.6 Actions of different production factors on yield and quality

Hurd yield is related to the stem diameter, together with the density and height of the plants that have not been smothered or fallen victim to disease. A long and problem-free growth period will favour these factors. Late flowering dioecious varieties are generally more productive than monoecious varieties, as the period of vegetative growth is longer.

The hemp seed yield depends on the duration of the flowering period, which itself is linked to the sowing date, to the sum of temperatures and the amount of water received during the flowering period. Monoecious varieties flower earlier than dioecious varieties and are therefore more productive.

Monoecious varieties

In Europe, almost all the hemp cultivated is of a monoecious variety whose yield is better the later it flowers, as this allows a longer vegetative growth phase. By contrast, early varieties grown in temperate climates produce a better hemp seed yield, as maturity is attained sufficiently early for the plant to benefit from favourable climatic conditions (summary of the trials on varieties from the FNPC, FDGEDA Aube and ITC).

We do not have enough experience to allow us to draw lessons for crops grown at lower latitudes. Hemp is a temperate climate crop, best suited to latitudes above 35 degrees.

The covering of the ground or 'canopy' improves the later flowering variety (trials FDGEDA Aube, 1999).

Density

The seeding rate is similar from one region to another for the same use, varying between 45 and 50 kg/ha for monoecious varieties. One of the objectives over and above the quality of the yield is to promote as rapid and as efficient coverage of the ground as possible in order to avoid the development of weeds.

Low densities favour plant height. This is particularly well illustrated by the 'border effect', where the plants growing at the edge of a plot are 20–30 cm taller and demonstrate more branching.

There is no correlation between the average height of a field of hemp and its hurd yield.

Low densities have a positive effect on hemp seed yield at the expense of the yield and quality of hurds.

Stem diameter reduces with increasing density.

Nitrogen

Nitrogen-rich manure must be spread so that it is available to the plants during the rapid growth of the vegetative phase. If fertilization is applied too late, it will not always meet the plant's needs during growth. The absorption of nitrogen at the end of the vegetative phase promotes the production of leaves, sustains the vegetative phase and delays the end of the cycle. This translates into difficulties in drying the hurds, which will also make fibre extraction more difficult.

An excess of nitrogen-rich manure increases the risk of plants lodging during the rapid growth phase.

Where nitrogen-rich manure is limited, achieving the same plant mass requires warmer growing conditions (FNPC trials, 2002).

The level of nitrogen-rich manure does not influence the fibre level in the hurds (summary of ITC trials, 2004).

As a general rule, nitrogen-rich manure applied at a rate of 100–120 U/ha allows the potential of the plot of land to be reached.[1]

6.3 Hemp Growing and Processing Innovation in Australia

6.3.1 Overview

The hemp fibre industry will succeed or fail on its ability to provide a constant supply of consistent quality fibre and hurds at a competitive price. While the technical attributes of both hemp outer long bast fibre and inner short fibre (hurds or shives) are considerable, some end-use manufacturers are reluctant to develop products that benefit from these attributes because there is no guarantee of sufficient or expanding supply or a competitive price.

For example, the current world price for Grade 2 fibre (suitable for non-woven textiles), set by European production, is 20% higher than that for most other similar fibres. If the price of Grade 2 hemp fibre was 20–25% less than the current world price and a consistent quality and supply assured, manufacturers could focus their processing and product development on the utilization of hemp fibres' unique characteristics, resulting in increased efficiency and further cost savings to the manufacturer.

For every unit of demand for fibre, one must grow, harvest, transport, process, store and sell four units of hurd. In simple terms, four-fifths (80%) of time, labour and energy is expended on hurds. Therefore, it is crucial that markets and products are developed to utilize hurds, and at a level that is relative, in volume, to the consumption of the outer long fibre. The key to a successful industry is how one manages and profits from the hurds and not just the fibre.

There are many sectors in present farm to finished product value chains that have both inefficiency and waste. The following approach is designed to reduce those inefficiencies but maintain or increase the profit within each sector. Obviously, a company with the appropriate technology will stand to benefit by maintaining/increasing profits and increased volume of production. This is where the greatest opportunity exists.

Therefore, the aim of this section is to outline ways in which one might achieve a 25–30% reduction in market price across the board, increase supply chain profits and increase the volume of supply to meet potential demand.

Areas capable of greater efficiencies exist across the entire value chain; potential targets are:

- 12% lower cost of raw material by achieving a 20% higher crop yield

- 15% higher proportion of fibre to hurds ratio in plant genetics
- 30% lower cost of production in harvesting and handling to mill
- 25% lower cost of Stage 1 processing 'classification' of raw materials.

Before these four points are dealt with in detail, some background into both hurd and fibre pricing and production will be given.

6.3.2 Markets and pricing

The current world price for Grade 2 fibre (suitable for non-woven textiles), set by European production, is €600/t and for clean dust-free hurds is €210/t (from the processing factory). This price has been consistent for several years.

The price for other fibres (flax, kenaf, sisal, jute) used in similar applications fluctuates significantly due to supply peaks and troughs; however, in general, prices would be similar to or 25% less than the price of Grade 2 hemp. Consequently, hemp makes up only a small portion of the total supply and is being blended with other fibres of different (lower or same) quality and price to meet existing demand, in the autoparts industry for example.

Blended fibres are not necessarily desirable, due to a number of quality inconsistencies, handling and preparation needs. Blended fibre composites require more weight of fibre in a product to make up for those inconsistencies than if a single source of fibre. The need for blending machines and fluctuation of supply and quality from all the fibre sources means that the products and methods of production are more expensive than if a single source of fibre was available.

If the price of Grade 2 hemp fibre was 20–25% less than the current world price (€480/t) and a consistent quality and supply was assured, manufacturers would no longer need to blend fibres to make a number of products. They could then afford to focus more of their processing and product development on the utilization of hemp fibres' unique characteristics, resulting in reduced equipment needs (blending and other machines) and increased efficiency of production, thus allowing further cost savings to the manufacturer.

It must be noted that the price paid to farmers is relative to the price they get for other crops grown in the same season for a similar production cost. Hemp has different input needs and competes with different crops in Europe to those in Australia. In so far as Europe is concerned, both agricultural production and processing attract substantial subsidies. This is in the order of €110/ha of crop production and €90/t of resulting fibre for processing.

In Canada, the present fibre production is, in effect, subsidized by the Hemp Food Industry. Growers in Canada receive a competitive crop income by growing hemp seed for the North American health food market. The remaining hemp hurds or stubble from the seed harvest (approximately 2.5 t/ha) can be bought by fibre and hurd processing companies at a relatively low price, as the main source of income is from the seed.

The present form of EU subsidies potentially will reduce over the next 2 years; however, other forms of subsidies probably will replace them. In Canada, the hemp food market is expected to expand at approximately 50% per annum, as it has done for the past 4–5 years. This will mean that an ongoing supply of seed crop stalk will be available, but only at a rate relative to the food market.

6.3.3 Understanding hurds

While the intention over time is to lower the cost of fibre to enable a greater market volume, the intention with hurds is to increase them by a similar percentage. This can be achieved by finding captivated high-volume markets with specific needs that require the technical attributes of hurds. Before going into this in detail, some volume figures one needs to appreciate about hurds are given below:

- Hurds make up 70–80% of the weight of stalk grown, harvested, transported, processed and sent to market.

- 300 ha of crop at 9.3 t/ha 2750 t of straw resulting in 1650 t of hurd.
- 1650 t of hurds is approximately 15,000 M^3 or a storage shed 50 m long × 30 m wide × 10 m high.
- A 40 ft container of hurds can hold approximately 75 M^3 or 8.25. A 40 ft container filled with fibre will weigh approx 19 t, just under the weight limit for a container of this size, which is approximately 20 t.
- If it costs $2000 to deliver a 40 ft container to a location, the transport cost to product weight for fibre is $0.10/kg and for hurds is $0.24/kg. The percentage cost of transport to product value for fibre is 10% and for hurds is 53%.

Clearly, the cost of transport and storage of hurds is a major limiting factor and must be made more efficient and cost-effective. This aspect is addressed only partially here; future papers will be written about the handling and transport of hurds, as well as specifically about the market development of hurd-based products.

6.3.4 Solutions and efficiencies

Potential efficiencies across the entire value chain are:

1. 12% lower cost of raw material by achieving a 20% higher crop yield.
2. 15% higher proportion of fibre to hurds ratio in plant genetics.
3. 30% lower cost of production in harvesting and handling to mill.
4. 25% lower cost of Stage 1 processing 'classification' of raw materials.

Higher crop yield (plant breeding and agronomic technologies)

As stated, farmers in the EU receive a €110 (AU$180) subsidy for every hectare of crop production. The price paid for bales of hemp delivered to the mill in the EU is approximately €150/t. Yields in the EU are approximately 6t/ha. However, comparing grower returns between different countries is somewhat irrelevant. The price paid in each region must be pitched at a level that is relative to what will entice the farmer to grow hemp as opposed to another crop.

In Australia, farmers are paid $250/t for semi-processed hurds when delivered to the mill, with an average yield of 10 t/ha or more. This rate makes hemp a reasonably attractive summer crop alternative compared to sorghum or maize, but not better than cotton or poppies. However, crop returns fluctuate: whereas sorghum was $150/t a year ago, demand and supply conditions have seen it rise recently to as high as $300/t. In the case of cotton, the price has gone the other way. The only way hemp can combat the competitive crop issue is to allow the price to fluctuate along with the other crops or forward contract crops and lock in a price with farmers in advance. If advance price fixing can be achieved, it will obviously be better for certainty in supply to manufacturers.

There is significant room for crop improvement, higher yields and lower production costs. With ongoing development in plant improvement and agronomic treatment, we could achieve a 25% increase in average yield from 10 t/ha to 12 t/ha. The mill price could then be reduced by 12% from $250/t to $220/t, yet still allow for a 5% increase in grower earnings. This would make hemp even more competitive as an alternative to traditional crops.

Overall higher yields, through plant breeding, could be achieved in approximately a 3–5 year time frame. Lowering production costs probably could be achieved even faster.

Yield increase efficiencies will be derived from:

- Plant breeding for higher biomass
- Better agronomic systems, planting, watering
- Targeted plant nutrient systems
- Reduced crop harvest loss/waste.

It will also be imperative that hemp production is promoted to farmers on its additional benefits such as:

- Low risk for crop loss
- Rotation benefits

- Stable fixed price (buffer to fluctuations)
- Low cost to grow crop (crop inputs)
- Potential for carbon credits for growing an industrial crop such as hemp.

Higher fibre/hurd proportion (plant breeding and planting systems)

Temperate varieties of hemp stalk grown in the EU and Canada have a ratio of fibre to hurds of approx 25:70%, whereas tropical varieties of hemp used by EIL are 18:77% (all include 5% loss). Using current market figures, temperate varieties have a higher net value than tropical hemp (i.e. temperate = AU$565/t, tropical = AU$527/t of stalk using a value of AU$1000/t fibre and AU$450/t hurd).

These figures obviously promote the need for a higher fibre content in all varieties of stalk, especially tropical varieties. A target of 35% stalk fibre content is achievable and some medium-yielding temperate varieties already have that. Breeding these characteristics into late flowering plant (tropical) varieties is very possible and would mean an increase in stalk value of AU$538–620/t at present pricing structures.

Separate to the issue of plant breeding and increasing the fibre content in stalk is the issue of the value of the two main products. To that end, it is entirely possible to attain an additional AU$150/t (+AU$450 = AU$600/t) for the hurds if product and market development is implemented sooner rather than later. Table 6.1 shows the effect of the fibre to hurd ratio in relation to potential prices for product. Five different price structures illustrate the point of higher bast yields. Tonnes yield per hectare does not enter into this equation but does play a major part in grower returns.

Lower cost of production in harvesting and handling to mill (PW System)

Before one can appreciate the efficiencies that the handling and processing system developed by Philip Warner (PW) can achieve, one must understand the farm-to-market processing and handling system currently used by most companies worldwide.

Table 6.1. The effect of the fibre to hurd ratio in relation to potential product prices.

Ratio%		Price structure 1–5						
		1	2	3	4	5	Low	High
5	Loss/t	$ –	$ –	$ –	$ –	$ –	$ –	$ –
77	Hurd/t	$ 450	$ 500	$ 550	$ 600	$ 650	$ 450	$ 650
18	Fibre/t	$ 1000	$ 950	$ 900	$ 850	$ 800	$ 800	$ 1000
100	**Stalk value**	**$ 527**	**$ 556**	**$ 586**	**$ 615**	**$ 645**	**$ 491**	**$ 681**
5	Loss/t	$ –	$ –	$ –	$ –	$ –	$ –	$ –
70	Hurd/t	$ 450	$ 500	$ 550	$ 600	$ 650	$ 450	$ 650
25	Fibre/t	$ 1000	$ 950	$ 900	$ 850	$ 800	$ 800	$ 1000
100	**Stalk value**	**$ 565**	**$ 588**	**$ 610**	**$ 633**	**$ 655**	**$ 515**	**$ 705**
5	Loss/t	$ –	$ –	$ –	$ –	$ –	$ –	$ –
65	Hurd/t	$ 450	$ 500	$ 550	$ 600	$ 650	$ 450	$ 650
30	Fibre/t	$ 1000	$ 950	$ 900	$ 850	$ 800	$ 800	$ 1000
100	**Stalk value**	**$ 593**	**$ 610**	**$ 628**	**$ 645**	**$ 663**	**$ 533**	**$ 723**
5	Loss/t	$ –	$ –	$ –	$ –	$ –	$ –	$ –
60	Hurd/t	$ 450	$ 500	$ 550	$ 600	$ 650	$ 450	$ 650
35	Fibre/t	$ 1000	$ 950	$ 900	$ 850	$ 800	$ 800	$ 1000
100	**Stalk value**	**$ 620**	**$ 633**	**$ 645**	**$ 658**	**$ 670**	**$ 550**	**$ 740**

While the complete PW System is not presently operational in Europe, China or North America, every segment has been proven and, to a large extent, those efficiencies documented. The PW System is not just a concept; it is a proven reality. Other new systems are being developed around the world; however, there is no 'complete' new system yet in existence where a comparison can be made.

The three main areas of inefficiency in the existing systems are in: (i) field harvesting; (ii) bulk handling and transport to mill; and (iii) Stage 1 feed into a mill processing system. All of these elements are addressed with the PW System and equate to a 30–35% reduction in production costs and a 20–25% reduction in the price of fibre and hurds to the market. While there is only a slight increase in field operation capital equipment costs (i) and (ii), there is considerable (40%) reduction in capital equipment at the processing and handling stage (iii).

EXISTING TECHNOLOGY AND PROCESSES USED. Existing technology to process bast fibres into the two distinctly different fibre fractions (bark and wood) effectively has not evolved over the past 50 years.

Field harvesting. Present harvesting systems for fibre require the crop to be cut down and dried before baling (Fig. 6.14a and b). A typical high-yielding crop will have stalk 4 m (12 ft) long with a density of over 80 plants/m² (10–12 t/ha). Because of the density and length, a special harvester is required to cut the stalk down and then chop it into approximately 0.5 m lengths, leaving it in a windrow for efficient drying. If the stalks are not cut into 0.5 m lengths, they will be too difficult to handle in the raking process. The windrow is raked two or more times to dry the stalk down to 12% or lower and dislodge all leaf matter before it is ready for baling. *This process means 3 machinery passes, fuel and labour.*

Bulk handling and transport. The windrows of dried stalk are compressed into large bales in the field (round or square), in a way similar to other straw or hay crops. These bales have to be picked up one at a time and put on to a trailer, which next moves to a storage location on the farm and is then covered with a tarpaulin or stored in a shed. *This process means 1–2 machinery passes, fuel and labour.*

Fig. 6.14a and b. Field harvesting with existing technology.

Fig. 6.15a and b. Bulk handling and transport with existing technology.

Later, these bales are loaded on to a truck by a tractor loader (1–2 at a time) (Fig. 6.15a and b). The truck transports the bales to the mill, where they are unloaded and again stored under cover until they are then loaded one at a time into the milling process. *To transport 15 t of bales to the point where they are being processed takes 6 h for two workers and associated equipment.*

Due to the fixed location of the traditional mill, crop production must be within 75 km of the processing location. The 'fixed mill' issue will be dealt with later but, needless to say, this and other issues dealing with distance to market or poor seasonal growing conditions have further implications on overall efficiency, commercial risk and extra costs.

Stage 1 feed into processing system. The first requirement in milling these fibres is to prepare the material by opening up the compressed bale or cutting it into smaller sections so that it may feed evenly and easily into the first stage of processing, or decortication. Feed inconsistencies occur at this stage, due mainly to the difficulty in untangling the compressed stalk in the bales. This has a knock-on effect at the first stage of processing, usually a hammer mill, where the feed rate alters significantly, resulting in stalks being over or under milled before going to the next phase. Currently, the world's largest mill has a maximum input of only 6 t/h before the material moves on to the primary processing phase.

Problems with the existing technology/approach above are:

1. Cost of harvesting, handling and processing, which translates to market price, expensive and not competitive.
2. Need for specialized harvester equipment with no alternative crop usage.
3. Harvested crop time lying in field: risk, a major factor where cutting, windrowing and

bailing are exposed to weather and potential spoiling or delays.

4. Bale double handling, farm loading, unloading and reloading and mill unloading and reloading into the Stage 1 process.

5. A centralized mill is exposed to supply problems, climatic and crop production changes or market location shifts.

6. Throughput capacity of mill is limited by the compressed bale form presented at the input stage.

7. Extra distribution costs by having to move the hurds (low-density material) from farm to mill, then mill to market.

THE PW SYSTEM BY COMPARISON.

Field harvesting. EIL harvests and processes to Stage 1 in the field. The crop is sprayed out with a registered and benign defoliant. The harvester is a typical forage machine that can be converted for hemp use in approx 4 h. This conversion enables the harvester to strip, hammer and cut standing crop hemp at a rate of 30 t/h. The stalk is decorticated to approximately 70% in this process and is cut into sections approximately 100–150 mm in length. This material is shot into a follow buggy drawn by a tractor, which when full (approximately 2–3 t) carries the material to a module packing machine on the side of the field. *This process means 2 machinery passes, fuel and labour.*

Bulk handling and transport. The follow buggy dumps the material into a module packing machine, which compresses the fibre and hurds into a 12 t block (solid haystack) in approximately 30–45 min (Fig. 6.16a and b). This machine is operated by one man. This system is similar to a cotton module. A 12 t module does not require special farm or mill storage and it can be left in the field until required. The level of fibre separation from the hurds and length of fibre is designed specifically to allow the module to remain as a solid block for up to 6 moves.

The module is loaded into a specialized 'chain bed' truck (Fig. 6.16c) in less than 10 min and can be offloaded at the mill by one person in the same amount of time. This process is equivalent to two moves. The module is stored outside on a specially prepared pad.

When the material is required for processing, a similar 'chain bed' machine picks up the module and feeds the (already partly processed) material into the Stage 1 process, which in this system is 'classification' of fibre, hurds and fines.

In (i) and (ii), there is a slight increase in capital outlay for equipment but, on balance with the traditional system, it is

(a)

(b)

(c)

Fig. 6.16a, b and c. Field harvesting, bulk handling and transport using the PW System.

minimal and is borne totally by the contractor not the grower.

Stage 1 feed into processing system. As the raw material is already partly processed (Fig. 6.17a and b), the input capacity is enhanced significantly to between 10–20 t/h and will achieve a consistent throughput compared to the traditional mill feeding system where feed rate varies. Also, as the primary process has been achieved, the amount of equipment (capital) and power required to drive it is reduced by approximately 50%, compared to traditional bale opening and hammer milling.

Table 6.2 shows in AU$ the traditional system as compared to the PW System from those points above.

Lower cost of Stage 1 'classification' of raw materials

Again, it will be necessary to compare the two different methods to appreciate the efficiencies

Fig. 6.17a and b. Processing the raw material in the PW System.

and equipment and supporting infrastructure costs. This next section will be broken down into three parts: (i) processing Stage 1 'classification'; (ii) processing Stage 2 'fibre refining'; and (iii) logistics, supporting infrastructure and costs.

The typical processing line in Europe for Grade 2 fibre (non-woven) is a mechanical process, braking, scutching and filtering as it progresses. These lines need to be fully enclosed and all dust extracted – as dust is a health hazard – by an air pressure system with a maximum 2 t/h capacity.

Other 6 t/h processing lines do exist but are mostly for paper pulp grade, which contains up to 20% hurds and is therefore less refined and not suitable for non-woven applications. In all EU mills, the 'classification' component and 'fibre refining' are incorporated into, or an extension of, 'the one line' in the same location. This means all the raw material has to make the journey to the mill.

In contrast, the PW System separates these two components for crop supply flexibility, transport and cost-efficiency reasons. The processing Stage 1 'classification' is undertaken as close to the supply of raw material as possible, even on the farm. This is achieved by having a relocatable process that can be erected and operational in 2 days. Because the raw material is already partly processed, the classification process simply removes, cleans and packs ready for transport approximately 50% of the raw material (most of the hurds, dust and fines). Ideally, all dust and fines are spread back on to the land as a soil amendment. The hurds are packed into shipping containers ready for direct delivery to the market. The fibre portion is remade into modules for transport to the second 'fibre refining' phase in a centralized location.

The next section deals with the classification part only of both the European 2 t/h system and the PW System, for ease of comparison.

PROCESSING STAGE 1 'CLASSIFICATION'
Conventional processing Stage 1 'classification'. These systems all rely on bales of hemp hurds. The first stage is to cut and then

Table 6.2. Farm to processing handling and transport comparisons of the traditional system and the PW System (in AU$).

Farm to processing handling and transport comparisons
(figures based on operating on a farm with a minimum of 100 ha and/or 1,000 tonnes hemp straw)

Conventional bale system Based on moving total 20 t 90 × 225 kg bales 2.4 × 0.9 × 0.8		Module system Based on 2 × 10 t module Total 20 t	
Action	$ per 20/t		$ per 20/t
Harvest crop into windrows Chopped length 0.6 m	120	Fibre crop spray defoliant Approx 10t/ha	30
Rake and windrow × 2 times Allow to rett stalk	60	Seed crop harvest, stalk remaining Strip fibre process standing crop	0 500
Rectangle bale 20 t Total area 2–3 ha @ 7–10 t/ha	450	into mobile basket. Total @ $25/t Cart chopped fibre to module	160
Pick up bales, cart to farm storage 90 bales @ $2/bale	180	20 t ($8/t) Make module	120
Unload bales and store 90 bales @ $1.5/bale	135	2 × 10 t ($60 per module 1/2 h) Cost of farm storage and tarp	100
Cost of farm storage 90 bales @ $0.50/bale	45	2 × $50 Pick up module on chain bed truck	150
Load flat 40 ft bed truck (1.5 trucks) 90 bales @ $2/bale	180	2 × 10 t ($75 per module) Transport to mill and unload	360
Transport to mill (1.5 trucks) Loading time/40 km/unload time	450	2 × 40 km @ $4.5/km Cost of module storage	10
Load bales into mill storage 90 bales @ $1.5/bale	135	Move module into processor for 1st stage classification	30
Cost of bale storage 90 bales @ $1.25/bale	112.5		
Move and load bales into processor 90 bales @ $2/bale	180		
Process straw to strip fibre level before clasification	40		
Total	2087.5	Total	1460
Rate per tonne	$ 104.38	Rate per tonne	$ 71.50

feed the hurds into the decorticator to separate the fibre and hurds. A stock conveyor serves to make the connection between the feeder and the bale opener.

The bale opener is equipped with a conveyor at the entrance and an exit conveyor to feed the line. At this stage, the conveyor under the bale opener and hopper feeder transports all the loose hurds, waste and straw, and the rocks and soil are extracted to a side line, where the hurds are aspirated and transported to a cleaning system.

Transportation is by air. An air separator puts the material in the cleaning system. This way of transportation allows the removal of stones and other foreign material while aspirating.

The material on the main line enters and is broken by roller drum crimpers and then passes through a first shaker and a first tooth drum. Then a conveyor and a shaker follow and afterwards the fibre passes through four sets of equipment, each set composed of a fine tooth drum, an inclined shaker and a horizontal shaker.

PW System processing Stage 1 'classification'. The module is picked up by an open chain bed-type machine called a module feeder buster. The module feeder buster carries the module of hemp to the first mill process stage and feeds the module into the first-stage mill processing equipment at a rate of 10 t/h, but cannot process up to 20 t/hr if the main processing line is made larger.

The classification mill. The first step in the mill is to remove the free shives, which reduces the flow-on material by 50%. This is done by feeding the material through a rotating screen. As the free hurds, dust and fines fall, the second step is to separate the remaining fibre and shives. The third step is to classify the remaining materials.

The end result of measures such as these will reduce labour and energy costs, capital costs on machinery and also reduce the unit processing cost by increasing throughput. This will mean a reduction of up to 25% of the traditional Stage 1 processing 'classification' cost.

At this moment in time, Stage 2 processing, or 'fibre refining', remains the same as the existing technology. This process needs further analysis. Techniques such as steam explosion, enzyme separation and even ultrasonic processing are being considered.

6.3.5 Summary

Regardless of the end processing technique, the area of production that presents the need for the greatest increase of efficiency is in the four points outlined. The methods illustrated are only one view as to how efficiency can be achieved and there may well be many other ways to achieve a similar outcome. Nevertheless, the fact remains that the present methods are not as efficient as they could or should be if bast fibre is to become one of the fibre sources to supply a world looking for more and more renewable resources.

6.4 The Harvesting Process Developed by the *Chanvriers de l'Est*

As part of the development of hemp cultivation and processing/transformation, a new organization was set up in the Lorraine region in the west of France. This project was named the *Chanvriers de l'Est* and started in 2005 with a few hectares of land.

At the end of their first year of production, the lessons learned were of two types:

1. An efficient hemp harvesting system is economically essential.
2. The pressing of high-density parallelipipedic bales, or 'square bales' as they are more commonly known, is a technical imperative.

The technique commonly used in France consisted of a first pass by the combine harvester in order to collect the seed, followed by the reaping of the hurds. This was very labour-intensive and had few supporters.

The farmers therefore approached an agricultural worker and together they visited the offices of Goetz, the company responsible for the modified combine harvester equipped with the Kemper maize beak.

The first machine purchased in 2008 demonstrated clear limitations and was modified dramatically for the 2009 season. It then gave very satisfactory results, allowing 250 ha to be harvested in excellent condition.

Another problem then arose. Impurity in the hemp seed was often found to be too high and this caused problems at the drying silo. A mobile cleaning system was therefore designed that allowed the impurity level to be reduced to under 5%.

A further line of improvement was also explored. The swaths of retting hemp needed turning in order to improve retting and encourage the hemp to dry uniformly. None of the machines trialled performed satisfactorily and so a prototype was constructed and used with great success during the harvest of 2009.

In order to ensure a good 2010 harvest of a crop covering some 500 ha, a new John Deere combine harvester was equipped and a number of 'swath turners' are in production.

In this way, it will be possible to harvest hemp seed while cutting the hurds at the same time. This will allow hurd pressing to be prepared efficiently, spending less than 20 min/ha, after which the hurds can be pressed into square bales.

The initial specifications are now filled.

Figures 6.18–6.21 illustrate the equipment used.

Fig. 6.18. Swath turners used by the *Chanvriers de l'Est.*

Fig. 6.19. Pressing of square bales. The weight of each bale varies between 350 and 550 kg. It varies as a function of the driver's skill and the size of the bale.

Fig. 6.20. Combine harvester with the 2009 equipment.

Fig. 6.21. View of the production process on site in 2009.

Note

[1] Different units of measurement can cause significant misunderstanding of results. European authors generally report total aboveground dry matter in tonnes/ha, where 1 metric ton = 1000 kg and 1 ha = 10,000 m^2. Those using the English system of measurement report yield in tons/acre, usually short tons (1 short ton = 2000 pounds) per acre (43,560 ft^2).

7 Legislative Controls on the Cultivation of Hemp

Sylvester Bertucelli

Fédération Nationale des Producteurs de Chanvre (FNPC), France

---

7.1 Introduction

In this chapter, we will be addressing three key areas:

1. French legislation arising from the *Code de la Santé Publique*.
2. European Union (EU) legislation governing agricultural subsidies.
3. The future of the Common Agricultural Policy (CAP) and EU subsidies.

7.2 French National Legislation

Within the EU, European Union law (historically termed European Community law) has primacy over national law, with a few particular exceptions, including legislation governing public health issues, on which national law retains its primacy.

In view of the biological similarities between industrial hemp and marijuana, a subject that has been dealt with thoroughly in this book, the cultivation and use of industrial hemp has been subject to national legislation under article R5132-86 of the *Code de la Santé Publique*:

The production, sale, use and consumption of the following are forbidden:

1. Cannabis: plants and resin as well as all preparations that contain or are derived from cannabis plants and resin.

2. Tetrahydrocannabinols: with the exception of synthetic delta 9 THC, its esters, ethers and salts and derivative preparations.

Derogations from these regulations may be accorded for the purposes of research and testing as well as for the production of authorised derivatives as agreed by the Director General of the "Agence française de sécurité sanitaire des produits de santé" (AFSSAPS).

The cultivation, importation, exportation and use (industrial and commercial) of cannabis devoid of narcotic properties may be authorised by the Director General of the agency with the necessary authorisation from the ministers of agriculture, customs, industry and health.

This article is the updated Article R5181 of the old *Code de la Santé Publique* that is commonly cited in the literature.

Article R5132-86 was introduced by the Order of 24 February 2004 and published in the *Journal Officiel de la République Française* on 21 March 2004. This article specifies the conditions under which industrial hemp may be used:

According to the article R5181 mentioned above, the cultivation, importation, exportation as well as the industrial and commercial exploitation of the seeds and fibre of varieties of *Cannabis sativa* L. are permitted providing the following criteria are met:

- The delta-9-THC concentration of these varieties must not exceed 0.20%.
- The collection of samples and determination of the delta-9-THC concentration are to be undertaken according to the EU methodology prescribed in the accompanying appendices.
- Applications to include a variety of *Cannabis sativa* L. amongst the varieties listed in article 2 must be accompanied by a report detailing the results of the tests conducted according to Procedure B and described in the Appendix of this Article. This must be accompanied by a description of the variety in question.

The Order of 24 February 2004 further specifies the maximum authorized concentration of THC in these varieties (0.2%), as well as the method of sampling and analysis.

The approved varieties, whose cultivation is authorized, are indicated in the following Orders published in the *Journal Officiel de la Republique Française*:

- 22 August 1990
- 27 May 1997
- 2 July 1999
- 24 February 2004
- 5 September 2005
- 21 February 2008

This list may, depending on how it is updated, vary from the list of hemp varieties authorized within the European Union and appearing in the *EU Common Catalogue of Varieties for Agricultural Plant Species*, which can be found at http://ec.europa.eu/food/plant/propagation/catalogues/agri2011/index_en.htm (accessed 17 June 2011).

In practice, and providing the seeds used are of an approved variety (and that this can be demonstrated), the cultivation of hemp for industrial purposes is perfectly legal.

A control system has been instituted by the French state to ensure the EU legislation is observed. The *Agence de Service et de Paiement* has been charged with policing this system. Random sampling and testing of 30% of the land under cultivation is undertaken, as permitted by the Order of 24 February 2004, in order to check *in situ* that the maximum authorized THC levels are being respected.

7.3 EU Legislation

The agricultural legislation and subsidy schemes will be presented here.

The Common Agricultural Policy (CAP) has evolved considerably since 1992. Initially, the aid distributed to farmers was individualized according to plant species. In order to conform to the economic rules of the World Trade Organization, the EU has worked to uncouple subsidies from specific crops. In its place, it has introduced a Single Payment Scheme (SPS) which provides a flat rate payment that may take into account historical factors. As of 2010, CAP subsidies are completely decoupled, that is to say they are no longer tied to specific crops. Payments under the SPS may be triggered for a wide range of crops, including hemp.

Industrial hemp attracts a specific 'transformation incentive' of €90/t pure fibre, providing the following conditions are met:

1. Processors in receipt of this aid must be approved by the *Agence de Service et de Paiement*.
2. Production on French soil must be on qualifying CAP land.
3. The relevant paperwork must be completed.

As of 2010, the French companies with the necessary authorizations are:

- Agrochanvre (Barrain group)
- Agrofibre (Euralis group)
- Câlin (CAVAC group)
- Eurochanvre (Interval group)
- Starthemp
- Terrachanvre

In addition to these companies, the German firm BAFA also cultivates hemp in France under a state licence.

Additional restrictions that must be satisfied are:

- An audit must be conducted of the decortication process to ensure that impurities do not exceed 25%.
- The recording process should be able to demonstrate that fibres are eligible.

The subsidies available for transformation/processing are planned for the 2010 and 2011 seasons, but may disappear in 2012.

7.3.1 Summary

Conditions for exploitation:

Example 1

Cultivation by a contracted farmer working with an approved processor:

- SPS triggered if the farmer applies.
- Incentive available for the processor for processed hemp at €90/t.
- Crop may be subjected to random testing for THC.

Example 2

Crop processed by an unregistered processor lacking, produced for own consumption or as an experimental crop:

- SPS triggered if available and applicable.
- Culture must be declared to the *Comité Économique Agricole de la Production de Chanvre* (Interior Ministry's Committee for the Production of Hemp).

7.4 Future of the CAP and Hemp Subsidies

At the time of publication, the future of the CAP after 2013 will probably be mapped out. Two possibilities are envisaged:

7.4.1 A liberal policy

- Creation of a single payment for each hectare within the EU, with a progressive reduction in this payment until it disappears.

Currently, this option is advocated by those who seek to shrink the agricultural budget. It threatens to produce significant changes to the agricultural landscape, may aggravate the volatility of market prices and weaken the agricultural industry, but could provide hemp with an unprecedented opportunity.

In effect, the SPS may herald a shift in payments from the central pillar of the CAP system (crop subsidies) towards a new priority, that of sustainable rural development. The growing importance of environmental considerations and land stewardship may shift community subsidies towards industrial crops such as hemp, which are more environmentally friendly.

- Carbon storage/capture in building materials.
- Cultivation without the need for herbicides.
- Longer crop rotation plans, leading to breaks in parasite cycles.

In this way, hemp production may benefit from the chaotic situation arising from the adoption of a liberal policy, assuming European agriculture survives as an industry.

7.4.2 A conservative policy

- Evolution and development of the CAP without upsetting the system.

This option is advocated by countries such as France, with a strong agricultural base, that historically have been the main beneficiaries of the CAP.

The need to secure and protect the European agricultural industry, thus realizing its potential in the face of opposition from those concerned more about free trade, is a key argument in favour of this option. In this case, there is likely to be a reinforcement of 'direct aid', while calling a halt on the standardization of subsidies. At the same time, a movement towards rewarding environmental added-value will be encouraged by modulating subsidies such that environmentally friendly products are preferentially weighted. In this case, hemp, with its intrinsic qualities, is again likely to benefit. This second option appears to be the most viable. There remains, however, the enormous question of the CAP budget.

8 The Agricultural Economics of Hemp

Pierre Bouloc

La Chanvriere de L'Aube (LCDA), France

8.1 Introduction

The primary objective of this chapter is to enable any reader with an interest in agricultural production to evaluate production costs, revenues and profitability.

The methodology presented here is not original, for it can be applied to almost any crop regardless of the continent on which it is grown. It involves an evaluation of the costs of production and revenue generated expressed per hectare of land cultivated and per tonne of crop harvested.

The proposed approach is to be commended, for the required data can be collected easily and lends itself easily to international comparisons.

In order to calculate margin per hectare, we will therefore address:

- the revenues resulting from crop production;
- the direct and indirect costs of production.

This approach will allow comparisons to be made between regions and will, most importantly, allow us to evaluate whether hemp production is profitable when compared with its standard competitors.

In order to be as complete as possible, the second part of this chapter will see a few international comparisons undertaken.

8.2 Information Sources and Documents Used

The information sources drawn on for this chapter include:

1. The OCERA (*Office de Comptabilité et d'Economie Rurale de l'Aube*).
2. A study undertaken by a French doctoral student, Thomas Bourguignon, at the University of Wageningen, Holland, and entitled 'Profitability Evaluation of Hemp Production in France and The Netherlands' (April 2004).
3. Data supplied by the Nova Institute.
4. A report entitled 'UK flax and hemp production: the impact of changes in support measures on the competitiveness and future potential of UK fibre production and industrial use', ADAS – Centre for Sustainable Crop Management, Cambridge, UK (2005).
5. Data supplied from the Eurostat New Cronos database (October 2004).
6. Data obtained by the author from a number of national and international experts.

8.3 Choice of Economic Model

8.3.1 Calculation methods

In order to tackle the basic question of the profitability of this crop, we need a simple model that can take into account the realities of production, whatever its size and complexity.

Proposed model:

Revenue per hectare – Cost per hectare
= Operating margin

8.3.2 Composition of revenues

The revenue generated from the cultivation of hemp within the 25 member states of the European Union is derived from three sources:

- the revenue from chopped straw;
- the revenue from hemp seed;
- the agricultural subsidy payment.

This is dependent on the following conditions being met: each producer must ensure that the straw and hemp seed supplied are of good quality, meet market standards and are free of foreign bodies.

Payments for straw

This is the average price paid to the farmer for the straw supplied by him that is wholesome and meets market standards. In reality, companies buy straw according to the amount and a number of qualitative variables including:

COLOUR OF THE STRAW. The straw must produce a fibre that is grey-white, yellow or light green in colour. These qualities will produce hurds, the broken pieces of the woody inner core of the hemp stem, which are white in colour.

These criteria are set by industry according to their needs. Accordingly, these types of straw will attract the highest payments.

Where the straw is green in colour, the plant is immature and the fibre produced will be dark green, as will the resulting hurds. The price paid for such straw is likely to suffer as a result. Where the straw has spent too much time on the ground, it will become grey or even black. This can occur where the weather conditions have not been favourable. In such cases, the fibre and the hurds will themselves be grey or black.

A large company has identified six colour grades, although it should be noted that these are visual characteristics that are designated by the company and approved by the owner/producer of the straw:

B1 white hurds – fibre grey white
B2 white hurds – fibre yellow
B3 white hurds – fibre light green
V1 green hurds – fibre dark green
G1 grey hurds – fibre grey
G2 grey hurds – fibre black

As mentioned above, the designation of straw colour is undertaken visually. This procedure is subjective, however, and the Fraser Institute in Bremen, Germany, has developed a chromatographic analysis that has the potential to avoid disagreements[2] if and when it is adopted.

HUMIDITY LEVELS. Generally speaking, it is accepted that the normal humidity of straw is approximately 15%. Where straw is supplied with a higher humidity, the suppliers are penalized. Where the humidity exceeds 19–20%, some companies may even reject the product.

By contrast, a humidity reading inferior to 14% may attract a bonus payment.

The humidity level is evaluated using a probe, which can be introduced into the heart of the bale. This testing is usually undertaken by the producer at the time of loading and then again on arrival at the factory.

WEIGHT OF STRAW BALES. In order to keep control of transport costs, the companies that retain responsibility for transport seek to fill their vehicles with as heavy a load as possible. Taking into account the density of the straw, it is in their interest to take the heaviest bales, as this will maximize the load on any given lorry. For this reason, suppliers are penalized for supplying bales that fail to meet a minimum weight. Any supplier delivering bales directly to the factory will also find it in their own interest to ensure that the bales are as heavy as possible.

PRESENCE OF STONES AND FOREIGN BODIES. Such foreign objects can increase transport costs and, more importantly, can interfere with the factory processing of the straw. Companies therefore penalize the supplier for a drop in the quality of the straw.

Companies working for the paper industry must ensure the straw is free from plastic (even string and small pieces of bag),[3] for plastic is strictly forbidden. Producers who have failed to clean their fields proactively in order to prevent such contamination will also be penalized.

STORAGE. Finally, certain companies reward farmers for storing the straw in their sheds.

These are the main ways in which straw production is rewarded.

Payments for hemp seed

Manufacturers pay by weight and for quality. It is for this reason that quality assays have been developed. Hemp seed is a fragile seed that must be aired and sorted quickly, to eliminate impurities, and dried, so that it keeps well. Payments vary according to quality and this is ascertained at the point of collection (silos) by taking samples from delivery batches.

This sample is subjected to sorting by a double passage through a ventilated separator. The humidity level is calculated on the 'good grains' collected from the separator.

The amount of impurities is calculated by weighing the 'waste' from the separator.

At the point where these samples are taken, the grain is also subjected to an evaluation of its oil content and quality.[4] This involves measuring the concentration of oleic acid.

PAYMENT. The payment received for the delivered hemp seed is a function of the quality, after sorting and ventilation, and the quantity delivered.

EU SUBSIDIES. EU subsidies, so often decried and poorly understood, were introduced at the time of the reform of the Common Agricultural Policy in 1992. This reform had two key objectives:

1. To lower the price of agricultural products substantially in order to reduce the costs of food materials and encourage exports. As an example, the price of wheat dropped from about €18 in 1990 to €10/t in 1995 (a rough estimate that does not reflect the fluctuations in price by region and according to the date of purchase of the product).

2. To compensate farmers partially for the resulting drop in their revenue by introducing the notorious Community Subsidy. These subsidies were calculated region by region in order to take into account local specificities. This is why subsidies can vary from one department (a French administrative area) to another.

NB The arrangement was modified in 2001 and again in 2003, with these changes coming into force in 2004. Changes were again made in 2007, at which point a new Common Agricultural Policy (CAP) came into force (since its origins in 1962, the CAP has been revised every 6 years). It is in order to take account of these variations that we are focusing on the period pre-2004 and that we will present at the end of this chapter a method of calculation that will allow anyone to make their own revised estimates, taking into account changes in the trading rates of each product and any changes to the subsidies received.

Note 1: the new method that is being introduced has been named 'ecoconditionality'. It entails, in principle, the subordination of a financial support for agricultural products in favour of environmental protection measures. Financial incentives become dependent on the adoption of agricultural practices that are environmentally friendly and respect the existing laws and regulations on the protection of the environment.

Note 2: at the time of the translation of this book (early 2011), a new Common Agricultural Policy was being prepared for implementation in 2013, the main elements of which can be found in the previous chapter.

In view of this, and in order to take into account the previous system, each farmer will receive a subsidy that differs from that of his neighbour. This will make comparisons difficult, especially if we fail to specify carefully the method of calculation and the measurements that remain in force.

8.3.3   Breakdown of costs relating to the industrial production of hemp

The costs requiring consideration are as follows:

- cost of the seed per hectare
- cost of fertilization
- cost of crop protection measures
- various equipment costs

- fixed equipment costs
- labour costs
- various other costs.

Cost of seed per hectare (expressed as €/ha)

This is the price per kg of treated, certified seed, including delivery and all taxes and charges relative to the amount sown per hectare. This figure is highly variable as, depending on the farmer, the sowing density can vary between 30 and 70 kg/ha.

Cost of fertilization (expressed as €/ha)

This is the cost of purchasing different fertilizers and delivering them to the farm multiplied by the amount used. Even where a crop does not require much fertilization, soil variety, and especially the soil's nitrogen, phosphate and calcium content, can introduce substantial variations in the global cost.

Cost of crop protection measures (expressed as €/ha)

Even where the crop does not require any particular treatments, as was discussed in the chapter detailing the results of the trials undertaken, this cost must be taken into account where it does arise.

Various costs and other charges

These include insurance costs, financial costs, energy bills, etc.

Variable equipment costs

These are, in the main, comprised of the rental cost for the specific pieces of equipment needed for hemp cultivation or the use of equipment from other businesses. In these cases, we must consider only the time needed or the tonnage processed.

In this way, the use of a combine harvester, supplied together with a driver, costs around €90/ha in France. These costs include labour costs.

Fixed equipment costs

This is made up of the cost of using materials and machinery that can be used for purposes other than the cultivation and harvesting of hemp. A tractor and trailer can be used in much the same way, regardless of the crop. The cost of this equipment will be calculated on the basis of the number of hours it is used in hemp production.

This hourly cost will include the price of depreciation and the running costs. All the multi-crop machinery used (tractor, trailer, sowing machinery, fertilizer, pulverizer and ground preparation equipment) needs to be considered.

Labour costs

Labour costs are one of the most significant where this particular crop is concerned. While the hemp farmer needs to visit his or her field only occasionally after sowing in order to check on the crop, the requirement for labour increases dramatically at harvest time and can reach 7 h/ha.

In the case of hemp, which is not harvested with a combine, the mowing, sheafing, baling and storage of the crop in a well-ventilated shed needs to be factored in.

Where hemp seed is being harvested, the costs of a combine harvester and seed collection must be considered.

8.3.4 Calculation of operating margins

Having collected the various constituent costs of hemp production, we now have:

- all the receipts (variable, due to the fact that cultivation subsidies have not been taken into account)
- all the costs (fixed and variable).

The difference between these two figures represents the cost of production. It should always be expressed per cultivated hectare or per tonne of straw or seed produced.

8.4 International Comparisons

8.4.1 Source of figures

France

The figures used in this chapter have been provided by the OCERA (*Office de comptabilité et d'économie rurale de l'Aube*). This organization is, as its name suggests, special-ized in the rural economy in its widest sense. It advises more than half of the farmers in the Aube. While respecting the anonymity of its clients, it produces a number of reports on the rural economy and agricultural production. It is recognized as a source of specialist advice by the authorities on the one hand and by farmers on the other. The results presented on the sub-ject of hemp production are significant as this *département* is an important hemp producer, with 6000 ha under cultivation (of a total of 9000 ha under cultivation in France).

Holland

The Dutch data are produced by the LEI. This is the main Dutch organization with an interest in economic and social research in the field of agri-culture. The organization is part of the University of Wageningen and its role is to provide solu-tions to the various problems arising from the integration of agriculture and the agroindustries into the economic and social environment. It is therefore a source of advice for farmers, dis-tributors, consumers and public authorities.

Germany

The source of the figures presented here is the nova-Institut GmbH, which, as far as hemp and other natural fibres are concerned, is a first-class source of data.

The work of the institute has ranged across the whole hemp economy, with a particular focus on certain markets for the fibre produced. It is in recognition of this that Chapters 11 and 12 of this book have been contributed by the Director General of the institute, Michael Carus, co-editor of the text *Der Hanfanbrau* with Bocsa and Lohmeyer. The institute's experts have studied both the industrial and the agricultural hemp economy.

UK

Data are taken from the ADAS report men-tioned earlier.

Australia

Data are from the work undertaken by the Ecofibre team, who have been responsible for the reintroduction of hemp in Australia.

Experts

Having consulted with national and interna-tional experts, the authors have regrouped and reorganized a large amount of data that allow us to produce an overview of the economic realities of hemp production which, while not exhaustive, is certainly well informed.

8.4.2 Designated production zones

Using the collected data, organized into territo-rial groups, we can compare the production of different regions.

France

The *Barrois* is a region of hills and plains in the south of the *département* of the Aube. A number of cereal crops typically are grown here, together with Champagne wine.

The *Champagne crayeuse* is a chalky plain located to the north of the town of Troyes. Polyculture is also practised here, although, despite the name, there is no champagne pro-duction. In its place we find sugarbeet, lucerne and potatoes.

Holland

The provinces of Groningen and Drenthe are in the north-east of Holland and are regions where cereals, sugarbeet and potatoes are grown along with hemp.

Germany

The data collected by the nova-Institut are aver-ages for the entire German production.

Australia

The data collected are derived from interviews with Australian experts working in the produc-tion and processing of hemp, both in Tasmania and in Queensland.

8.4.3 The situation in France

We will present the situation in France as seen in two different, but neighbouring, regions.

Both regions are important producers and are responsible for more than 60% of the land under production in France (see above).

Calculations were based on 3 years of production, in order to take into account the changes in methods and calculation introduced by the new CAP in 2001, and on large samples representing more than half of the registered producers.

The Barrois

Table 8.1. Revenue generated from hemp production in the Barrois.

Variables	Average 2000–2004	Observations
Straw yield (t/ha)	5.38	
Seed yield (t/ha)	0.92	
Sale price of straw (€/t)	83.50	
Sale price of seed (€/t)	313.33	
Revenue from straw (€/ha)	449.51	
Revenue from seed (€/ha)	289.84	
Total revenue (€)	739.35	Equivalent to an income of €137.42/t of straw

Table 8.2. Costs incurred by hemp production in the Barrois.

Variables	Average 2000–2004	Observations
Fertilizer costs (€/ha)	116.00	Nitrogen: 110 kg/ha Phosphorus: 00 kg/ha Potassium: 100 kg/ha
Seed costs (€/ha)	140.00	Sown at 48 kg/ha
Plant health costs (€/ha)	2.00	Generally, pesticides are not used on hemp
Other costs and variable charges (€/ha)	51.00	
Total of all variables for the crop (€/ha)	309.00	
Cost of harvesting (€/ha)	90.00	Equivalent to 1 ha at €90/ha
Cost of mowing and sheafing (€/ha)	52.00	Equivalent to 1 ha at €52/ha
Cost of baling (€/ha)	53.80	Equivalent to 5.38 t/ha at €16/t for baling and transport
Transport costs (€/ha)	32.28	Equivalent to 5.38 t/ha at €16/t for pressing and transport
Total of all variables for harvesting (€/ha)	228.08	
Fixed mechanical costs	160.00	
Variable labour costs (€/ha)	80.00	
Total costs	777.00	Equivalent to a spend of €144.43/t of straw

Table 8.3. Operating margins for hemp cultivation in the Barrois.

Total revenue (€)	739.35	Excluding any subsidies
Total costs (€)	777.08	
Operating margin	−€37.73/ha	Equivalent to a margin of −€7.01/t of straw

The Champagne

Table 8.4. Revenue generated from hemp production in the Champagne region.

Variables	Average 2000–2004	Observations
Straw yield (t/ha)	8.72	
Seed yield (t/ha)	1.07	
Sale price of straw (€/t)	83.50	
Sale price of seed (€/t)	313.33	
Revenue from straw (€/ha)	728.12	
Revenue from seed (€/ha)	334.22	
Total revenue	*1062.34*	*Equivalent to an income of*
		−€121.83/t of straw

Table 8.5. Costs incurred by hemp production in the Champagne region.

Variables	Average 2000–2004	Observations
Fertilizer costs (€/ha)	137.33	Nitrogen: 130 kg/ha
		Phosphorus: 100 kg/ha
		Potassium: 110 kg/ha
Seed costs (€/ha)	140.00	Sown at 48 kg/ha
Plant health costs (€/ha)	2.00	Generally, pesticides are not
		used on hemp
Other costs and variable charges (€/ha)	51.00	
Total of all variables for the crop (€/ha)	*330.33*	
Cost of harvesting (€/ha)	90.00	Equivalent to 1 ha at €90/ha
Cost of mowing and sheafing (€/ha)	52.00	Equivalent to 1 ha at €52/ha
Cost of baling (€/ha)	87.20	Equivalent to 8.72 t/ha at €16/t
		for baling and transport
Transport costs (€/ha)	52.32	Equivalent to 8.72 t/ ha at €16/t
		for pressing and transport
Total of all variables for harvesting (€/ha)	*281.52*	
Fixed mechanical costs	160.00	
Variable labour costs (€/ha)	80.00	
Total costs	*851.85*	*Equivalent to a spend of*
		€97.69/t of straw

Table 8.6. Operating margins for hemp cultivation in the Champagne region.

Total revenue (€)	1062.34	Excluding any subsidies
Total costs (€)	851.85	
Operating margin	+€210.49/ha	Equivalent to a margin of
		€24.14/t of straw

Discussion

INCOME. The income generated is the average of that realized over the 3-year period 2001–2003. This period was used because the data are comparable, particularly with regard to the cost of straw and the amount of EU subsidy.

Before 2001, the system used by the EU differed, both in nature and in terms of the amount paid. It influenced prices markedly. It is for this reason that we have chosen to use information from this period. Information from earlier periods has been retained and is available to readers.

Costs

COST OF FERTILIZER. Chapter 5 discusses the amount of fertilizer usually needed for hemp production. During the period under consideration (2001–2003), the price of fertilizer fell. These prices correlate with the price of oil and this should be borne in mind for future estimates.

COST OF SEED. By comparison, the price of seed moved in the opposite direction and increased during this period. This is due to the:

- Increase in contributions and taxes related to seed production;
- A general increase in costs which is not mitigated by an increase in seed production.

OTHER COSTS. These remain more or less constant for the period under study.

VARIABLE MECHANICAL COSTS. The cost of €90/ha is the price paid to an outside business for harvesting. Were the farmer to use his or her own equipment or work as part of a cooperative, he or she would be able to reduce these costs.

The variable mechanical costs can vary significantly between areas. This is due to the difference in yield per hectare and the resulting differences in the denominator. As for fixed costs, these are identical between regions, as the method of production is the same.

LABOUR COSTS. The time spent labouring is based on the information provided by farmers. In general, the time required is of the order of 7.5 h/ha.

8.4.4 The situation in Germany

Table 8.7. Revenue generated from hemp production in Germany.

Variables	Average 1996–2002	Observations
Straw yield (t/ha)	6.22	
Seed yield (t/ha)	0.800	This figure relates to a small number of producers only and is therefore given for information
Sale price of straw (€/t)	110.00	Including the cost of transport to the factory
Sale price of seed (€/t)	313.33	In the absence of German information, the French figures from the same period have been used here
Revenue from straw (€/ha)	684.27	
Revenue from seed (€/ha)	250.66	
Total revenue	*934.93*	*Equivalent to an income of −€150.31/t of straw*

Table 8.8. Costs incurred by hemp production in Germany.

Variables	Average 1996–2002	Observations
Fertilizer costs (€/ha)	109.11	Nitrogen: 100 kg/ha Phosphorus: 80 kg/ha Potassium: 150 kg/ha Calcium: 120 kg/ha
Seed costs (€/ha)	122.71	
Plant health costs (€/ha)	0.00	Generally, pesticides are not used on hemp
Other costs and variable charges (€/ha)	104.30	Including the cost of renting the land
Total of all variables for the crop (€/ha)	*336.12*	
Cost of harvesting (€/ha)	132.94	By an outside contractor and expressed at 1 ha/h
Cost of mowing and sheafing (€/ha)	56.79	
Cost of baling (€/ha)	122.71	
Transport costs (€/ha)	11.52	

Continued

Table 8.8. Continued.

Variables	Average 1996–2002	Observations
Total of all variables for harvesting (€/ha)	*323.96*	
Fixed mechanical costs	190.32	
Variable labour costs (€/ha)	65.75	Equivalent to 6.43 h/ha at €10.23/h
Total costs	*916.15*	*Equivalent to a spend of €147.29/t of straw*

Table 8.9. Operating margins for hemp cultivation in Germany.

Total revenue	934.93	Excluding any subsidies
Total costs	916.15	
Operating margin	+€18.78/ha	Equivalent to a margin of €3.02/t of straw

8.4.5 Presentation of data from Holland

Preliminary remark: in view of the economic and other difficulties encountered in the exploitation and processing of industrial hemp, the Dutch company exploiting crops in excess of 2000 ha has ceased trading. It has not resumed trading and the figures provided here, though exact and validated by a reliable source, only cover the period prior to 2004.

Table 8.10. Revenue generated from hemp production in Holland.

Variables	Average 2000–2004	Observations
Straw yield (t/ha)	8.00	
Seed yield (t/ha)	0.00	No seed harvested during this period
Sale price of straw (€/t)	100.00	Including the cost of transport to the factory
Sale price of seed (€/t)	0.00	
Revenue from straw (€/ha)	800.00	
Revenue from seed (€/ha)	0.00	
Total revenue	*800.00*	*Excluding subsidies*

Table 8.11. Costs incurred by hemp production in Holland.

Variables	Average 2000–2004	Observations
Fertilizer costs (€/ha)	95.40	Nitrogen: 70 kg/ha Phosphorus: 20 kg/ha Potassium: 150 kg/ha
Seed costs (€/ha)	111.30	
Plant health costs (€/ha)	0.00	
Other costs and variable charges (€/ha)	28.00	
Total of all variables for the crop (€/ha)	*234.70*	
Cost of harvesting (€/ha)	113.00	Equivalent to 1 ha at €113/ha
Cost of mowing and sheafing (€/ha)	36.30	Equivalent to 1 ha at €36.30/ha
Cost of baling (€/ha)	214.00	Equivalent to €214/ha or €26.75/t for baling

Continued

Table 8.11. Continued.

Variables	Average 2000–2004	Observations
Transport costs (€/ha)	90.76	Transport from the fields to the farm for (and including) storage
Total of all variables for harvesting (€/ha)	*454.06*	
Fixed mechanical costs (€/ha)	124.75	Tractor: 5.5 h at €66/ha Preparation of the soil: 3.70 h at €42.55/ha Sowing: 0.9 h at €10/ha Fertilization: 0.9 h at €8/ha
Variable labour costs (€/ha)	99.00	5.5 h/ha at €18/h
Total costs	*912.51*	

Table 8.12. Operating margins for hemp cultivation in Holland.

Total revenue	800.00	Without subsidies
Total costs	912.51	
Operating margin	−€112.51/ha	Equivalent to a margin of −€14.06/t of straw

Revenues

In this region, the revenue from hemp does not include payments for hemp seed. In fact, due to maturation problems, the seeds are not harvested (i.e. the crop is not 'combined').

Costs

The variable costs of mechanization appear to be higher in Holland. This is because mowing, sheafing, baling and transport are all undertaken by a manufacturer using equipment that has been developed for the purpose. The manufacturer bills the farmers accordingly.

The resulting cost is greater than if the farmer had undertaken the same work with his or her own equipment.

Labour costs are lower than those of other operators and include transportation costs.

8.4.6 Presentation of data from Australia

Two sets of tables allow us to consider both irrigated and non-irrigated cultivations. The difference in yields and in costs is spectacular, but the margin per hectare is not very different.

Non-irrigated hemp

Table 8.13. Revenue generated for non-irrigated hemp production in Australia.

Variables	Non-irrigated average	Observations
Straw yield (t/ha)	5.00	Non-irrigated
Seed yield (t/ha)	0.00	
Sale price of straw (€/t)	151.65	Price includes delivery to the factory
Sale price of seed (€/t)	0.00	
Revenue from straw (€/ha)	758.28	
Revenue from seed (€/ha)	0.00	
Total revenue	*758.28*	*Equivalent to an income of €151.65/t of straw*

Table 8.14. Costs incurred by non-irrigated hemp production in Australia.

Variables	Non-irrigated average	Observations
Fertilizer costs (€/ha)	139.27	Nitrogen: 200 kg/ha Phosphorus: 40–60 kg/ha Potassium: 80 kg/ha
Seed costs (€/ha)	126.85	
Plant health costs (€/ha)	22.19	
Other costs and variable charges (€/ha)	39.32	
Total of all variables for the crop (€/ha)	*327.63*	
Cost of harvesting (€/ha)	198.08	Equivalent to €39.62/t
Cost of mowing and sheafing (€/ha)		Included in the costs of harvesting.
Cost of baling (€/ha)		Included in the costs of harvesting
Transport costs (€/ha)	71.18	Equivalent to €14.23/t
Total of all variables for harvesting (€/ha)	*269.26*	
Fixed mechanical costs (€/ha)		
Variable labour costs (€/ha)		
Total costs	*596.89*	*Equivalent to a spend of €119.37/t of straw*

Table 8.15. Operating margins for non-irrigated hemp cultivation in Australia.

Total revenue	758.28	Without subsidies
Total costs	596.89	
Operating margin	+€161.39/ha	Equivalent to a margin of €32.27/t of straw

Irrigated hemp

Table 8.16. Revenue generated from irrigated hemp production in Australia.

Variables	Average when irrigated	Observations
Straw yield (t/ha)	8.00	Non-irrigated
Seed yield (t/ha)	0.00	Crops were not combined
Sale price of straw (€/t)	155.40	Price including delivery to the factory
Sale price of seed (€/t)	0.00	
Revenue from straw (€/ha)	1243.20	
Revenue from seed (€/ha)	0.00	
Total revenue	*1243.20*	*Equivalent to an income of €155.40/t of straw*

Table 8.17. Costs incurred by irrigated hemp production in Australia.

Variables	Average when irrigated	Observations
Fertilizer costs (€/ha)	222.84	Nitrogen: 200 kg/ha Phosphorus: 40–60 kg/ha Potassium: 80 kg/ha
Seed costs (€/ha)	139.27	
Plant health costs (€/ha)	22.19	

Continued

Table 8.17. Continued.

Variables	Average when irrigated	Observations
Other costs and variable charges (€/ha)	39.32	
	+190.27	Irrigation costs
Total of all variables for the crop (€/ha)	*613.89*	
Cost of harvesting (€/ha)	198.08	Equivalent to €24.76/t × 8 t
Cost of mowing and sheafing (€/ha)		Included in the harvesting costs
Cost of baling (€/ha)		Included in the harvesting costs
Transport costs (€/ha)	131.20	Equivalent to €16.40/t × 8 t
Total of all variables for harvesting (€/ha)	*329.28*	
Fixed mechanical costs (€/ha)		
Variable labour costs (€/ha)		
Total costs	*943.17*	*Equivalent to a spend of €117.89/t of straw*

Table 8.18. Operating margins for irrigated hemp cultivation in Australia.

Total revenue	1243.20	Without subsidies
Total costs	943.17	
Operating margin	+€300.03/ha	Equivalent to a margin of €37.50/t of straw

8.4.7 Presentation of data from the UK

At the time of writing, there is only one company (Hemcore Ltd) in the UK that processes industrial hemp. The data presented below are drawn from the ADAS report cited earlier, itself based on data provided by this company.

Unfortunately, this information is somewhat incomplete, as we do not have data for the fixed mechanical costs or the labour costs. The figures obtained from these data therefore provide us with a rough margin of production rather than the more accurate operating margin provided for the previous examples.

Table 8.19. Revenue generated for hemp production in the UK.

Variables	Average 2000–2004	Observations
Straw yield (t/ha)	5.5	
Seed yield (t/ha)	0.00	Crops were not combined
Sale price of straw (€/t)	147.21	
Sale price of seed (€/t)	0	
Revenue from straw (€/ha)	809.65	
Revenue from seed (€/ha)	0	
Total revenue	*809.65*	*Equivalent to an income of €147.20/t of straw*

Note: These calculations were undertaken using the average exchange rate for December 2005, where € = £0.6793.

Table 8.20. Costs incurred by hemp production in the UK.

Variables	Average 2000–2004	Observations
Fertilizer costs (€/ha)	95.69	Nitrogen: 110 kg/ha Phosphorus: 60 kg/ha Potassium: 60 kg/ha
Seed costs (€/ha)	173.70	37 kg/ha
Plant health costs (€/ha)	0.00	Included for information only as pesticides are not usually used on hemp
Other costs and variable charges (€/ha)	?	
Total of all variables for the crop (€/ha)	*269.39*	
Cost of harvesting (€/ha)		
Cost of mowing and sheafing (€/ha)	36.80	Equivalent to £25/ha for mowing and £9/ha for baling
Cost of baling (€/ha)	58.88	Equivalent to baling at £7.27/t
Transport costs (€/ha)	97.15	Transport equivalent to 5.50 t/ha at £12/t
Total of all variables for harvesting (€/ha)	*192.75*	
Fixed mechanical costs (€/ha)		
Variable labour costs (€/ha)		
Total costs	*379.32*	*Equivalent to a spend of €60.57/t of straw*

Table 8.21. Gross margins for irrigated hemp cultivation in the UK.

Total revenue	809.65	Excluding subsidies
Total costs	379.32	
Gross margin	€430.33/ha	Equivalent to a margin of €78.24/t of straw

8.5 Comparative Study of the Costs Across the Seven Areas Under Consideration

8.5.1 Synoptic presentation of the data from the seven zones under study

Table 8.22. Summary of revenue.

Variables	Champagne – Ardennes	Barrois	Holland	Germany	Australia (non-irrigated)	Australia (irrigated)	UK
Straw yield (t/ha)	8.72	5.48	8.00	6.22	5.00	8.00	5.5
Seed yield (t/ha)	1.07	0.92	0.00	0.800	0.00	0.00	0.00
Sale price of straw (€/t)	83.50	83.50	100.00	110.00	151.65	155.40	147.21
Sale price of seed (€/t)	313.33	313.33	0.00	313.33	0.00	0.00	0.0
Revenue from straw (€/ha)	728.12	449.51	800.00	684.27	758.28	1243.20	809.65
Revenue from seed (€/ha)	334.22	289.84	0.00	250.66	0.00	0.00	0.0
Total revenue/ha	*1062.34*	*739.35*	*800.00*	*934.93*	*758.28*	*1243.20*	*809.65*
Revenue (€/t of straw)	*121.83*	*134.92*	*100.00*	*150.31*	*151.65*	*155.40*	*147.21*

Table 8.23. Summary of costs.

Variables	Champagne – Ardennes	Barrois	Holland	Germany	Australia (non-irrigated)	Australia (irrigated)	UK
Fertilizer costs (€/ha)	137.33	116.00	95.40	109.11	139.27	222.84	95.69
Seed costs (€/ha)	140.00	140.00	111.30	122.71	126.85	139.27	173.70
Plant health costs (€/ha)	2.00	2.00	0.00	0.00	22.19	22.19	0.00
Other costs and variable charges (€/ha)	51.00	51.00	28.00	104.30	39.32	39.32 + 190.27	?
Total of all variables for the crop (€/ha)	330.33	309.00	234.70	336.12	327.63	613.89	269.39
Cost of harvesting (€/ha)	90.00	90.00	113.00	132.94	198.08	198.08	
Cost of mowing and sheafing (€/ha)	52.00	52.00	36.30	56.79			36.80
Cost of baling (€/ha)	87.20	53.80	214.00	122.71			58.88
Transport costs (€/ha)	52.32	32.28	90.76	11.52	71.18	131.20	97.15
Total of all variables for harvesting (€/ha)	281.52	228.08	454.06	323.96	269.26	329.28	192.75
Fixed mechanical costs (€/ha)	160.00	160.00	124.75	190.32			
Variable labour costs (€/ha)	80.00	80.00	99.00	65.75			
Total costs	851.85	777.08	912.51	916.15	569.89	943.17	379.32
Costs in €/t of straw	97.69	143.37	114.60	147.29	119.37	117.89	

Table 8.24. Summary of operating margins.

Calculation of margin	Champagne– Ardennes	Barrois	Holland	Germany	Australia (non-irrigated)	Australia (irrigated)	UK
Total revenue	1062.34	739.35	800.00	934.93	758.28	1243.20	809.65
Total costs	851.85	777.08	912.51	916.15	596.89	943.17	379.32
Operating margin	+€210.49 /ha	−€37.73 /ha	−€112.51 /ha	+€18.78 /ha	+€161.39 /ha	+€300.03 /ha	€430.33 /ha
Equivalent to a margin of €/t of straw	+24.14	−6.88	−14.07	+3.02	+32.27	+37.50	+78.24

Note: by way of comparison, in 2002, French experts provided figures that compared well with those presented above, despite being in a somewhat different form (Table 8.25).

Finally, at the time of completing this book, we are able to provide data from the 2005 harvest, as supplied by the OCERA (Table 8.26).

8.5.2 Discussion

It is not surprising that we can identify differences in profitability from one area to another. It is significant that the French, in the *département* of the Aube, produce both hurds and

hemp seed, thus creating two revenue streams, whereas other producers do not exploit this possibility. This is often due to climatic constraints that do not allow the seed to mature adequately. This affects the price of the straw.

It is also to be noted that the price of the straw varies markedly, with a 25% difference between France, Holland and Germany. This arises because, in the French case, the price quoted is ex farm, whereas in the other cases, the price quoted is delivered to the factory.

8.5.3 Costs

Input

The variations in costs can be explained by variations in the amount and dose of fertilizers used and variations between different countries in the price of any one product. It is to be noted that the French use more phosphate than the Dutch. By contrast, the Dutch use 50% more nitrogen, while the Australians use a large amount of nitrogen (200 kg/ha).

The products used to protect plant health are recorded here only to remind the reader that hemp, as a population, is not subject to significant fungal and insect damage and therefore requires little in the way of treatments. At present, such attacks are exceptional. The cost of the seed varies according to the rate applied

Table 8.25. Expenses for hemp production (2002).

Expenses	In €/t of straw
Input	251.00
Cost of cultivation and harvesting:	
– Using dedicated materials	302.00
– Using other materials	108.00
	410.00
Cost of transport andstorage	61.00
Other costs	54.00
Total	776.00

Table 8.26. The economics of hemp production in Barrois and Champagne Crayeuse (OCERA, 2005).

€/ha		Champagne Crayeuse				Barrois			
		Hemp		Rape		Hemp		Rape	
Product	Straw	9.5 × 85	808	4.1 × 195	800	6 × 85	510	3.4 × 195	663
	Hemp seed	11.5 × 29	334		–	10.5 x 29	305		–
	Subsidy		pm		pm		pm		pm
			1142		800		815		663
Costs	Supply		310		380		280		300
	Mechanical		300[a]		270		300[a]		270
	Other		50		35		40		35
	Straw harvesting	€16/t × 9.5	152[a]		–	€16/t × 6	96[a]		–
Semi-net margins (rounded up)			*330*		*115*		*99*		*58*

Notes: [a]Material specific to CUMA. The semi-net margin thus calculated is, on average, superior to that for rape.

and the original cost. The original cost is more or less restricted by various taxes and other fiscal charges, depending on the country concerned.

Machinery charges

Given that, plus or minus a few euros, the costs of harvesting (mowing) are more or less the same, it is in the baling and handling of the hemp – as well as during its transport – that differences in costs are seen between Holland, on the one hand, and France and Germany, on the other.

In Holland, harvesting is undertaken by the manufacturer.

In France, a number of farmers have banded together to form a cooperative (CUMA: *Coopérative d'utilisation de matériel agricole*). This cooperative has a very flexible and geometrically varied structure that allows it to purchase, use and maintain all the machinery that its members believe to be worth sharing. This way of organizing users is extremely popular in France (numbering 13,100, with a total membership of 233,000) and allows the cost of using farm equipment to be kept very low.

Another phenomenon must also be taken into consideration when explaining these differences in prices and that is the average surface area devoted to hemp production.

The technical choice of baling equipment also explains some of these differences.

The Dutch manufacturer makes use of an automated machine, which produces parallelipipedic (six-faced) bales, that is only used in hemp production. The French simply make use of round balers, which they also use to bale hay and straw. The costs of acquiring these machines are very different. The classical balers are inexpensive and receive many more hours of use. This makes the cost of using them considerably lower.

Finally, the manner in which transportation is organized, as already mentioned, is completely different. These differences in the way the straw is processed clearly make comparisons difficult, but it is not unreasonable to attribute a farm-to-factory transport cost of around €15/t.

Labour costs

Where labour costs are concerned, the choices made by the farmer (hired help or working themselves) can impact significantly on costs.

Similarly, an increasing number of farmers practise a technique called *sans labour*, or 'without labour', in which an intensive/deep preparation of the ground is replaced by a superficial preparation with the selective use of herbicides. This latter technique results in significant economies in terms of time and labour.

In Australia, labour costs are not distinguished from the total costs.

8.6 Modelling Table

To round up this presentation on the economics of hemp production, a table has been prepared (Table 8.27) that allows us to model hemp production in order to compare it with other crops that could be grown on the same land.

Table 8.27. Modelling.

	Hemp		
Product harvested	Straw	Seed	Seed and straw
Yield per hectare			
Seed			
Straw			
Amount of work required per hectare excluding work undertaken by contractors/companies:			
– labour, preparation of the soil, sowing			
– fertilizer spreading, etc.			
Specified costs:			
Fertilizer			
Seed			

Continued

Table 8.27. Continued.

Product harvested	Hemp		
	Straw	Seed	Seed and straw
Herbicides			
Insurance against hailstorms			
Amount of work required per hectare, excluding			
work undertaken by contractors/companies:			
– combining			
– mowing and sheafing			
– baling			
– transport			
Cost of machinery and traction (combining,			
mowing, sheafing, baling):			
Machinery, traction in €/ha			
Costs involved in receiving and drying hemp seed			
(transport and handling)			
Labour costs:			
Hourly rate (× €/h)			
Total expenditure			
Direct payments:			
Area contribution			
Contribution for straw T × €/t			
Contribution for seed T × €/t			
Total receipts			
Yield for commercialized products:			
Seeds			
Straw			
Operating margin			
Price required for hemp products in order for			
them to be as profitable as wheat			
Seed			
Straw			

Notes

[1] In the context of globalization, everyone should try to know what hemp competitors are and it should not remain the 'plant in the field'.

[2] For the industrial utilization of the fibres, different materials and methods have been developed and used to evaluate the colour and dimensions of the fibres. Readers are innted to refer to chapter 9.

[3] The presence of foreign bodies, in particular plastic pieces, however small, result in the formation of holes in the paper or variations in quality of the paper.

[4] The composition of hemp seed is explained in chapters 3 and 11.

9 The Industrial Hemp Economy

Pierre Bouloc
La Chanvriere de L'Aube (LCDA), France

9.1 Introduction: What is the Hemp Industry?

Hemp and its constituent parts can be used in a great variety of ways. As such, to describe the composition of hemp in each case would require a magic wand. It is therefore appropriate to consider the uses that hemp components will be put to. To start with, it is essential that we first have an understanding of the objectives of the manufacturer and the expectations and requirements of the market(s) that the manufacturer is seeking to satisfy.

In reality, hemp straw is not used as it is.

It is appropriate to remind the reader of the different components of hemp and their usual proportions (Fig. 9.1).

All the uses of the fibre and the other components of hemp are presented in Chapter 11 and are therefore summarized only briefly in Table 9.1.

These various uses require well-differentiated raw products. With this in mind, the role of the hemp processor can be summarized as follows. Hemp processing consists of 'the decortication of a complex plant into as many constituent parts as possible'.

The processor's work consists of separating all the components and collecting them in such a way that they can be exploited to their full potential. This equates, more simplistically, to the separation of the hurds, their refinement

and the recovery of the powder produced as a by-product.

This process of 'breaking up the straw' has existed since hemp was first used in the ancient Yunnan province of China. It still exists to this day, although the procedures have changed and the source of power is no longer dependent on human power. The fundamental steps, however, remain unchanged.

9.1.1 Preliminaries requiring consideration

Before presenting the different procedures used to transform straw into usable and saleable components, the problem must first be placed in context, thereby illustrating the complexity of the process.

1. Depending on the final use to which the fibre is put, the industrial processes can be completely different. There is, therefore, no standard factory, even though the basic principles are and will remain the same:

- Separate the fibre from the hurds.
- The fibres must be cleaned according to their end usage.
- The powder produced in this process must be recovered.

2. The second difficulty to overcome relates to the abrasive nature of the plant. This can result in increased wear of the machinery used and relatively significant maintenance costs.

3. The process of fibre removal from hemp is naturally polluting unless precautions are taken to collect the powder that is produced by the decortication of the straw.

4. The fourth difficulty arises from the significant investment required and the ongoing maintenance costs. This will be discussed later on in this chapter, but we can say at this point that the investment required to establish an entire factory exceeds the annual turnover of that factory.

5. Finally, it should be pointed out that working with hemp creates considerable logistical problems.

A certain amount of basic data needs to be brought to the reader's attention at this point in order to provide an idea of the extent and difficulties posed by these problems in terms of volume, surface area and tonnage (Boxes 9.1. 9.2, 9.3 and 9.4).

By volume, taking into account the density:

It is clear from these figures that a significant part of the investment required in the hemp industry is to provide the storage facilities necessary to house the raw materials and the materials produced when it undergoes refinement.

Finally, and far from being one of the most insignificant difficulties, we are working with a product that has very little unitary value. In France, hurds are sold by the farmer at €80/t.

Consequently, the industrial economy of hemp will be affected by the differing investment costs of materials and property, which will be discussed below.

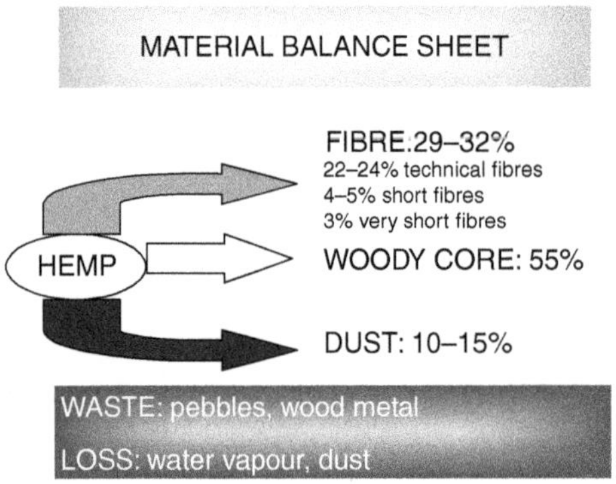

Fig. 9.1. Industrial hemp: a breakdown of straw's derivative products.

9.2 Presentation of the Hemp Decorticating Operation

This presentation will start with fibre and will then discuss from an industrial perspective the hurds, powder and hemp seed. First, however, the essential operation of retting is discussed.

9.2.1 Retting

The ultimate use of the fibre, that is to say, the industrial process, starts with retting.

Table 9.1. Principal uses of the constituent parts of hemp.

Fibre	Hurds	Powder	Hemp seed
Paper	Animal litter	Granulated animal bedding	Human food
Used industrially together with polymers to produce composites	Chipboard	Fuel	Animal food
	Mulch	Soil improver	Cosmetic products
	Insulation	Etc.	Cosmetic oils
Hemp wool used as an insulator	Hemp cement		Oils destined for use in industry
Felt	Etc.		Etc.
Geotextiles			
Textiles			
String, yarn and rope			
Etc.			

Box 9.1. Value by weight.

1 ha of hemp produces:
- − 4–12 t of dried straw with a humidity of 15%
- − 0.6–1.5 t of hemp seed

1 t of straw can be broken down into:
- − 0.55–0.65 t of hurds
- − 0.2–0.3 t of fibre
- − 0.15 t of powder

1 ha of hemp therefore produces (based on a yield of 8 t/ha):
- − 4.8 t of hurds
- − 2 t of fibre
- − 1.2 t of powder

Box 9.2. Volumes required when storing the derivative constituents of hemp.

1 t of straw pressed into round bales occupies approximately 6.5 m³
1 t of fibre pressed at high density occupies approximately 1.9 m³
1 t of hurds occupies approximately 5.5 m³
1 t of powder occupies 0.5 m³

Box 9.3. Volumes necessary to store the material produced by 1 ha of hemp.

Straw: 6.5 m³ × 8 t = 52 m³

As hurds:
  As fibre
  As powder

Box 9.4. Surface area needed to store the material produced by 1 ha of hemp.

Straw:
1 ha producing 8 t of straw/year, stored as round bales with a diameter of 1.65 m
Weighing 380 kg/bale, i.e. 21 bales/ha
Ground surface area required for one bale = 2.14 m²
Weight per m²/bale: 380 kg/2.14 m² = 0.177 t/m²
Storage over five levels: 0.380 t × 5 = 1.9 t
Weight per m² = 0.89 t/m²
Surface area needed for 1 ha: 4.2 m² (out of circulation) with piles of 5 bales

Fibre:
Weight of a bale of fibre = 400 kg
Volume = 1.5 × 0.8 × 0.8 = approximately 1 m³
Ground surface area required for 1 bale = 1.2 m²
Weight per m²/bale = 0.330 t/m²
Storage over 6 levels = 1.98 t/m²
Surface area needed for 1 ha = approximately 0.5 m² (out of circulation)

- to facilitate the removal of the lignified component.

Principle

Retting produces a partial breakdown of pectins, together with a degradation of hemicelluloses as well as the interpolymer bonds of the parietal system. This phenomenon is attributable to the activities of pectinases and hemicellulases, as well as to enzymes that have yet to be characterized fully (disconnecting enzymes, oxidases, etc.).

Finally, it would appear that the extraction and trapping of calcium by organic acids produced by microorganisms also make a significant contribution to the retting process and help destabilize pectins.

Obviously, these catalytic processes are under the control of microorganisms, or a population of microorganisms, according to modalities that remain relatively unknown to this day and still remain to be determined.

Attempts to mimic this biological process using enzymes or specific chemical agents fail to reproduce the natural phenomenon. These imperfect artificial processes may, however, be adequate for an industry that can use these techniques to control the fibre-removing process and refining of hemp.[2]

The French verb 'to rett' is *rouir*. It has its origins in the 18th century and describes the isolation of textile fibres (from flax and hemp) by the destruction of the sticky matrix that binds them together. This is undertaken by macerating them in water or by any other procedure (translation of the definition appearing in the *Petit Robert* 1984 edition).[1]

Retting is defined as the fibres being bound together and to other parts of the plant by pectins that are insoluble in water: these substances must be broken down by a fermentation process, termed 'retting':

- in order to free the fibres of the stem and attack the lignified part;

The retting process

At the start of the retting process, the viscous epidermis presents an impermeable protective barrier and prevents anything from happening.

During the first few hours, water creeps through tiny gaps in the bark and is absorbed by the internal cells, causing them to swell up. The absorption of water increases with time until the epidermis ruptures. Certain parts of the straw break up: this is the washing phase.

The protopectins that bind the fibres to each other are the most exposed and serve as substrate for bacteria. The primary organism is the anaerobe 'plectridium', together with the bacillus 'amilobacter'. This work is undertaken by numerous bacteria and fungi.

Note that research undertaken into the retting of flax, in 1965, by J. Arthaud of the INRA, demonstrated the presence of:

- *Clasporium herbarum*
- *Alternaria (A. tenuis, A. tenissima, A. consortiale)*
- *Epicoccum*
- *Pullularia (P. pullularis)*
- *Stemphylium*

Recent work undertaken by Tamburini *et al.* has emphasized the role and the activity of *Clostridium felsineum* and *Bacillus subtilis*, together with *C. acetobutylicum*, *C. sacchar-obutylicum* and *B. pumilus*.

Whether retting is undertaken on the ground, 'dew retted', or in water, as was done in the past and is still done today in certain parts of Vojvodina, Serbia, bacteria and soil-dwelling fungi are entrusted with the retting process.

To summarize, these organisms:

- colonize straw where the humidity and temperature levels are favourable[3] (normal temperature >6°C);
- secrete exogenous enzymes which break down the cement that binds the fibres together, resulting in a loss of tissue cohesion.

Retting, like fermentation, is therefore one of the oldest known biotechnologies.

It is, however, an empirical process that is largely dependent on the local environment, including the hygrometry and temperature.

The enzyme's attack on the plant material must proceed sufficiently to allow the separation of the fibres, but must stop before the microorganisms are able to damage the fibres.

This microbiological degradation of straw thus enables the fibrous bodies to be released from within the stem and from each other during a fibre-stripping procedure that is termed 'de-cortication'. This process, which we will describe in detail, consists of the mechanical separation of:

- fibres
- wood
- the epidermis and cement that will form the powder.

The degree of retting affects the ease, or indeed the difficulty, of the decortication process. It will especially affect the quality of the final product, taking into account the equipment available.

The particular case of enzymatic retting

Humans have sought to reproduce and improve on the biotechnological process that nature has invented for the retting of fibres using existing bacterial flora. Their ambition is legitimate and understandable, given that so many other natural phenomena have been adapted and improved following their initial observation. A prime example is that of fermentation, which humans now use in the production of alcohol (wine, beer, cider, etc.), having mastered natural fermenting agents.

We will describe quickly the steps of this technique, drawing on an article by Drager *et al.* that appeared in the *Journal of Industrial Hemp* (Volume 7, 2002). This article compared the respective merits of fibre extraction using enzymatic and steam-explosion methods.

This research made use of industrially created enzymes to extract the fibres in an effort to produce textile-grade hemp.

In order to do this, they made use of several formulations and procedures and, in particular:

- MPG (macerating polygalacturonase at a dose of 1,220,000 PGU/mg, where PGU is measured by the viscometric degradation of a solution of pectin at 30°C and at a pH of 3.9);

• PE (pectinmethyl esterase at a dose of 135,780 PE/g).

The preparations tested were essentially pectinolytic and hemicellulolytic enzymes that acted by dissolving the cell walls to release the fibres. These enzymes are produced by microorganisms during fermentation, including a 'bacillus' species and a culture of *Aspergillus niger*.

Different varieties of enzymes with various capacities for breaking down pectin were tested. The filtrates of these stabilized, and concentrated enzyme cultures were then dissolved in an aqueous solution at a pH of 4.0 for the pectinases PE and PGU, and at a pH of 8.0 for BioPrep 3000 L, and then placed in contact with the fibres.

The samples obtained were shaken mechanically for between 0.5 and 2 h at 40°C for the pectinases PE and PGU and at 60°C for the BioPrep 3000 L. They were then neutralized and rinsed for 2 min in cold water.

In principle, this procedure does not appear overly complex. How complex is it in practice?

The authors concluded their paper indicating that the research conducted on the enzymatic separation of fibres demonstrated their viable potential as a means of producing fine fibres with high-tension resistance. The BioPrep 3000 L was described as the best performing enzyme. They qualify their findings by adding that these experiments need to be repeated in an industrial setting.

At the time of writing, despite numerous experiments being conducted in China, Italy and Germany, a procedure suitable for industrial use does not appear to have been found.

Results of retting

There are four principal results of retting:

1. Ease of fibre extraction. The more the straw is retted, that is to say the longer and more pronounced the enzymatic attack, the easier will be the fibre extraction. Thus, if we are looking to make the work easier, it makes sense to allow retting to take longer. This was what the ancient hemp farmers were doing when they left hemp to rot in waterways or retting baths for 3 weeks or more.[4]

2. Straw colour. The longer retting takes, the more the straw, and therefore the fibre, goes brown, or even black, before it eventually becomes completely rotten.

3. The more the fibre is grey or black, the more the hurds will be grey or black.

4. Loss of material. As soon as the bacteria attack, they start to consume plant material. This is particularly pronounced if the straw is manipulated, because when the stems are broken, the hurds are released. In the past, the hemp farmer would allow the hurds to rot away completely until they were washed away. Hurds that were collected during decortication were used to stoke the fires that dried the hemp before it was processed further. Today, the manufacturer seeks to make use of all parts of the plant, including the hurds. It is therefore not in the manufacturer's interest to lose this part. Hence, the manufacturer will therefore seek a weak to moderate degree of retting in order to minimize this loss.

Sought-after fibre qualities[5]

The analysis of fibre quality is based on three criteria: purity, colour and fibre position.

PURITY. The degree of purity (or of refinement) that is sought is expressed as a proportion of the hurds.[9] For industrial use (for technical fibres), the purity sought has a level of hurds/fibre ≤2%. For paper production, the level can be as high as 35%. For the production of yarns and textile threads, string or rope, the level must be close to 0%.

COLOUR. Although today this is not a deciding criterion of choice, the preferred sought-after colour is white or blonde, or at the very least, a light colour.

NB The issue of colour will be revisited when it comes to hurds, for the users of hurds, whether it be as animal bedding, in hemp cement or plaster, all require white hurds.

FIBRE POSITION. The main users of hemp fibre (industry and paper manufacturers) do not need the fibres to be organized. What matters to them is their homogeneity. By contrast, where thread is being produced, it is essential that the fibres be aligned in parallel. This therefore requires a very particular production organization and equipment.

These qualitative characteristics will guide the conceptualization of the industrial tool(s).

9.3 The Different Techniques for Fibre Preparation

Fibres can have different end uses corresponding to particular technologies. We will discuss in order:

1. Fibres for use in textile manufacturing.
2. Fibres for use in paper manufacturing.
3. Fibres for technical uses.

9.3.1 Preparation of threads for yarn, textile and rope, etc.

Quality demanded for the manufacture of textile fibres

At the end of the processing phase, the fibres extracted must be as long as possible, organized longitudinally and perfectly clean, without any residue of hurds.

Retting

In order to obtain the quality of product described above, hemp must be retted for a long period of 2–3 weeks.

Today, this phenomenon cannot be seen in France and it is necessary, in the 21st century, to visit facilities in Serbia or Ukraine in order to get a good idea of what was still practised in Europe as recently as a few decades ago.

In Ukraine, hemp is cut by a reaping machine (using a Russian machine, the design of which resembles that of a reaper-binder), then gathered into sheaves after a period of time spent retting on the ground. The sheaves are then placed in bundles of 5–8. In this way, the straw continues to ret while drying. Once

dried, the straw is stored in stacks before being sent to the mill.

In Serbia, we have observed the ancient technique of immersion in which the straw is plunged into a retting bath. At the end of this process, the sheaves are taken out and placed to dry in bundles, as described above.

The retting achieved by this technique is both deep and intense, the straw is easy to break and fibres are easy to extract.

Fibre extraction

The fibre extraction techniques currently used to produce textile fibres are similar to those used for flax fibres and rely on the same principle:

- mechanical crushing using fluted rollers;
- the chain is fed manually or mechanically, ensuring that an even layer of stems is fed into the rollers;
- the crushing rollers break up the fibre and the hurds.

Once the fibre has gone through two sets of rollers, it is collected by an operator, who creates a bundle weighing between 1 and 3 kg.

The hurds are recovered from below the rollers using a pneumatic or chain-driven system. Figure 9.2 illustrates the principles of this operation and the machines involved. These machines are still based on a tool designed at the end of the 19th century (Renouard, 1909).

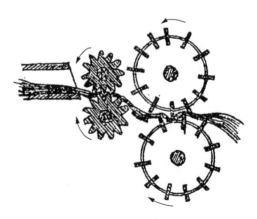

Fig. 9.2. Principle of the crushing machine.

Fibre cleaning

The bundle of fibres is taken by the operator and presented to a beating machine. This machine consists of a helix with a diameter of 1.3 m made from large wooden blades spaced 30 cm apart. These, when turned, hit the bundle and remove any remaining hurds. This technique resembles in all respects that used for flax. Figure 9.3 illustrates the principle of this machine. It is based on a tool invented in the 19th century that is still used today in Serbia and Ukraine.

Results obtained

These procedures are outdated in terms of cost and security, but are extremely efficient and are able to remove all traces of impurities completely. The resulting fibre will be suitable for use in the procedures described below, leading to the production of a thread for weaving.

9.3.2   Preparation of fibres for paper manufacturing

Result required

Tho papor induotry uoeo hemp fibreo ao well ao other natural fibres (including flax, jute, cotton and sisal) because it requires material that is rich in cellulose and hemicelluloses, as explained by Bernard Brochier in Chapter 13, in order to provide the paper with structure from the long, resistant fibres.

The paper manufacturers want long fibres (≥10 mm) in order to produce a fine and porous paper without losing strength. At the same time, the bundles of fibres must not be too long (<20 mm), in order to ensure their manoeuvrability during the operations preceding cooking. Levels of residual hurds up to 30% can be accepted. The humidity level must be less than 15%. It should be noted that the fibres must be free of all foreign material (stones, metal, plastic waste), as their presence can cause problems during preparation and cooking and will result in faults in the paper (holes, irregularities in thickness, etc.).

Industrial paper production requires that the straw is retted for no more than 4–5 days. This is for several reasons:

1. Clear straw will produce white hurds. The main use for hurds is in the animal bedding industry. The manufacturers demand a short retting for the straw.
2. The paper manufacturer seeking to produce a white paste prefers the fibre to be as clear as possible.
3. The farmer prefers to free up his fields as quickly as possible after harvesting in order to start the autumnal work of ploughing and sowing.

For these three reasons, hemp is left in the field only long enough for it to dry, 4–5 days. This represents enough time to start the retting process and will render fibre extraction easier.

Fibre extraction and fibre cleaning

The separation of the three different components of hemp traditionally is undertaken by two procedures, which are summarized in Fig. 9.4.

In essence, three principal methods of fibre crushing exist in the market:

* crushing with hammers
* crushing with rollers
* beating.

These are illustrated in Fig. 9.5.

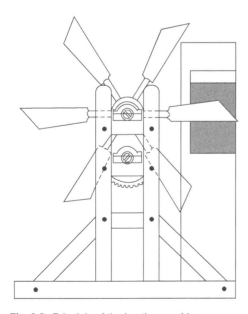

Fig. 9.3. Principle of the beating machine.

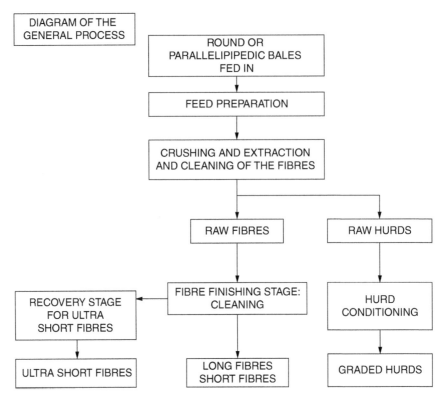

Fig. 9.4. Diagram of the general process of decortication and fibre extraction.

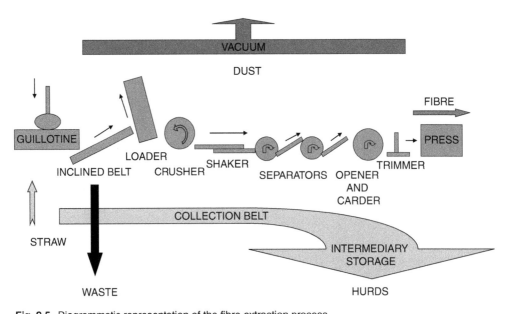

Fig. 9.5. Diagrammatic representation of the fibre extraction process.

We will also present two further fibre extraction processes: the turbine process and extraction by steam explosion.

Having summarized the main steps of the fibre extraction process, and in order to illustrate the industrial production line, we will present an example, that of Van Dommele-Laroche (Fig. 9.6).

This illustration will now serve as the starting point for a description of the most common fibre extraction process.

GUILLOTINE. The hay bale (with a diameter of 1.65 m and a weight of approximately 380 kg) is placed on the belt and conveyed to the 'guillotine'. This consists of a horizontal, hydraulically powered blade that cuts the straw into pieces 30–50 cm in length (for other methods of cutting straw, see box 9.5).

LOADER OR PRE-OPENER. The chopped straw falls on to an inclined belt, which takes it to a loader. The loader's role is to:

- eliminate by gravity, stones and other undesirable foreign objects;
- spread out the straw into a bed that is as homogeneous and as even in thickness as

possible. A blower or aspirator may be used to eliminate further dust and foreign objects.

METAL DETECTOR. In general, at this stage of straw processing, some sort of metal detection is used. Most commonly, the elimination of metal must be undertaken manually, but systems also exist that allow the suspect item to be redirected towards an outside stock.

CRUSHERS. This is the heart of the fibre extraction process. Three main devices are used. As shown in Fig. 9.6, two hammer crushers are installed in order to ensure that the fibre is cleaned as thoroughly as possible.

HAMMER CRUSHER. In order to reproduce the mechanism of the old crushing machinery, whose function was to crush the straw, a system was designed that projected the straw pieces on to a fast revolving cylinder, equipped with mobile hammers.

The projected straw is caught in the space between the cage and the cylinder. The hammers are brought down on to the straw by

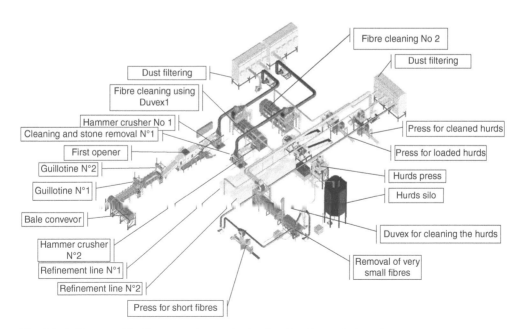

Fig. 9.6. A diagram of the fibre extraction production line, Van Dommele-Laroche.

Box 9.5. Other methods for cutting straw before fibre extraction.

The advantage of these alternative techniques lies in the elimination of the sectioning procedure at the start of the extraction process.

1. The straw is cut in the field by a Roto-Cut system. This automatic machine cuts the stem at ground level and then, as in a silage maker, an internal mechanism cuts it into pieces of 30–50 cm in length. These pieces are then swathed for drying (see Chapter 6 for a description of the German method – the BAFA method). After drying, the straw is compacted into cylindrical bales. This procedure is used mainly in Germany.

2. Another method, described in Chapter 8 uses a straw or fodder press equipped with an integral cutting device. This consists of blades that can cut the straw in advance, before it enters the compression chamber.

3. Finally, a bale-grinder system (using the sort of technology used for grinding household waste) is used to chop up the bales and turn them into short pieces of straw (Trommel Systems, ECP-CPI, ARTECH).

These different technologies, while interesting in principle, still need further improvements.

---

centrifugal force and break up the straw on contact. At the end of this operation, a mix of hurds, fibre and dust is produced. This then needs to be subjected to a refining, or sorting, operation.

CRUSHER WITH ROLLERS. In a similar procedure to that described above for the production of fibres for the textile industry, the straw is passed between two serrated rolling tables turning in opposite directions in order to move the straw forward. The serrations get progressively smaller and the upper and lower rollers increasingly closer to each other as they move towards the end of the table. In this way, the straw is subjected to an increasing onslaught as it progresses through the rollers.

It is possible to have three or four successive beds of rollers, and the machinery can have more than 30 pairs of rollers in total.

After a first passage, as with the hammer crusher, a mix of hurds, fibre and dust is produced, which must then be sorted and refined.

BEATING PROCESS. The working principles are much the same for this process: the straw is subjected to a mechanical treatment that separates the fibre from the hurds. In this case, a beater or thresher is used that takes its inspiration from the threshing arms of a combine harvester.

As in the process described for the hammer crusher, a high-speed tangential blow must

be applied to the straw; this shock will provoke the rupture of the stem.

TURBINE PROCESS. In this process – which is, to our knowledge, used in only one factory in Germany – the principle of wood decorticating machinery has been applied. The straw is cut into pieces measuring 30 cm or less and these are passed between two vertical plates with a diameter of 1.3 m, one turning very fast, the other immobile.

The straw passes into the gap and literally explodes. The resulting fibre is very fine. This technique is both sophisticated and noisy, requires a great deal of precise control and appears to consume a great deal of energy.

All of these procedures, while relatively easy to describe, are extremely precise and are very difficult to regulate. It is necessary to find the balance between the volume of straw, the speed of rotation and the gap between the two plates.

FIBRE EXTRACTION BY STEAM EXPLOSION. Technologies will be described which are still at the experimental or laboratory, rather than industrial, stage.

The technique of fibre extraction by steam explosion can be used with all sorts of fibrous plants and has been used to decorticate hemp and flax. It would appear to have its beginnings in Germany during the interwar years. In 1933, the German government, preparing for the war effort and anticipating that the maritime conflict would interrupt its supply of cotton, restarted hemp production.

It was at this time that the process of steam explosion was used on a grand scale, as it was of particular interest for the production of textile fibres. Where thread was concerned, it allowed the production of single fibres (as opposed to fibres in bundles), which could then be used to produce thread, and ultimately cloth. The technique used needed to reproduce on an industrial scale all the characteristics and qualities of the fibre: length, tension resistance, suppleness and fineness, etc.

This process will be described briefly, for, given the cost of exploitation, it is now used for laboratory research only[6] (Kessler and Kohler, 1996).

The machinery consists of a compression chamber and an extrusion system. The fibre (bundles of single fibres that are still attached to each other), having already been subjected to a mechanical fibre extraction process, is placed in the reactor at a pressure of 2.9 mPa of saturated vapour and a temperature of 150–220°C. The fibres are impregnated with caustic soda in order to break down the pectins. The fibre stays in the reactor for 5–15 min, during which time the pressure starts to break up the fibres. At the end of the treatment, the material is sent to a cyclone, where it literally explodes, allowing the fibres to be separated individually. This mechanical and chemical treatment produces extremely fine fibres.

The experiment reported shows that the key point is the diffusion of steam within the bundles of fibres at the start of the process. A minimum time of 5–10 min gives even better results.

This time guarantees a uniform penetration of the material and yields fibres of excellent quality. By contrast, longer treatments (>20 min) at high temperatures break down the cellulose components of the fibre.

It would appear that with this procedure, and the resulting adaptation of the chemical treatment, the qualitative differences found in the raw materials could be reduced, thereby producing fibre that corresponds to the specifications of the end user. For this to happen, the length of steam treatment and the dose of soda must be adjusted according to the material being used.

In this way, a uniform batch of fibres can be produced, even though the raw material is far from uniform.

SIFTING AND CLEANING. A number of methods are used to separate the fibre from the hurds. Essentially, there are two common processes: shakers and rotating sieves (most commonly made by Duvex). These will knock the separated hurds from the fibres; the hurds fall into a drainage channel that delivers them to storage.

RECRUSHING – PASSAGE THROUGH A SECOND CRUSHING DEVICE. After passing through the first crusher, and regardless of the extent of retting, a large amount of hurds will still be attached to the fibres. In order to complete the breakdown of these two elements, a second mechanical treatment is required. This follows the same principles as the first, will often use the same machinery and subjects the straw to a new trituration.

CLEANING AND PURIFICATION. In the same way and with the same technologies as described earlier, the fibre is subjected to a second sieving. Separators, shakers and sieves, moving vertically, laterally and rotationally, are all added to the mix. The combination of these movements deposits the hurds gradually on the belt that takes it away to storage.

OPENER CARDER. At the end of this process, an opener carder is used to arrange the cleaned fibres into a uniform bed.

TRIMMING. Where the fibre needs to respect a dimensional standard, it is fed continually through a trimmer.

CONDITIONING. Finally, the fibre is conditioned by high-density presses into bales weighing in excess of 400 kg, with dimensions of the order of 1.5 × 0.8 × 0.8 m.

Results obtained

At the end of this process – a process that lasts only a matter of minutes from the time that the bale enters the system to the time the bales of processed fibre emerge – fibre is obtained containing between 5 and 35% of hurds.

A proportional amount of hurds and powder is therefore also produced.

9.3.3 Preparation of the fibres for technical uses

'Technical fibres' is the professional term used for fibres that are destined for technical or industrial uses, as distinct from the fibres discussed previously, which generally are destined for use in paper manufacturing.

Their main uses are:

1. Non-woven products, including mats and felts of different thicknesses;
2. Hemp wool;
3. Inclusion in compound plastics;
4. Geotextiles.

These different uses each have their own particular demands and requirements.

*Expected characteristics
of technical fibres*

FIBRE QUALITY FOR USE IN THE PLASTICS INDUSTRY: CLEANLINESS AND DEGREE OF RETTING. Fibre length is not an important criterion, the fibres having been cut before they are processed. By contrast, they must be totally clean (ideally with 0% hurds), because the presence of such contamination of the fibre, and therefore of the compound, can render the end product fragile and increase the number of points at which it is prone to break. Hurds can, however, be used in plastics manufacturing where the intention is to produce a particular aesthetic effect (in flowerpots, for example). In these cases, they are added by the manufacturer according to a selected percentage, and is therefore a matter of choice.

The physical properties of the end product are not as good if the fibres have been subjected to a prolonged retting process. Fibres that have been retted excessively are not accepted by plastic manufacturers as they are likely to cause problems, affecting both the colour and resistance of the final product. It is therefore necessary to find the optimal degree of retting in order to produce clean fibres and materials with the right properties. Studies are

being undertaken to determine the optimal amount of retting (cf. the work of the INRA undertaken by Bernard Kurek).

The automobile and construction industries demand products that have constant and reproducible characteristics. The plastics industry therefore needs hemp fibres that always present the same technical characteristics, that is to say 'standard hemp'. The technical processes and the varieties need to be defined precisely according to the final product. There is also a need for traceability.

The interested reader is referred to the explanations provided by Dr Joërg Müssig on the subject of the desiderata of plastic technicians (see Chapter 10) and how to qualify any natural fibre easily.

FIBRE QUALITY FOR NON-WOVEN TISSUES: LENGTH AND DEGREE OF RETTING. Where non-woven materials are concerned, manufacturers need long fibres (optimum length 20 cm). These must be very clean in order to avoid clogging up the needles when the material is finally carded.

A light retting improves the final quality of the product by facilitating fibre extraction. Degradation of long fibres into shorter fibres is reduced.

QUALITY OF FIBRES DESTINED FOR USE AS INSULATING MATERIAL (HEMP WOOL): LENGTH, ABSENCE OF ALLERGENIC SUBSTANCES AND COMPRESSION RESISTANCE. Here again, manufacturers need very clean, long fibres. Two points are essential: the hygiene aspect and the compression resistance of the material.

These fibres are destined to be used in the home and must therefore be free of anything that could impact negatively on human health. The extent of retting can have an influence on this process and can introduce potentially allergenic microorganisms into the fibres. As an insulating material, the compression resistance of this wool is also important. Retting, and therefore the solidity of the fibres, together with the date of harvesting are two important factors. As with the fibres destined for use in the plastics industry, research is ongoing to determine the optimum maturity at harvest and the degree of retting required.

COMMON QUALITIES FOR ALL TECHNICAL FIBRES. Generally speaking, the quality of a technical fibre can be defined according to a number of physical and chemical measurements, as well as 'functionality' criteria. The ideal fibre is described globally by technicians in Table 9.2.

These characteristics are dependent on the environmental context of the culture and, to a lesser extent, on genetic factors.

Generally speaking, the users of technical fibres not only need to know the factors that influence the ease of fibre extraction but also those factors that will impact on the properties of the end product.

On this subject, the work undertaken by Dr Joërg Müssig is particularly useful.

The necessary installations

In order to manufacture technical fibres, straw decortication relies on the same technical principles as were described for the production of fibre for the paper industry. The main difference lies in the mode and extent of refinement required.

Among the architectural drawings on public file by construction firms, it is possible to find diagrams for proposed installations. These installations can be adapted easily for the processing of flax or other natural fibres.

To illustrate this, the production line TEMAFA, used by the German hemp producer BAFA, is presented in Fig. 9.7.

An original process for sorting fibres

An additional variable that can be introduced stems from the choice of method and technology for sorting fibres.

At the end of the sorting and refining process, the fibres are trimmed in order to meet the maximal length specifications.

These fibres then move along a belt, where they are subjected to a vacuum cleaning, the speed and volume of which is controlled carefully.

Each piece of aspiration equipment is calibrated according to the type of fibre. By using three to five 'aspirators', three to five types of fibres are obtained at the end of the chain.

This highly sophisticated procedure takes a while to regulate. It gives excellent results and allows the value of the fibres to be realized better by taking into account precise requirements.

9.3.4 Presentation of some original fibre extraction systems

Fibre extraction in the field

Several constructors have attempted to develop fibre extraction machines that could be used in the field where the industrial hemp is produced.

The objective is to:

* Avoid transport costs by only removing the fibre from the field once it has been separated from the hurds.
* To leave behind the by-products of the hurds. These will rot down and contribute organic matter to the soil.

Table 9.2. Selection criteria for technical fibres.

Physical criteria	Chemical criteria	Desirable functionalities
Resistant to rupture	Maximal amount of dry matter	Technological performance (ability to improve the properties of the supporting material)
Capacity for elongation good	Lignin levels low/minimal	Industrial performance (suitable for factory use)
Fine	Hemicellulose levels low/minimal	Form and appearance
Clear fibres	Cellulose levels high/maximal	Uniform
Fibres silky/smooth	Crystallinity high/maximal, favouring a capacity to synergize with the matrix in which it is to be integrated	

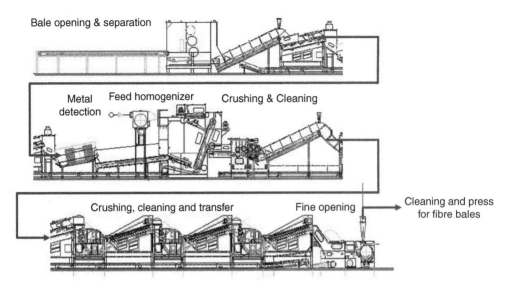

Fig. 9.7. Diagrammatic representation of the general fibre extraction line produced by TEMAFA and used by BAFA.

Working on this idea, a Dutch[7] company has designed a prototype which, at the time of writing, appears to have treated only a few hundred hectares.

A Canadian company[8] from Ontario has marketed a mobile piece of machinery for the decortication of fibrous plants using 14 pairs of rollers organized in the horizontal plane. This machine moves from one storage place of straw to another and processes the material in the field. The concept appears to be ingenious, but performance appears to be poor. The theoretical yield is 1 t/h using this model, but it is unclear what actual yield this corresponds to or how many machines have been sold.

These machines may constitute a means of reducing the cost of production in areas that are considering starting hemp production but that do not want to invest heavily in a costly and uncertain sector. That said, the performance of these machines appears to be poor.

Australian fibre extraction processes

Hemp production was relicensed in Australia in 1998. Over the past few years, Australia has developed a large-scale production programme – a complete production process from field to factory.

Very large fields are farmed in Australia, with areas for hemp production typically measuring some 100 ha. Due to this, they have adapted slightly different means and methods.

1. Hemp is cut by a machine that churns out straw chopped into 1 m pieces and spreads it, without overlap, in order to facilitate drying.
2. A first fibre extraction is undertaken in the field by a machine that takes in the straw continuously and decorticates it by 60–80% before loading it into a container.
3. This container is then emptied into a press that produces a parallelipipedic rectangular bale, weighing 14 t, called a *ballot* in French. The *ballot* is covered with tarpaulins in order to ensure its preservation until it is processed at the mill.
4. The mill, when ready, sends for the *ballot*.
5. At the mill, the *ballot* is fed into the system, which processes straw at a rate of 10 t/h.
6. The first decorticating process in the factory reduces hurds by 50%.
7. The second decortication removes the rest of the hurds, bringing the amount of residual hemp down to 15%.

The total cost of this procedure, including collection, is estimated at AU$275. Using the exchange rate current in October 2005, this equates to €162/t.

9.4 Treatment of Hurds

9.4.1 Constraints imposed by the product

It should be remembered that hurds in bulk have a density of 0.12–0.3, equating to 12–30 kg/m^3. Bearing in mind this physical characteristic, it is inconceivable that this material be transported without compressing it severely.

It should be noted also that this material is not very fluid and its rheological behaviour is unpredictable and varies with humidity levels and ambient temperature.

Finally, 1 t of straw generates between 600 and 700 kg of hurds, occupying a volume of 5.5 m^3.

The objective is to pack, as cheaply as possible, the maximum amount by weight into the smallest possible volume.

9.4.2 General layout of the industrial plant

To achieve the objective of maximal compression, a single technique is used.

The existing installations all adopt the same design:

- Hurds are delivered on a belt or via pneumatic transport into intermediary storage.
- Hydraulic pressure (of the order of 150–200 bars) is used to compress the hurds.
- The compressed material is evacuated, generally as a parallelipipedic rectangle packed in a PVC bag. These bags are preformed. 'Wrap around' techniques are also used. As their name suggests, these consist of the creation of a covering by wrapping it around the compressed material while it is still under pressure.

NB These techniques are in no way original and are also used for the bagging up of products such as wood chipping, cut straw and hay, etc.

9.4.3 Cost of the installations and running costs

Costs vary, making it difficult to provide an accurate figure.

9.5 Treatment of the Powder

9.5.1 Characteristics of the powder

First, the characteristics of the powder are given:

- The decortication of 1 t of straw produces 150 kg of powder.
- This equates to a volume of 0.5 m^3 with a density of between 0.2 and 0.3.
- The product is abrasive.

9.5.2 Objective

Considering the density and pulverized nature of the product, it requires compacting for transport and use. One exception is that, where the product is to be used as an enriching agent, it can be transported in bulk, providing the fields on which it is to be used are not far away. In such cases, covered lorries or tankers with a pneumatic loading system must be used in order to avoid losses during transportation.

9.5.3 Necessary installations

Taking into account the market value of the product, procedures need to be kept cheap and simple. An appropriate method is to use the granulating equipment used in factories that desiccate lucerne and other livestock feeds.

The powder is subjected to pressure to force it through channels; this causes the temperature of the product to rise (up to around 60°C), resulting in compaction.

The end product is a pellet shaped like a cork, with a diameter, depending on the intended use, varying between 0.5 and 7 cm and a length varying between 1.5 and 10 cm. The density ranges from 0.2–0.3 to 0.6–0.7.

The product can then be bagged (for granulated animal litter and soil enrichment) and transported.

It is also possible to manufacture bricks that are suitable for burning in stoves and other domestic heating systems.

9.6 Treatment of Hemp Seed

The harvesting of hemp seed by combine harvester yields a mixture of dry, ripe seeds, together with seeds that are only partially ripe or even completely green. Additionally, this material can also contain bits of leaf material and pieces of greenery from the inflorescence. The humidity levels observed are 20%. This is one of the reasons farmers are asked not to harvest until humidity levels are below this level.

It is well known that the more green material the harvested crop contains, the higher the risk of overheating, which itself is accompanied by risks to the material. Hemp, as with other oil-rich seeds, is very sensitive to hydrolysis and oxidation of fatty acids. It is therefore necessary for the drying process to be undertaken quickly.

Second requirement: dry slowly!

In order to produce a marketable product, the hygrometry level must be reduced to 9%.

Given the quantity of material processed, this can be achieved only in silos of a sufficient capacity and equipped with driers. The seeds should have first been sorted and aired in order to remove all non-ripe seeds.

Drying is undertaken over a relatively long period in order to avoid damaging the product. The drying temperature must not exceed 40°C.

Finally, given the friable nature of the outer integument, new impurities can appear during the treatment process. There is therefore a need for another cleaning, by aspiration, before the seed can then be stored.

The store itself must be well aired and monitored carefully.

9.6.1 Harvesting

Farmers eager to harvest hemp seed must ensure that it is ripe. This sounds obvious, but it must be emphasized every year to over-hasty farmers.

This is not as simple as it may sound, however: a stand of hemp does not mature evenly. Additionally, the lower seeds reach maturity before those higher up the plant.

This gives rise to two basic common-sense rules that should be respected. Harvesting should take place when:

- the highest leaves start to turn yellow;
- the seeds at the base of the inflorescence are ready to fall off on 75–80% of plants.

In terms of the harvesting equipment, combine harvesters with axial fan blades are used at a moderate rotational speed (~300 turns/min).

9.6.2 Installations and costs

In France, all the hemp seed collected and destined for sale transits through the silos of companies that store cereals. Only they have the capacity to deal with and store 4000–5000 t of seed/year.

Taking into account that which was explained above, the costs of drying, based on figures from January 2005, are of the order of €15–18/t.

9.7 Industrial Costs: Investment and Running Costs

The informed reader will understand readily that it is not easy within this text to provide definitive and intangible values for investment and production costs. Much depends on the degree of sophistication of the installation, its size, volume of material processed, rate of throughput and the nature of the end product (degree of refinement). An attempt will be made to present and apply a methodology and to produce a rough estimate for an industrial set-up and its equipment destined for the production of fibre and the conditioning of hurds.

9.7.1 Investments

The cost of the fibre extraction production line is strictly a function of the hourly volumes that one aims to process. The following rough estimates are provided.

Theoretical production: 1.5 t/h (equivalent to 4000–6000 t of straw/year).

The nova-Institut estimates an investment required at more than €3,000,000 (based on 2005 figures) simply for fibre extraction. Others estimate a figure of €5 million for a capacity of 5 t/h.

The true difference between these investment costs is due to:

- the degree of refinement that is sought (higher levels of fibre purity require more decorticating equipment – for example: an increase in the number of rollers);
- the degree of sophistication of the equipment used to treat the dust and waste produced in order to avoid pollution;
- the degree of sophistication of the equipment used to treat and condition the hurds and powder;
- the surface area and volume of storage provided for storing the straw and the end products.

Depending on the choices made, the amount of investment required can double.

9.7.2 Operating costs

The values provided in Table 9.3 are averages obtained from observations and exchanges of information with European hemp producers. The following rough figures are offered:

- Numbers: one person is required for the fibre extraction of 2000 t of straw/year.
- Energy: 100–150 kWh are required per tonne of straw (including the conditioning of the hurds).

Table 9.3. Operating costs.

Cost (2004 values) in €/t	Source: Experts	Source: Nova-Institute
Salaries and social charges	43.56	47.00
Repayments	38.71	54.00
Energy	10.67	11.40
Financial charges	11.10	Included with other charges
Maintenance	8.40	9.00
Other costs	24.83	27.40
Total	*137.27*	*148.80*

- Maintenance: 2 qualified workers per team.
- From this it would appear that the cost of treating 1 t of straw lies between €120 and €150/t.

As a result, we are able to propose two approaches to the calculation of production costs (these are average values and exclude the cost of purchasing straw).

9.8 Conclusion

This chapter has demonstrated and explained that the refining of industrial hemp into useable products is both difficult and costly. The biggest challenge lies in defining clearly the end use of the product and making a sale.

Notes

[1] Other authors indicate the term may come from the Latin Rivus or Ros or from the German Rozen.
[2] For a scientific explanation of retting, one should refer to Chapter 3 where B. Chabbert and M. Kurek describe the chemical composition of hemp.
[3] In water at 20°C, the process of decomposition lasts 8 days.
[4] This practice polluted water and the atmosphere leading the authorities to ban the technique.
[5] The reader is referred to the contribution of Dr Jöerg Müssig, in Chapter 10 on the methods used to characterise the natural fibres.
[6] An explanation of this technology can be found in Chapter 10 by Dr Jöerg Müssig.
[7] Dunagro.
[8] Hill Agra Machinery.

10 Integrated Quality Management for Bast Fibres in Technical Applications

Jörg Müssig,[1] Gabriel Cescutti[2] and Holger Fischer[3]

[1]*Hochschule Bremen – University of Applied Sciences, Bremen, Germany;*
[2]*European Patent Office, Rijswijk, The Netherlands;* [3]*Faserinstitut Bremen e.V., Bremen, Germany*

10.1 Introduction

In recent years, natural fibres have become increasingly popular for use in industrial applications, for example as a reinforcement for plastics. Thus, since 1996 the nova-Institut GmbH has been documenting the rise in the volume of natural fibres used by the European automotive industry (Karus *et al.*, 2006). Priced at approximately 60 cents/kg, such fibres are also economically attractive as reinforcement for plastics. In car interiors, natural fibres have already become established for applications such as trim panels (with a market share of approximately 40%), seat shells and rear shelves (Gassan, 2003).

Due to environmental influences, the properties of natural fibres show much greater variation than synthetic fibres. This poses a problem for software-based industrial design processes that require reliable material data as an essential prerequisite for numerical simulation. According to Harig and Müssig (1999), the conditions that material tests on natural fibres have to meet in order to yield useful results are:

- the supply of fibres with reproducible qualities, based on a quality management system that has been put in place along the whole value-added chain, with reliable proof of origin;
- the objective determination of accurate fibre properties for calculating the properties of the finished product.

The main argument against the industrial use of natural fibre reinforced plastics is that the quality of the fibres depends on the year in which they are grown. It is, nevertheless, possible to obtain fibres of consistent quality, as well as reliable data enhancing the predictability of the properties of natural fibre reinforced plastics by using a quality management system that starts at the cultivation stage and which is based on reproducible proof of origin and harvesting parameters (Cescutti and Müssig, 2005).

To demonstrate the possibilities offered by quality control in agriculture, the results of cultivation trials performed by the Chamber of Agriculture Weser-Ems in Germany will be presented here. The test location lies in the Wehnen district, in Bad Zwischenahn, near Oldenburg in the north of Germany, close to the North Sea. The experiments were carried out in 1997 and 1998 (detailed information on the experimental arrangement can be found in Müssig and Martens, 2003). The experiments were carried out with hemp variety Fedrina

74 at a rate of 200 seeds/m². Sowing date was 16 April in 1997 and 8 May in 1998. The soil was fertilized with 100 kg N/ha with CAN (calcium ammonium nitrate).

It is important to start the concept of integrated quality management at a very early stage. Even at the harvesting stage, the choice of an unsuitable harvesting process can cause damage to the fibres. In cooperation with the Chamber of Agriculture Weser-Ems in Germany, different techniques were tested for the harvesting of hemp. The techniques used can be divided according to the following system:

- Harvesting and baling of the stems in non-oriented position
  – whole stem
  non-squashed: rotor disk mower, double-knife cutter bar
  squashed: swath mower System Fortschritt
  – part of the stem
  non-squashed: 2-step double-knife cutter bar, HempFlax harvester
- Separation of wooden part of stems and fibres in the field
  – part of the stem
  field decortication: Nölke harvesting system.

Bassetti *et al.* (1998) give a detailed overview of the harvesting technologies for hemp. The systems used in our experiments are described in their work.

As shown in Fig. 10.1, the harvesting techniques for the category 'Harvesting and baling of the stems in non-oriented position' had no destroying impact on fibre strength, whereas use of the intensive harvester had a bad influence on fibre properties and led to a dramatic decrease in fibre strength.

When discussing the advantages and disadvantages of retting, the risk of damaging the fibres by over-retting has to be taken into account. The complete harvest could even be lost in the case of bad weather conditions during retting. Therefore, in cooperation with the Chamber of Agriculture Weser-Ems, we investigated the influence of retting on fibre quality. To measure the influence of retting on collective strength, trials were organized in 1996, 1997 and 1998. Samples were taken each week, starting with the harvesting date 17 September, named 'Date 1', to 'Date 4'. Because of extremely bad weather conditions in 1998, the stems could not be stored and still lay in the field in November. The last sample of this experiment was taken on

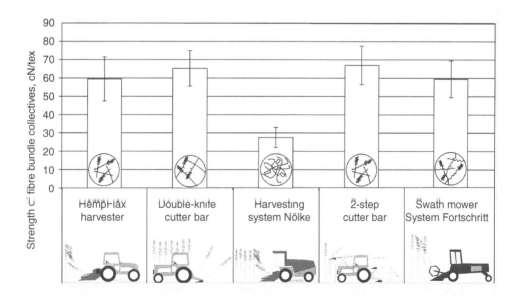

Fig. 10.1. Influence of the harvesting technique on collective strength (Stelometer/80 measurements per sample) (Müssig and Harig, 2000).

10 November as 'Date 5'. In the first step, the stems were separated with a decortication unit and, in the second step, with a coarse separator. A detailed description is given by Müssig and Martens (2003). After separation, the tensile strength of the fibre bundles was determined by testing collectives with the Stelometer device (for a description, see the section on collective strength/Stelometer); the results are presented in Fig. 10.2. In 1996 and 1997, the progress of retting was very smooth and the stems could be baled and stored without any problems. As far as strength values were concerned, no damage to the samples could be found after 1 month of retting. The extreme decrease of strength observed later in 1998 was related to the high amount of precipitation during the first days of October. In 1998, a high amount of precipitation occurred after cutting the stems. After constant retting in the beginning, the water and temperature conditions changed after testing Date 3. Wet conditions led to high microorganism activity, which damaged the fibres.

With a view to producing high-quality hemp fibres for industry, it is important to control the strength and fineness of the fibres along the whole production chain. The effect of crop management decisions such as cultivar choice, sowing date, plant density and harvesting date on these parameters is important. To achieve good fibre properties, quality management must start with cultivation, harvesting, harvest date and, last but not least, the retting process and duration. Considering all these issues, hemp (*Cannabis sativa* L.) can be a profitable crop with the right profile to fit into sustainable farming systems (Müssig and Martens, 2003).

Internal quality systems have been developed by fibre suppliers to ensure the origin and quality of the raw material. In recent years, adapted or new testing methods for bast and leaf fibres have been developed. While international standards for testing bast fibres are available for the textile industry (ISO 2370:1980 Textiles – Determination of Fineness of Flax Fibres – Permeametric methods, ASTM D6961-03 Test Method for Color Measurement of Flax Fibre and ASTM D7025-04 Test Method for Assessing Clean Flax Fibre Fineness), there are no international standards for the testing of natural fibres for technical applications like reinforced plastics. Various authors have presented the properties of several natural fibres and have collected the results of numerous investigations (Hearle and Peters 1963; Haudek and Viti, 1980; Batra, 1998). But, the comparison of more than one fibre

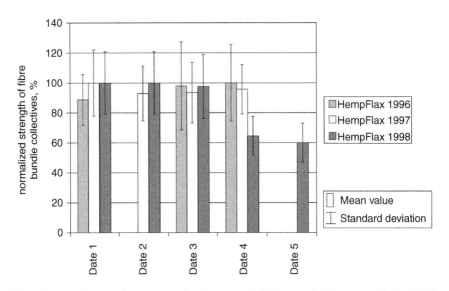

Fig. 10.2. Influence of retting duration on collective strength (Stelometer) (Müssig and Harig, 2000).

property, especially by comparing the data of different authors, is often difficult because fibres with unknown history and different testing methods have been used. For this reason, the following sections focus on making the various properties of different fibres comparable by using identical samples and testing methods. Furthermore, explanations of the economic significance and the suitability as well as the description of testing methods should help the reader in choosing test methods for measuring fibres.

10.2 Aspects of Fibre Quality Management

10.2.1 Economic and technical significance of natural fibre tests

For continuous production, it is important that the supplied fibres have been tested systematically, allowing a reliable determination of deviations occurring in the key properties. The main fibre quality characteristics relevant for the use of natural fibres in the reinforcement of plastics are shown in Table 10.1.

Morphological properties
and mechanical data

For use in textiles, the quality of the yarn is determined by the short fibre content, fineness, fibre–fibre friction and strength. For use in composite materials, length, fineness and mechanical properties like Young's modulus (stiffness) and the strength of the fibre influence greatly the mechanical performance of the reinforced plastics (Kaw, 1997). To predict the properties of composite materials, several analytical and semi-analytical models for laminates, micromechanical models and failure criteria have been developed. Apart from the mechanical properties, these models require knowledge of the fibre morphology. For a numerical simulation, knowledge of the length of the fibres, their fineness and their mechanical data is essential.

Colour

In retting control and in trade, colour plays an important role. It can be measured objectively with a colour system like the CIE (L*a*b*) colour system.

Chemical composition,
accompanying substances

For car-interior applications, the chemical composition of the fibres and the resulting properties such as fogging and odour have to be taken into account. The share of volatile components determines whether or not the materials will be approved for use in vehicle construction. The same holds true for the use of NF-PP[1] in end-user applications like car interiors, which is ruled out if there is any intensive or unpleasant odour.

Table 10.1. Key fibre properties and their industrial significance.

Category	Fibre property	Industrial relevance/application example
Morphological properties	Fineness	Processing
	Length	Simulation
		Mechanical behaviour
Mechanical data	Strength (Young's modulus)	
Colour	LAB or CIE values	Design application
Chemical composition, accompanying substances	Fogging behaviour	Automotive industry Rail vehicle construction
	Odour emission	Car interior applications
Harvesting, mechanical separation	Wooden stem parts	Mechanical behaviour, processing, design
	Degree of retting	Mechanical behaviour, processing

Harvesting, mechanical separation

Harvesting parameters play an important role for further processing steps by the compounder or the needle felt manufacturer. A high proportion of wooden stem parts, for example, can interrupt processing and can also create mechanical weak points in the component. With an optimum retting time, the relationship between fibre strength and fibre fineness (as regards the transmission of forces between the matrix and the fibres) can be improved. This also optimizes the mechanical properties of the finished part.

10.2.2 Suitability of the measurement method

In the course of history, different measuring systems and testing specifications for cotton, wool, bast fibres and synthetic fibres have come into use in the textile industry. Each method has been developed to determine the morphological properties of one kind of fibre. Thus, it cannot simply be applied to other fibres. We have carried out an evaluation of the present measuring methods with regard to their usability for quality control in mass production, comprising the following topics:

- objectiveness
- suitability for incoming control tests and/ or numerical simulation of properties
- flexibility and comparability of the results
- economic aspects like cost per measurement or cost of the testing device.

Care has been taken to recommend only flexible methods that are in widespread use. Table 10.2 gives an overview of the methods evaluated.

10.2.3 Recommended methods

As an orientation aid for industry in selecting test methods, recommendations have been set up as part of the N-FibreBase project: www.n-fibrebase.net. These test recommendations are listed in Table 10.3. They were developed in

close coordination with the working group on natural fibre reinforced polymers of the German Federation for Composite Materials (AVK-TV).

Fineness

According to the detailed analysis of possible fineness measurement methods suitable for very different natural fibres, we decided to concentrate on two methods. Both these methods have also been favoured by the N-FibreBase proposal (www.n-fibrebase.net) and AVK-TV group. In the following, the gravimetric fineness measurement method and the image analysis system, *Fibreshape*, in combination with a high-resolution scanner will be described in detail.

GRAVIMETRIC FINENESS MEASUREMENT. After 24 h of conditioning (20°C, 65% relative humidity of air), the fibres and fibre bundles are combed very softly without losing fibres on any of the two sides of the prepared sample. Very important for this preparation step is not to separate the bundles into finer bundles. After this preparation, the sample is cut to a length of 20 mm. The fibres are separated using tweezers. Each fibre or fibre bundle with a length of 20 mm is counted and stored for subsequent weighing of the complete samples. For each sample, 500 fibres or fibre bundles are counted and weighed. The gravimetric fineness gF in tex is calculated by using the following equation:

$$gF = \frac{\dfrac{M}{mg}}{N_B \cdot \dfrac{l_B}{mm}} \cdot \underbrace{10^{-3} \cdot 10^{6}}_{=1000}$$

where N_B is the number of counted fibres, M the mass of the collected fibres in mg and l_B the length of the cut fibres in mm (here 20 mm).

FIBRE WIDTH DISTRIBUTION WITH FIBRESHAPE. Fibre fineness is an important fibre property because it describes the quality of the separation process. An image analysis system called

Table 10.2. Qualitative evaluation of testing methods for measuring physical and mechanical fibre properties.

Fibre property	Equipment (method)	Remarks	Equipment cost[a]	Measurement cost[a]	Usual method?[a]	Suitable for[b]
Strength and Young's modulus	Tensile tester (single-element test)		-	○	++	NF
	Stelometer (collective test)	No extension/ force diagram No Young's modulus	+	++	+	Co, Wo, B
	Diastron, Textechno Favimat (single-element test)		-	○	-	NF
	HVI (collective test)	Developed for Co, not adequate for B	−	++	+	Co
Fineness	Scanner/ FIBRESHAPE (single-element test)	Quick analysis, not yet widespread	++	+	-	NF
	Laboratory balance (collective test)	High time demand	++	-	++	NF
	OFDA/Laserscan (single-element test)	Quick analysis, well reproducible. Problems with thick fibres/ bundles	-	+	○	Wo (B)
	Airflow	Indirect measurement (results not fully comparable)	+	++	+	NF
Length	ALMETER (collective test)	Maximum length: 250 mm	-	++	-	Co, Wo, (B)
	Tweezers, scale (single-element test)	Extremely high time demand	++	−	+	NF
	Imaging system, software (single-element test)	High time demand, not yet fully automatic	-	-	○	NF

Notes: [a] Evaluation scheme (−, -, ○, +, ++); [b] abbreviations: Co = cotton fibres; B = bast fibre bundles; Wo = wool fibres; NF = all kinds of natural fibres.

Fibreshape has been developed for the quality control of fibre particle mixtures and the needle density of felts. It determines width distribution and the orientation of fibres (Müssig and Schmid, 2004).

Measurement of the width distribution of the fibres and fibre bundles is done by preparing the elements between two glass slides for colour positives (Company Gepe, Zug, Switzerland, type 69 01) after 24 h conditioning at 20°C

Table 10.3. Selected measurement methods for fineness, length and strength.

Fibre/bundle property	Method alternative	Equipment (method)	Measurements	Type of standard or instructions for testing	Test information	Data suitable for numerical simulation	Incoming control
Fineness	1	Scanner/ FIBRESHAPE	Width (µm) (complete distribution)	Internal test method (defined measurement mask, etc.)	Resolution (dpi) number of measured elements	X	X
	2	Gravimetric fineness/ balance	Gravimetric measurements (dtex) (mean value)	EN ISO 1973, 1995	Staple length, number of measured elements		X
Length	1	Capacitive measurement/ ALMETER	Length (mm) (distribution cross-section related)	Wool testing standard (IWTO 17:1985), cotton (manufacturer's data), bast (internal test specification)	Cross-section related, preparation (A-barbe), number of measurements	X	X
	2	Manual test process (tweezers process)	Length (mm) (distribution)	DIN 53808	Number of tested elements	X	
Strength	1	Single-element test	Strength (cN/tex) (mean value, standard deviation), Young's modulus (cN/tex), conversion to N/ mm² possible	DIN EN ISO 5079	Clamping length, test speed, test structure, material used for clamping material	X	
	2	Collective tensile test/ Stelometer	Strength (cN/tex) (maximum value, standard deviation)	ISO 3060, 1974	Clamping material, clamping length, mass of tested collective, number index		X

and 65% relative humidity of air. The slides are scanned by a Minolta Dimage Scan MultiPro at 4800 dpi (Langenhagen, Germany). Scanner software Dimage Scan 1.0 in combination with Photoshop 5.0 LE (Adobe Systems GmbH, Munich, Germany) is used to process the data. An adapted set-up has been developed to measure the width of the prepared fibres and fibre bundles.

Length

As shown in Table 10.3, two methods were chosen to test the fibre and fibre bundle length.

1. A manual method, which allows a cheap and easy way to characterize the length distribution of a sample.
2. An automatic system, for a quick determination of the length distribution of fibres and fibre bundles up to a length of 260 mm.

LENGTH MEASUREMENT USING TWEEZERS. In this single-element testing method, the length of the whole element is measured in straightened state. The crimped fibre or fibre bundle has to be straightened with an adequate force, avoiding elongation. The instrumental set-up of the measurement is displayed in Fig. 10.3. The edge on the left side is the starting point for measurement, where the fibre end is fixed by the weight. The fibre is then drawn by using the tweezers with low force, to avoid elongation. The length of the fibre can now be read

from the length class lines painted on the ground plate. Based on the amount of fibres in each length class (Fig. 10.4.), the distribution can be calculated.

LENGTH MEASUREMENT USING AN ALMETER. The Almeter (Siegfried Peyer AG, Switzerland) is an electronic apparatus for measuring length and has been developed to test wool fibres (Fig. 10.5). The device works with a specimen of fibres aligned in parallel using a preparatory machine called a 'Fibroliner'. Detailed information has been published by Grignet (1981), where he stresses the importance of sample preparation by the Fibroliner. The Fibroliner has to be modified for bast fibre bundles like hemp. The number of needles per comb must be changed from 153 for wool to 75 for bast.

The Fibroliner prepares a sample by rearranging the fibres or bundles, with all having one of their ends situated approximately on a line perpendicular to the direction of the fibres.

As shown in Fig. 10.5, the Almeter itself consists of two parts assembled in the same housing: a device for measuring the local sum of the cross sectional areas of the specimen and an electronic unit that calculates the length distribution automatically during measurement.

The device for measuring the local sum of the cross-sectional areas consists of a specially designed rectangular condenser with dimensions

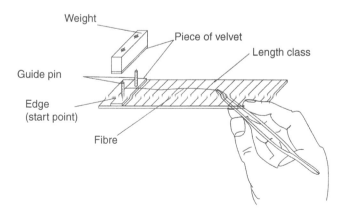

Fig. 10.3. Length measurement using tweezers.

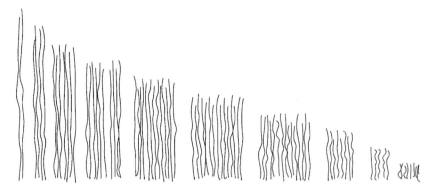

Fig. 10.4. Fibres, separated into length classes.

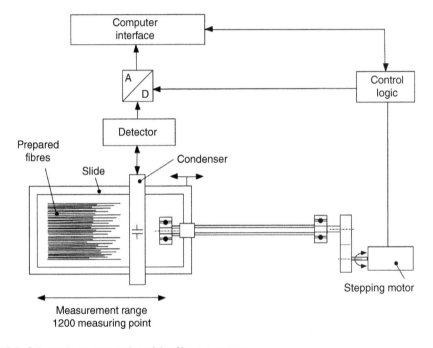

Fig. 10.5. Schematic representation of the Almeter system.

1.8 mm by 175 mm. The very short dimension of the condenser in the direction of the fibres or bundles allows detailed scanning of the local sum of the cross-sectional areas in the longitudinal direction of the elements. The test specimen is drawn with constant speed between the electrodes of the measuring condenser.

Before measurement, the samples had to be conditioned at 20°C ± 2°C and relative humidity 65% ± 2%. For each sample, three measurements had to be taken using approximately 0.6 g of fibres or fibre bundles. Each test specimen was held with one end in each hand and straightened by slight tension and by subjecting it to gentle shaking for preparing a so-called 'A-beard' using the Fibroliner. This A-beard was transferred into the sample slide of the Almeter by using the specimen holder of the Fibroliner and then measured.

LENGTH MEASUREMENT USING FIBRESHAPE. Fibre or bundle fineness improves the tensile properties of composites. As well as fibre length, fineness is a decisive property for the quality of injection-moulded parts. A modified set-up has been developed for the image analysis system 'Fibreshape' to measure the width and length of fibres and fibre bundles extracted from fibre-polypropylene (PP) compounds.

Measurement of fibre length and fineness distribution was done by preparing the extracted fibres as described above. The measured values were used for the numerical simulation of injection-moulded flax/PP samples. With this method, it is possible to show the differences in fibre length reduction that result from using different compounding systems. Further details have been published by Bos *et al.* (2006).

Strength

Tests for the tensile strength of hemp fibres are not standardized and very different preparations and methods have been used in the literature. A comparison of the values found in the literature is difficult, if not impossible, to undertake. The different characteristics of hemp fibre and fibre bundles used in tests are shown in Fig. 10.6.

According to the N-FibreBase proposal (Table 10.3), two methods were chosen to the tensile properties. The collective test was useful as a fast method to compare the quality of fibres with nearly the same fineness. The single-element test was chosen for testing the exact mechanical properties.

COLLECTIVE STRENGTH/STELOMETER. The different samples were measured with the collec-

tive test by the Stelometer. This test requires the sample to be conditioned for 24 h at 20°C and 65% relative humidity. Samples were then clamped in a Pressley clamp (Fig. 10.7) with Plexiglas jaws at a gauge length of 3.2 mm. The Stelometer was adjusted according to ASTM D 1445. To obtain a representative set of results, more than 20 collectives had to be tested. The strength of fibre and fibre bundle collectives in cN/tex could be calculated from the mass of the bundle collective tested (kg) divided by its mass-related fineness (tex). The mass of the collective is measured with an accuracy of 0.01 mg.

SINGLE-ELEMENT STRENGTH/DIA-STRON. The influence of fibre stiffness and the force elongation characteristics on composite behaviour is clear. Reproducible methods are required for the determination of the fibre properties providing data best suited to composite material development (Müssig *et al.*, 2005). According to Nechwatal *et al.* (2003), the single-element test is of particular importance in determining the tensile properties of fibres. Problems in the implementation of such tests lie particularly in:

- the influence of the clamping mechanism and of fibre slip in the clamp;
- various gauge lengths and whether this influence is taken into account;
- the determination of the fibre or fibre bundle cross-section surfaces;
- calculation of the stiffness, e.g. Young's modulus.

In order to solve the problems mentioned above and to reduce the number of possible influences on the testing result, an appropriate testing instrument was obtained from Dia-Stron Ltd, UK, and adapted to the requirements in an intensive exchange with this company.

In this method, the individual elements to be tested are no longer clamped but glued, in order to reduce the influence of clamping, and may be tested at gauge lengths of 30, 20, 10, 5 and 3.2 mm. The compliance of the system, 99% of which is due to the load cell, is corrected directly in the analysis. The cross-sectional surface area of each element is measured by

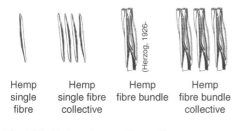

| Hemp single fibre | Hemp single fibre collective | Hemp fibre bundle | Hemp fibre bundle collective |

(Herzog, 1926)

Fig. 10.6. Various forms of hemp fibres (Müssig, 2001a).

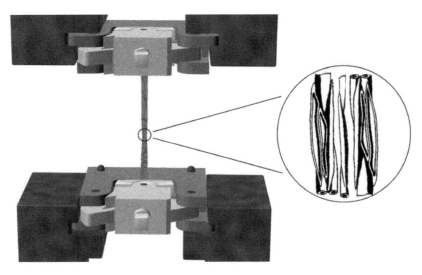

Fig. 10.7. A collection of hemp fibre bundle fastened to the Pressley clamp.

means of a laser beam in several places. The sample is then transferred automatically to the tensile testing system. After the tensile test with the desired pre-tension, the sample containers are removed and the next samples are tested automatically. Forty-five individual elements may be prepared in advance and inserted in the system automatically by autosampler. The software allows both the determination of surfaces (N/mm²) and of fineness-related values (cN/tex). The analysis program allows an extensive evaluation of the data, such as the determination of the true zero point, elongation correction, extensive possibilities for module determination, determination of breaking energy and also the energy required for the removal of crimp and the cross-section determination of the single elements.

Before measurement, the fibres and fibre bundles were conditioned for 24 h at 20°C and 65% relative humidity. Ninety elements were tested per sample. The free clamping length was 3.2 mm and the test speed was 2 mm/min (Müssig *et al.*, 2005).

Density

For measuring the specific density of cellulosic fibres, it is important that the material is dried carefully. This is necessary because moisture would cause a swelling process, thus affecting the specific density. The sample is divided in three subsamples and each of these is immersed in carbon tetrachloride (CCl_4), taking care that no air bubbles remain on the sample surface. Due to the high density of the liquid (1.59 g/cm^{-3}), the sample stays at the surface. In the next step, the sample is transferred into a mixture of 90% CCl_4/10% xylene (density 1.52 g/cm^{-3}). If it stays at the surface, it is again transferred into a mixture of 80% CCl_4/20% xylene (density 1.45 g/cm^{-3}) and so on, until the density of the liquid is low enough to allow the sample to sink. When the sample is at the bottom, the density of the liquid is increased by adding CCl_4 drop by drop, until the sample is able to float in the middle of the beaker. In this state, the liquid and the sample have identical densities and the density of the CCl_4/xylene mixture can be measured easily by means of a pycnometer, with an accuracy of ±0.005 g/cm^{-3}.

Odour determination

The evaluation of odour is still commonly performed by a panel of testers. Unfortunately, there are hugely different standards established in the various industrial sectors. Interior parts for cars in Germany are, for example, analysed according to the VDA 270 standard, using an inconsistent scale for evaluating the samples, starting with objective criteria for low-odour samples and changing to hedonic criteria for high-odour samples. This is even in contradiction to the basic rules of sensors, resulting in less reproducible results.

Where the odour of natural fibres is concerned, a two-step evaluation that, in the first step, uses an objective scale and, in the second step, a hedonic one has been shown to be optimal (Fischer *et al.*, 2004b). In Fig. 10.8, the results of identical samples evaluated by (i) VDA 270 and (ii) the FIBRE method (evaluation by objective intensity scale) are compared, displaying clearly the lower standard deviation obtained by the FIBRE method.

Another interesting result of this study is the reduction of the odour potential of hemp

by enzymatic treatment in order to remove the pectins. When hemp is exposed to low or medium temperatures (<150°C) during processing, the odour of hemp treated enzymatically is much lower than the odour of raw hemp (sample GDE02, see the section below on tested fibres for details), which contains more pectins (Fig. 10.9). A general decomposition of the material starts at higher temperatures (200°C), resulting in nearly identical odour intensity for both variants.

10.2.4 Differences between hemp and flax

As can be seen in Table 10.4, hemp and flax differ in lignin and pectin content, causing different behaviour in several processes. Both

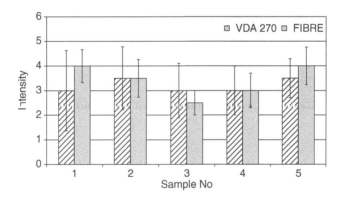

Fig. 10.8. Different results of odour evaluation by VDA 270 and the FIBRE method.

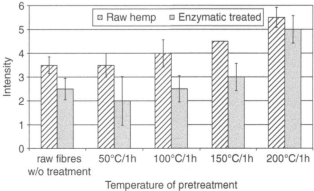

Fig. 10.9. Influence of enzymatic treatment on the odour behaviour of hemp.

Table 10.4. Composition (in %) of natural cellulose fibres at 10% moisture content (Krässig *et al.*, 1989).

	Cotton	Jute	Flax	Hemp	Ramie	Sisal	Abaca
Cellulose	92.7	64.4	62.1	67.0	68.8	65.8	63.2
Hemicellulose	5.7	12.0	16.7	16.1	13.1	12.0	19.6
Pectin	0	0.2	1.8	0.8	1.9	0.8	0.5
Lignin	0	11.8	2.0	3.3	0.6	9.9	5.1
Water solubles	1.0	1.1	3.9	2.1	5.5	1.2	1.4
Wax	0.6	0.5	1.5	0.7	0.3	0.3	0

types of fibres contain more than 60% cellulose, about 16% hemicellulose, but 0.8 resp. 1.8% pectin and 3.3 resp. 2.0% lignin. Unfortunately, these fibres are often compared as cellulosic fibres to cotton, which is a nearly pure cellulose fibre containing neither pectin nor lignin.

In practice, there are differences observed between hemp and flax, for example, in mechanical processing and in the production of composites. For mechanical processing, different equipment has been established. Recent work on the enzymatic treatment of hemp in order to produce very fine fibre bundles has shown that the processing parameters from, for example, flax treatment cannot be transferred easily to hemp, since the structure of the adhesive components between the fibres is different (Fischer *et al.*, 2004a, 2006). This is a foreseeable effect caused by the very specific type of enzyme action, but these differences in pectin structure even cause a different behaviour in chemical treatment, for example, in EDTA solutions.

10.3   Properties of Natural and Man-made Cellulose Fibres

In the following sections, the testing methods described are applied exemplarily to typical natural fibres used as reinforcement for plastics.

10.3.1   Tested fibres

The following plant fibres have been used as test specimens:

- Jute: the jute (*Corchorus olitorius* L.) fibre bundles (Schilgen) without any

  avivages on the surface were provided by the company NAFGO GmbH, Neerstedt, Germany, in 2001.
- Sisal: the sisal (*Agave sisalana* Perr. ex Engelm.) fibre bundles were provided by the company NAFGO GmbH, Neerstedt, Germany, in 2001.
- Flax A: the flax (*Linum usitatissimum* L.) fibre bundles (HL_04_01a) were provided by the company Holstein Flachs GmbH, Mielsdorf, Germany, in 2004. The flax was grown near Mielsdorf in 2003 and retted only briefly. The fibre bundles were separated from the briefly retted long flax during the scutching process in the separation plant at Holstein Flachs GmbH.
- Flax B: the flax (*Linum usitatissimum* L.) fibre bundles (HL_04_01b) were provided by the company Holstein Flachs GmbH, Mielsdorf, Germany, in 2004. The flax was grown near Mielsdorf in 2003. After harvesting, the stems were field retted and baled. The fibre bundles were separated from the very homogeneously retted long flax during the scutching process in the separation plant at Holstein Flachs GmbH.
- Hemp: the hemp for the trials (*Cannabis sativa* L.) was provided by the company NAFGO GmbH, Neerstedt, Germany. Quality management has been set up for their products which allows tracing of the fibre origin back to the grower. Through fine-tuned growing, harvesting and separation management, constant fibre qualities can be provided for technical applications. The fibre bundles used are described in detail in the following. The hemp variety *Fedora* was grown near Neerstedt (Oldenburg region, Germany) in 2002 and briefly field retted (A1000-value: 1.08 ± 0.13). The stems were coarsely by the

company NAFGO GmbH (Neerstedt, Germany) using a DEMTEC® line with four drums (Demaitre B.V., Belgium). A detailed description of the industrial equipment has been published by Müssig (2001b). The resulting raw fibres were labelled as GDE-02.

- Ramie: the fibres for the experiments were made available by the company Buckmann, Bremen, Germany. In this case, chemically separated ramie fibres (*Boehmeria nivea* H. and A.) from China that had been processed into card slivers were used (delivered 2000).

- Cotton: in the experiments, the following cotton fibre was used: cotton (*Gossypium barbadense* L.) species: US Pima; mean length by number (Almeter) Ø 25.13 mm; fineness (gravimetric) Ø 1.452 dtex (FIBRE, 1994).

- Lyocell: lyocell (Lyo… from Greek: lyein = dissolve, cell from cellulose), on the basis that cellulose belongs to the group of industrially created fibres as do viscose or cupro (Schnegelsberg, 1999). Lyocell fibres are gained from a solution of cellulose in an organic solvent (N-methyl-morpholine-N-oxide) by regenerating the cellulose to fibre form with the aid of the NMMO procedure and thus consist of almost 100% pure cellulose. The special characteristics of lyocell fibres such as high strength and comparatively low elongation are due mainly to a highly crystalline structure and the orientation of the molecules. Application areas of lyocell fibres are, according to Koch (1997), particularly in the clothing sector, but also in the field of technical textiles such as filters and special types of paper. Even though the lyocell fibre does not belong to the group of natural fibres, it is included in

the studies presented here because of its high cellulose proportion. Lyocell fibres with a fineness of 6.7 dtex were made available by Lenzing AG, Austria.

These fibres and fibre bundles were tested with the above-described testing methods. The results are described in detail in this chapter. Because of the extreme values reached by flax fibres, two very different flax varieties were tested.

10.3.2 Density

Natural fibres have attracted attention in the last decade as a possible substitute for glass fibres in composite components. Especially for lightweight applications, natural fibres offer a significant lower density (>40%) compared to the weight of glass fibres (2.7 g/cm³). This can be confirmed in the results of the true density measurements in Fig. 10.10. The observed density ranges from 1.30 g/cm³ for sisal fibres to densities of over 1.5 g/cm for cotton, ramie and lyocell. The true densities depend more on the structure than on the chemical composition of the fibre. Bobeth (1993) has pointed out that various factors, like the presence of impurities on the surface following a chemical treatment, the conditioning state of the samples and the presence of closed cavities, may influence the results and thus affect the measurement. Batra (1998) also noticed a wide scattering when comparing values from several authors.

Due to the porous structure of natural fibres, differences can appear between the apparent density and the true density. Comparative studies (in 1971 and 1974) reported by Batra (1998) have measured differences up to 53% for banana fibres. The lowest difference

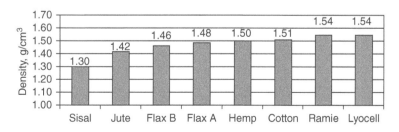

Fig. 10.10. Density of different natural fibres and lyocell using the suspension method.

was found for a ramie sample (10.7%), while for *Crotalaria* ('sunn hemp') a difference of 12% was observed.

10.3.3 Using density values for strength calculation

Well-documented studies of the mechanical properties of natural fibre reinforced composites lead to the conclusion that it is very important to know the exact density, because then it is possible to calculate precisely the fibre volume content or the cross-sectional area of the fibres and, subsequently, the composite properties (Cichocki and Thomason, 2002; Madsen and Lilholt, 2003). Globally, mechanical notations for engineered parts are related mainly to the surface; a common unit is N/mm^2 or MPa. The measurement of fineness in the textile industry is usually expressed in tex. This measurement is convenient for fibre bundles like bast fibres that do not have circular or nearly circular cross-sectional areas. The gravimetric fineness measurement method described above can be carried out with simple equipment. Generally, a conversion is made using a standard density value taken from the literature. The disadvantage is that no reliable data for calculation can be achieved.

An accurate measurement can be obtained by analysing the density of the sample. In this case, the volume of the lumen and porosities will not be considered. This allows a proper conversion from a mass-related to a surface-related mechanical property. This can be confirmed by the results of single-element tensile tests realized with the automated miniature tensile tester Dia-Stron plotted in Fig. 10.11. The mechanical values and the cross section were measured for each fibre type. The measured density was used to calculate the mechanical strength in cN/tex. The line shows the calculated mass-related density values using a standard literature density of 1.45 g/cm^3. A proper conversion from mass-related values into surface-related ones is not possible if only such standard density values are available. For materials with densities >1.45 g/cm^3 (e.g. lyocell, ramie), the surface-related

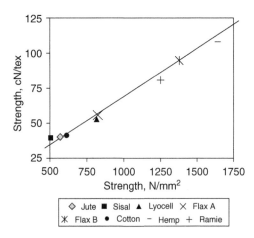

Fig. 10.11. Fibre tensile strength in terms of linear density (cN/tex) versus a surface-related strength (N/mm^2)/single-element strength tests using Dia-Stron.

mechanical properties will be underestimated. For materials with lower densities (e.g. sisal), the opposite is true.

10.3.4 Mechanical properties of natural and man-made cellulose fibres

It has been shown that the mechanical properties of natural fibres are competitive with those of glass fibres, especially as short fibre reinforcement in composites (Schwill et al., 2004). Table 10.5 summarizes the mechanical values of the tested fibres and fibre bundles. At first glance, the wide distribution of the mechanical values does not seem to allow any calculation of the mechanical properties of, for example, composites. In case of the bast fibre samples, flax and hemp, the standard deviations (not shown here) of strength and Young's modulus are equal to or bigger than the mean values.

The reason for the wide strength distribution lies in various natural defects and property variation. Ruys et al. (2002) observed in plant fibres natural defects like cracks, holes, regions of reduced cross-sectional area, dislocation bands and miscellaneous irregularities in shape. They point out that the breaking operation results in a high level of defects in individual fibres. The presence of these defects

Table 10.5. Mechanical properties and cross-sectional areas of the fibre tested/single-element strength tests with Dia-Stron.

Fibre/number of tested elements		Corrected break strain (%)	Break load (N)	Cross-sectional area (µm²)	Gravimetric fineness[a] (tex)	Strength (N/mm²)	Strength (cN/tex)	Young's modulus (N/mm²)
Sisal[b]	Mean:	11.4	8.0	20,518	26.6	428	33.0	4,575
84	Median:	11.3	6.5	15,820	20.5	424	32.6	4,387
Flax B[b]	Mean:	6.6	6.0	8,254	12.0	874	59.9	14,583
84	Median:	6.5	5.5	7,020	10.2	790	54.2	12,840
Hemp[b]	Mean:	8.0	8.3	15,368	23.1	827	55.0	12,984
66	Median:	7.8	7.9	13,927	20.9	588	39.1	8,285
Flax A[b]	Mean:	4.5	3.3	6,291.5	9.3	787	53.0	18,968
85	Median	4.3	2.8	5,071.3	7.5	521	35.1	13,250
Jute[b]	Mean:	3.2	1.2	2,388	3.4	571	40.3	17,339
93	Median:	3.1	1.1	2,040	2.9	540	38.1	16,350
Cotton[c]	Mean:	7.3	0.1	121.9	0.18	618	40.9	11,844
117	Median:	6.9	0.1	107.2	0.16	561	37.1	10,490
Lyocell[c]	Mean:	16.7	0.2	308.3	0.48	815	52.8	8,633
93	Median:	16.1	0.2	300.5	0.46	837	54.2	8,273
Ramie[c]	Mean:	3.8	0.5	474.9	0.73	1,250	81.0	35,958
158	Median:	3.7	0.5	461.1	0.71	1,271	82.4	35,285

Notes: [a]Calculated values using measured density values (compare Fig. 10.10); [b]fibre bundle; [c]single fibre.

results in a dependence of single fibre strength on the degree of damage. There is also a link between fibre tensile strength and fibre fineness. This relationship is well known for glass fibres. Flemming and Roth (2003) report the observations from Griffith (1920) on the increase of strength following a decrease of fibre diameter. Regarding the four scatterplot diagrams grouped in Fig. 10.12, which display the single-element tensile strength versus the cross-sectional area, the same statement can be made for natural fibres. The hemp sample (as an example of stem fibres plotted in Fig. 10.12b) shows a wide distribution in the cross-sectional area. This can be explained by the combination of single fibres in the low cross-sectional area and fibre bundles in the high cross-sectional area. Tensile strength decrease is linear below 2000 µm, while in the upper cross-sectional area nearly constant tensile strength is observed. Both flax samples (not displayed here) showed a similar dependence. The tensile strength of sisal leaf fibres is plotted in Fig. 10.12d. Behaviour similar to that of the hemp sample can be observed. For comparison, the tensile strength–fineness relationships for single fibres like the seed hair fibre–cotton and the bast fibre–ramie are

plotted in Fig. 10.12a and c. For both of these, only a continuous decrease of tensile strength is observed.

10.3.5 Length

The results of length measurement with tweezers are shown in Figs 10.13 and 10.14. The lyocell fibres and the very long flax B fibre bundles are not included in the diagram. Figure 10.13 shows the wide spread of length values for ramie, jute and sisal, while cotton, flax and hemp are narrow in their length distributions.

In yarn production, the short-fibre content is an important measure of fibre loss at spinning. For the production of fibre-reinforced composites, fibre length also must not be below a critical fibre length. For needle felt, production fibres of >50 mm mean length are necessary. If such felts are used for composite production, the fibre length is always above the critical length. All fibres discussed in this section fulfil this condition. The fibre bundles of hemp, in particular, allow a large variety of fibre lengths, which can be controlled by modifying process parameters in mechanical separation.

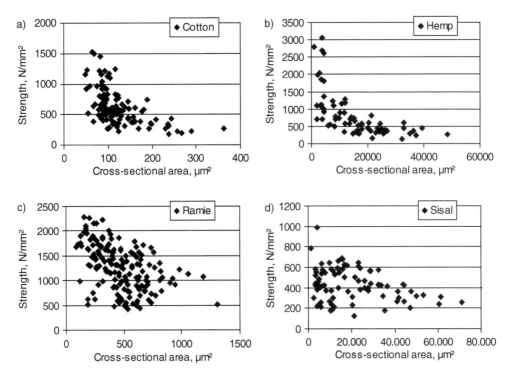

Fig. 10.12. Young's modulus versus cross-sectional area for (a) cotton, (b) hemp, (c) flax and (d) sisal.

In contrast to this, the length of reinforcing elements in the injection moulding process must not exceed a few millimetres, but must still be above the critical length. Moreover, natural fibres undergo a considerable amount of thermal and mechanical degradation during injection moulding. Joseph *et al.* (2003) noticed a fibre shortening of up to 60% for melt-mixed short sisal fibres in a matrix of polypropylene.

10.3.6 Fineness (fibre width)

The results of fineness measurement by means of the Fibreshape system are presented in Table 10.6 and Fig. 10.15. The factors influencing the fineness of natural fibres are, for example, the variety, the method and the degree of separation. Especially in the case of bast fibre bundles, fineness can vary depending greatly on the degree of retting, as the flax varieties in Table 10.6 demonstrate quite clearly. The distribution of the values for hemp and flax A indicates an uneven separation degree of

the fibre bundles, while the distribution of the more strongly retted flax B is narrower. For cotton, the fineness is very close to a normal distribution, indicating measurement of mainly single fibres.

Fineness plays an important role for fibre-reinforced materials, since the surface in contact with the matrix increases with finer fibres or fibre bundles. By increasing the contact surface between the fibres and the matrix, the stress transfer into the fibres can be improved. In contrast to natural fibres constituted of single fibres like cotton, bast fibre bundles can be opened during processing due to shear effects. For this reason, the fineness of bast fibre bundles after processing can decrease down to the fineness distribution of single fibres.

10.4 Integrated Quality Management for Hemp

Integrated quality management for hemp fibres is discussed to describe the fibre bundle

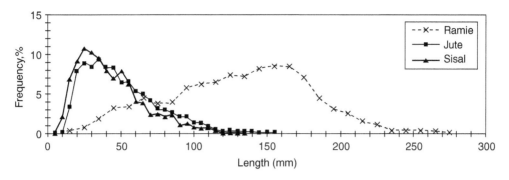

Fig. 10.13. Fibre length distribution, length measurement using tweezers with at least 1000 fibres per sample.

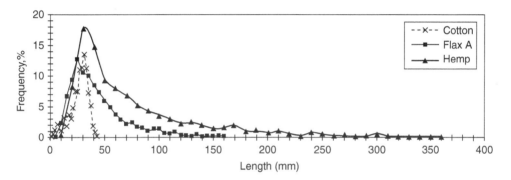

Fig. 10.14. Fibre length distribution, length measurement using tweezers with at least 1000 fibres per sample

Table 10.6. Measurement of the fibre width/optical system, Fibreshape.

		Sisal	Jute	Hemp	Flax B	Flax A	Ramie	Cotton
Number of elements tested		1,011	3,492	5,074	6,974	18,894	5,172	18,029
Centile $X_{0.90}$:	µm	162.6	62.8	76.3	69.5	32.8	31.4	20.4
Quartile $X_{0.75}$:	µm	121.1	50.3	48.7	31.3	22.4	24.7	18.0
Median $X_{0.50}$:	µm	85.3	34.8	26.8	19.0	16.9	18.5	15.4
Quartile $X_{0.25}$:	µm	27.7	12.5	13.4	12.3	12.2	13.5	12.8
Centile $X_{0.10}$:	µm	11.6	6.0	7.6	7.3	8.2	9.8	10.2
Mean value:	µm	**84.4**	**34.8**	**37.6**	**30.7**	**20.2**	**19.8**	**15.5**
Standard deviation	µm	59.9	23.5	39.3	36.2	15.5	8.5	4.4
Coeff. of variation	%	70.9	67.4	104.5	117.7	76.9	42.9	28.2
Level of confidence 95%/5%	µm	3.7	0.8	1.5	1.1	0.3	0.2	0.1

properties after different steps of separation for trial K1 (GA – GADO). Because of their structure, hemp fibre bundles can be modified within certain limits, in particular with regard to fineness and length. This offers a very broad range of applications. Coarse hemp fibre bundles extracted by low-cost processes can be used in engineering applications such as soil and hydraulic engineering or automobile manufacturing. By more intensive (mechanical, chemical,

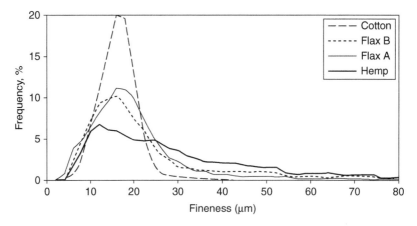

Fig. 10.15. Example of fibre fineness distributions for cotton and bast fibre bundles.

chemical-physical or enzymatic) separation, the property profile of the fibre bundles can be varied in accordance with product requirements.

To demonstrate the adaptability of the proposed testing methods, hemp fibre bundles were tested after different separation steps. For this trial, hemp (*Cannabis sativa* L.) variety Felina 34, grown in Klagenfurt, Austria, was used. A detailed description of the harvesting process of the hemp used is given in Müssig (2001a). The thin and well-retted hemp stems were separated in a separation plant, as shown in Fig. 10.16.

Samples were taken at different positions of the line. The peeled bast of the thin stems was labelled as (RS). Coarsely separated (GA), medium (MA) and finely separated (FA) fibre bundles were produced. After the coarse separator, a part of the material was removed from the line to be separated in a steam explosion process (GAD) and afterwards opened with an opener (GADO). The objective of this trial was to demonstrate the wide range of possible fibre properties that could be produced based on hemp. The results of the measurements are presented in the following sections.

10.4.1   Fineness (GA – GADO)

Gravimetric fineness (GA-GADO)

Refinement of hemp bundles by the separation process is shown in Fig. 10.16, as determined by the gravimetric fineness measurement method. The results given in Fig. 10.17

illustrate how large the variations in fineness are at different separation steps.

As described previously, the gravimetric fineness measurement test is very time-consuming, but the need for special testing equipment is low. A disadvantage of this process is that no fineness distribution is available based on the measurements of one sample set (here, 370 tested fibre or fibre bundles). This leads to missing statistical data like the standard deviation and the coefficient of variation. In this context, Simor (1965) mentioned the work of Fröter and Zienkiewicz (1952), in which a level of confidence of ±3% at 95% security level was achieved only after 5000 gravimetric measurements on flax. With a view to fast measurements, the gravimetric fineness measurement method is not the best choice.

OFDA

The fineness of fibre bundles was examined by the OFDA (optical fibre fineness analyser) method. The optical fibre fineness analyser was developed for measuring the diameter of wool fibre (Baxter *et al.*, 1992). This apparatus measures the width distribution of bast fibre bundles efficiently and results have been found to correlate well with those of other methods. Because of the large number of measurements taken, our results could be reproduced well. This method is very rapid, as well as being highly reproducible (Drieling *et al.*, 1999).

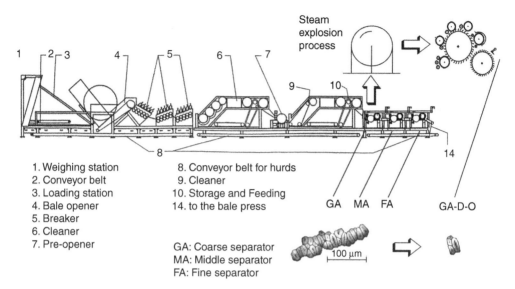

1. Weighing station
2. Conveyor belt
3. Loading station
4. Bale opener
5. Breaker
6. Cleaner
7. Pre-opener

8. Conveyor belt for hurds
9. Cleaner
10. Storage and Feeding
14. to the bale press

GA: Coarse separator
MA: Middle separator
FA: Fine separator

Fig. 10.16. From hemp stems via mechanically to chemically separated hemp fibre bundles – separation trial K1 (Müssig, 2001a).

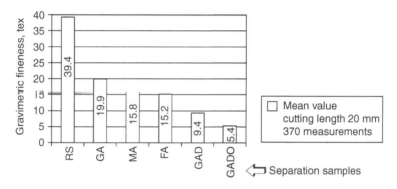

Fig. 10.17. Gravimetric fineness values (Müssig, 2001a).

Prior to the experiment, the hemp samples were conditioned for 24 h at 20°C and 65% relative humidity. The OFDA values presented in Table 10.7 show clearly that the distribution of the fibre bundle width of hemp samples is readily measurable by this method. It has to be emphasized that all samples have a very good level of confidence.

One can observe that there is a significant refinement of the fibre bundles in the stems (sample RS) by the coarse separator (Table 10.7). The subsequent treatment by medium separator MA and fine separator FA causes no significant additional refinement of the fibre bundles. For the hemp sample presented here,

the use of a steam explosion process leads to a distinct refinement of the fibre bundles to a mean width <20 μm.

It has to be mentioned that OFDA is not able to detect very coarse fibre bundles (Table 10.9), as in general – if the sample properties are not very well known – the use of systems like Fibreshape is recommended.

10.4.2 Length

Almeter

The length of the samples was determined by the Almeter method. The results are presented in Table 10.8 and Fig. 10.18. A significant shortening by the medium separator MA can be seen, while the influence of the fine separator FA on length is nearly negligible. As with the width distribution, the steam explosion causes a drastic shortening of the fibre bundles.

10.4.3 Strength

Fibre samples were taken at different positions in the separation line and tested for their strength properties. The fibre bundles were tested as collectives and as single elements. The results are presented in the following sections.

Collective test

The results of the collective test with the Stelometer device are presented in Fig. 10.19. From the results presented in Fig. 10.19b, it can be seen that the collective strength values for mean and median are equal and the measured values are almost normally distributed.

A general advantage of the collective test, in comparison with the single-element test, is that it yields a high statistical accuracy based on a relatively small number of tested samples. For cotton, Warrier and Munshi (1982) showed in a single-element test that 300 measurements were needed to achieve a level of accuracy of 5% at a significance level of 95%.

The same level was achieved in the hemp collective tests on sample K1/RS after 60 and for sample K1/GADO after 35 measurements. The mean number of necessary tests was 80. By decreasing the level of accuracy down to 10% at a significance level of 95%, the number of collectives to be tested was 22 on average.

As shown in Fig. 10.19a, the collective strength decreases with each separation step from RS to GADO. It has to be investigated whether this decrease is influenced by damage to the fibres or can be explained by differences in the number of elements in the collectives. This effect is described by Suh et al. (1994) for cotton: the collective strength decreases dramatically with an increasing amount of cotton fibres in a collective.

As explained by Suh et al. (1994), the number of elements in a collective has an important impact on the results. To make the results more comparable, it is necessary to know the number of single elements in a

Table 10.7. OFDA results from the samples K1/RS – K1/GADO.

Sample	Number	Mean (μm)	Standard dev. (%)	Coefficient of variation (%)	Level of confidence 95% (μm)	Median (μm)	Amount <30 μm (%)	Amount >100 μm (%)	Width at 1% of the measured elements (μm)	x_{50} (LNV) (μm)
RS	12,477	54.3	59.4	109.3	2.6	30.2	49.9	15.9	266.1	32.8
GA	22,470	47.0	49.4	108.8	1.5	26.9	53.6	12.0	233.2	30.4
MA	20,777	47.3	49.0	104.9	1.6	27.2	53.2	12.2	231.4	30.7
FA	22,745	46.8	47.7	102.9	1.3	27.9	52.5	11.6	228.4	30.7
GAD	44,941	16.8	12.5	74.5	0.3	13.5	93.7	0.4	69.1	14.5
GADO	62,166	16.2	10.2	63.0	0.2	13.5	94.8	0.2	57.7	14.3

Table 10.8. Almeter values of the tested hemp fibre bundles.

Sample	Mean (Q) (mm)	CV (Q) (%)	L(Q) <25 mm (%)	L(Q) 50% (mm)	L(Q) 1% (mm)
GA	47.8	78.4	37.7	36.0	180.8
MA	40.9	68.4	37.3	31.9	123.5
FA	38.2	80.2	45.0	27.7	156.5
GAD	13.0	66.4	92.8	10.6	45.5
GADO	13.1	65.8	93.4	10.8	39.8

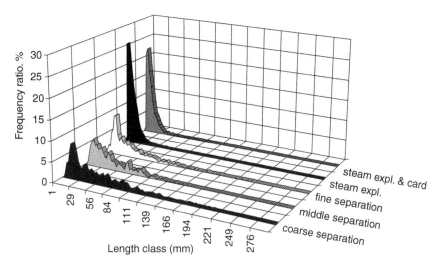

Fig. 10.18. Almeter length values.

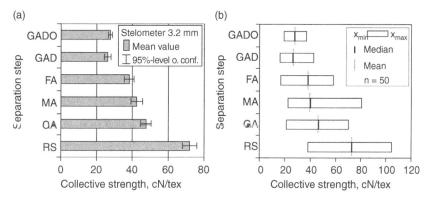

Fig. 10.19. Stelometer values (Müssig, 2001a).

collective. While it is not possible to count the elements in a collective after each Stelometer test, a concept was worked out to calculate a so-called 'number index' (Müssig, 2001a). Using this number index, it is possible to compare Stelometer results in a much better way. The number of elements in the collective can be estimated by the following equation:

Number index = (mean fineness of the tested collectives in tex)/(gravimetric fineness of the sample in tex)

The results (Stelometer collective strength, gravimetric fineness and calculated number index) are given in Fig. 10.20.

    The calculated mean number of elements in the tested collectives increases from a low level (1.7 for K1-RS) to 19.6 for hemp sample K1-GADO. The increasing number of elements leads to the effect described by Suh *et al.* (1994) and reduces the collective strength. How large this effect is and whether there is an overlapping effect from the general damage to the fibre bundles from the separation process can only be answered by measuring the fibres and bundles in a single-element test. The results achieved by single-element tests are given in the following section.

Single-element test

INSTRON UNIVERSAL TENSILE TESTER. Samples RS, GA, FA and GADO from trial K1 were tested with an Instron universal tensile tester

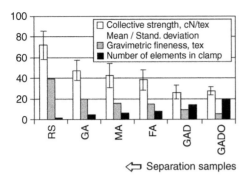

<- Separation samples

Fig. 10.20. Influence of the number of tested elements on the collective strength (Müssig, 2001a).

(Instron, Pfungstadt, Germany). For these trials, Pressley clamps covered with Plexiglas were used. The clamping length was 3.2 mm and the testing speed 2 mm/min. The results are given in Fig. 10.21. Beside the mean value, the median and the maximum and minimum values are shown graphically.

Based on the results shown in Fig. 10.21, the differences in strength when using the single-element method are much smaller compared to the collective test. It should be mentioned that the values are not normally distributed and the resulting values are left-skewed for all samples. It seems that there is still a correlation between strength decrease and the separation step.

Therefore, two important aspects should be introduced into the discussion: (i) what is the influence of the calculation of a mass-related strength using a mean mass measured for all the tested samples; and (ii) is 50 measurements enough to give more than just a trend?

Problems with the implementation of such tests (mass-related, single-element strength with clamped samples) relate particularly to:

- the influence of the clamping mechanism and of fibre slip in the clamp;
- the determination of the mass of the tested elements;
- the calculation of the fibre mass-related tensile strength by using the mass of the elements as a mean value.

In order to solve the problems mentioned above and to reduce the number of possible

influences on the test result, the samples were tested with an appropriate testing instrument from Dia-Stron Ltd (Andover, UK).

DIA-STRON. Samples GA, MA, FA, GAD and GADO from trial K1 were tested with a Dia-Stron system. The device and the method are explained in an earlier section. Table 10.9 gives detailed information about the measured values. In addition to the cross-sectional area and the area-related strength, the gravimetric fineness and the mass-related strength (both measured in tex) were calculated using a mean density of 1.5 g/cm³. Comparing the calculated gravimetric fineness with the measured values given in Table 10.9, a good correlation can be found. The values identified whether coarser (e.g. sample FA) or finer (e.g. GAD) fibre bundles were used in the tensile test in comparison to the mean gravimetric fineness.

From the values given in Table 10.9, it can be concluded that the damage to fibre bundles in the separation process is small to negligible. These results underline the need for modified testing methods for bast fibres to obtain exact physical properties. Exact fibre data are extremely important if used in numerical simulation software to calculate the properties of fibre-reinforced plastics (Müssig et al., 2006).

Strength summary

The design of a testing method (collective or single-element test) has a fundamental influence on the results.

If the values of hemp fibre bundles tested in a collective test were compared with other results, the number of tested elements should be nearly identical, otherwise a comparison would not be possible. The 'number index' gives a better understanding of differences between the values. Our proposal is to add the information about fineness of the tested collectives and the number index to the tensile strength values of a collective. Having these data in mind, it is easier to decide if a smaller collective strength value is caused by damage to the fibre bundles or by an increase in fibre bundle fineness.

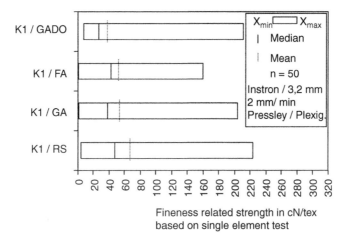

Fig. 10.21. Fineness-related strength measured with the Instron universal tensile-tester by testing single elements (Müssig, 2001a).

Table 10.9. Dia-Stron values of the tested hemp (single-element test).

Fibre/tested elements		Corrected break strain (%)	Break load (N)	Cross-sectional area (µm²)	Gravimetric fineness* (tex)	Strength (N/mm²)	Strength (cN/tex)*	Young's modulus (N/mm²)
K1/GA	Mean	7.6	7.7	13,299	19.9	735.0	49.0	11,481
82	Median	7.1	6.6	11,565	17.3	627.0	41.8	8,202
K1/MA	Mean	6.5	6.7	13,044	19.6	672.0	44.8	11,345
70	Median	6.5	6.0	10,036	15.0	572.6	38.2	9,539
K1/FA	Mean	7.8	8.6	15,927	23.9	611.7	40.8	9,013
88	Median	8.0	8.0	14,786	22.2	561.0	37.4	7,812
K1/GAD	Mean	4.5	1.7	3,318	5.0	666.8	44.5	15,989
80	Median	4.2	1.3	2,343	3.5	596.2	39.8	13,470
K1/GADO	Mean	5.0	1.9	3,863	5.8	696.9	46.5	14,968
73	Median	4.9	1.6	2,810	4.2	446.4	37.6	11,440

The collective test is the first choice for a quality check to compare the strength of hemp with a comparable fibre bundle fineness. Because of the small number of necessary tests, this method is very suitable for production quality control, for example in a mechanical separation plant, or, as described by Dreyer et al. (2002), to compare hemp produced in an enzymatic separation process.

To get physically precise data, for example for a numerical simulation of the behaviour of a natural fibre reinforced plastic, it is necessary to use the single-element test. With the Dia-Stron system, it is possible to get area-related strength and Young's modulus values. Using this method, new knowledge can be generated regarding the mechanical properties of fibres.

10.5   Conclusion and Outlook

From the results presented here, it is obvious that even the process of harvesting has significant influence on fibre quality. Processing freshly harvested stems causes greater damage than processing dried stems. Subsequent carding of damaged fibres causes enhanced loss of material due to the higher number of short fibre bundles.

The measurements necessary for quality control can be undertaken easily by using the methods presented here. These methods are adapted and optimized for bast fibre bundle testing and even if testing only a small number of collectives produces reliable tensile strength data, assuming fineness (i.e. the number of fibre bundles in the collective) is taken into account. A collective consisting of many fine fibre bundles has lower tensile strength than a collective of a few coarse fibre bundles. To obtain precise data for a numerical simulation of the behaviour of a natural fibre reinforced plastic, it is necessary to use the single-element test. With the Dia-Stron system, it is possible to obtain area-related strength and Young's modulus values.

For fine bast fibre bundles, testing fineness is possible by the OFDA method or by Laserscan. Coarser samples can cause errors, because OFDA and Laserscan have been developed for wool, where very coarse fibres are not present. Complete analysis of fibre and fibre bundle morphology (including fineness) is best done by the Fibreshape system.

Knowledge of fineness is essential for the selection of optimal fibre lots for special products. The tensile strength of, for example, needle felts is much higher if they are produced from fine fibre bundles. This is due to greater surface area and thus better sticking of the fine fibre bundles to each other. Similar effects are observed in the production of natural fibre reinforced thermosets. If the felt structure is penetrated completely by the monomer, the mechanical properties are enhanced significantly. On the other hand, the small intermediate spaces of fine fibre bundles can sometimes prevent complete penetration by the monomer. In such special cases, use of coarser fibre bundles – obtained by less intensive mechanical separation – can help. This pathway can also be useful for the cheap production of low-performance composites by spray technology.

In each case, the steps of harvesting and retting have to be considered in a quality management system, because they have significant influence on the fineness of the fibre bundles. Decortication is possibly better and separation leads to finer fibre bundles – if retting has been well controlled. Use of unretted or badly retted stems increases fibre loss, produces shorter bundles and leaves more residual hurds in the subsequent separation and carding processes.

In fibre bundle separation, the changes in fineness and length are the most important parameters to be controlled by a quality control system. In normal production with fibre bundles <250 mm, length distribution can be controlled easily by an Almeter, combined with a modified preparation unit. This is also valid for the subsequent carding step, because fibre bundle length influences the needle felt properties directly.

The application of such a reliable quality control system can be a big advantage for the industry and enhance the use of domestic fibre plants. This would be not only of ecological significance but also economically advantageous, as it would offer short transport distances in the whole value-added chain, as well as the possibility of making natural resources available in constant qualities.

Note

[1] Natural fibre reinforced polypropylene.

11 Use of Natural Fibres in Composites for German Automotive Production from 1999 to 2005

Michael Karus

Industrial Hemp Association (EIHA), c/o nova-Institut GmbH, Hürth, Germany

11.1 Introduction – Methodology and Data Correction

Since 1996, the nova-Institut, Hürth, Germany, has been surveying data on the use of natural fibres (NF) in German automotive production. In a comprehensive investigation by means of e-mail questionnaires and telephone interviews, data for 2004 and 2005 were surveyed in the summer half-year of 2006. As in previous years, the data of suppliers active in Germany were focused on and could be surveyed almost entirely. Additional exemplary interviews of employees of automotive companies, NF mat producers, machine manufacturers and raw material suppliers served the purpose of ensuring further data.

Almost all data proved to be consistent with the surveys of previous years. However, the data on the amounts of NF composites had to be revised considerably: the 45,000 t for 2003, as published so far, proved, in retrospect, to be wrong; it was not before 2005 that 30,000 t could actually be achieved – and this at a continuously increasing use.

The reason for the miscalculation back then was the conversion of the NF amount into the composite amount. Up until then – in coordination with branch representatives – an average NF share of 40% had been presumed. But, because for this year's survey for the first time not only could the NF amount be surveyed but also the composite amount, the NF share for thermoset and thermoplastic techniques could now be calculated precisely – and it was clearly beyond 40% (see below). In addition, for the first time also edge trim was taken into consideration, with a presumed average loss of about 20% in the course of the moulding process.

On the whole, given consistent data for the use of NF, this resulted in newly calculated data for the respective composites. This correction was done retroactively for 1999 until 2005, in order to obtain a new, coherent database.

11.2 Results and Their Interpretation

Figure 11.1 shows that the use of NF in German automotive production has increased even further in 2004 and 2005 – even with slower growth rates of less than 3%. This growth is based primarily on the rising use of the press flow moulding and injection moulding techniques (both new to natural fibres), while established compression moulding is stagnating.

11.3 Natural Fibres in Motor Cars

In 2005, for the first time, 19,000 t of NF (not including wood and cotton) were used in

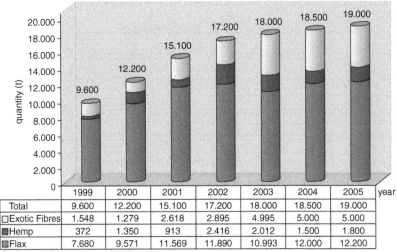

	1999	2000	2001	2002	2003	2004	2005
Total	9.600	12.200	15.100	17.200	18.000	18.500	19.000
☐Exotic Fibres	1.548	1.279	2.618	2.895	4.995	5.000	5.000
■Hemp	372	1.350	913	2.416	2.012	1.500	1.800
☐Flax	7.680	9.571	11.569	11.890	10.993	12.000	12.200

* without wood and cotton

Fig. 11.1. Use of natural fibres for composites (not including wood or cotton) in the German automotive industry, 1999–2005 (nova-Institut, November 2006).

automotive composites. At the same time, the proportion of NF used has changed. While exotic natural fibres – jute and kenaf, sisal, coir and abaca – increased substantially between 2000 and 2004, both on a percentage basis and absolutely, there has been a stagnation ever since. This was linked directly with the prices of European flax fibres, which were quite high in the same period and have been decreasing again since as recently as 2004; simultaneously, in recent years there have been significant price increases for jute and kenaf on the world market. Accordingly, flax could expand its market position again in 2004 and 2005. Hemp's market share is determined mainly by its short supply. Due to the failure of a large producer, its use decreased in 2004, then recovered again.

11.4 Current Market Shares of Different Natural Fibres

Figure 11.2 shows the current shares of different NF for 2005 in the form of a pie chart.

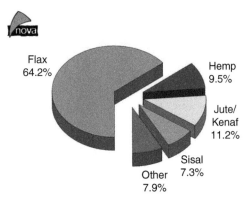

Fig. 11.2. Use of natural fibres for composites in the German automotive industry, 2005 (nova-Institut, November 2006).

The predominance of flax fibres (market share of almost 65%) that are produced almost exclusively in Europe becomes clear, in most cases as a by-product of textile long fibre production. Hemp fibres, also almost exclusively from European production, currently show a market share of just under 10%. Larger shares are not possible until further processing capacities are established or the hemp insulation material market decreases.

'Exotic natural fibres' can be itemized for 2005, which previously had not been possible due to lack of the respective data. The most important exotic fibres are jute and kenaf with 11%, followed by sisal with 7%.

While jute is by far the fibre with the highest turnover worldwide, thus making it the 'leading fibre' among technical NF, insufficient data are available for kenaf. In the trade sector, jute and kenaf are often not differentiated from one another properly. For this reason, these two Asian fibres are always listed together. Sisal is the second most important technical NF worldwide, coming mainly from Africa and South America.

Other exotic fibres, particularly coir from Southern Asia, are used in the first exterior part of the framework in the press flow moulding process. A couple of other natural fibres can be used for composites.

11.5 Shares of Different Production Techniques

Figure 11.3 shows the range of different production techniques for NF composites. As in recent years, compression moulding is dominant, though slightly less than previously.

Compression moulding, which has been used more than 99% in recent years, has since declined to 95%.

For the first time – in the use of natural fibres – two new techniques can be observed: press flow moulding and injection moulding. For both techniques, considerable increases are possible in the coming years, while compression moulding seems to have entered a phase of saturation. This is due partly to the fact that its main field of application is high-class interior parts of the medium and luxury class, where it is hard to achieve further growth (see also section 11.11).

Currently, it is often said that NF compression moulding has passed its peak, already being on a downward swing. Our survey cannot confirm this, but merely detects a stagnation. However, a shift among suppliers is noticeable, which could explain the impression mentioned above: while, indeed, the production of NF compression moulding parts is decreasing quantitatively among many small and medium suppliers, production accordingly is increasing among a few large suppliers, thus compensating for the decrease among the smaller suppliers. When asking all suppliers, one necessarily gets the impression that compression moulding is declining for the most part – although it remains qualitatively constant.

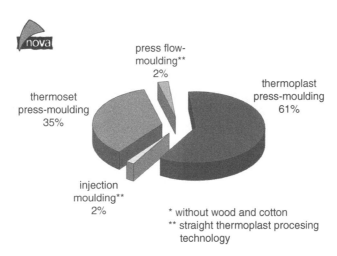

Fig. 11.3. Shares of different processing technologies of natural fibre composites (not including wood or cotton) in the German automotive industry in 2005 (nova-Institut, November 2006).

A closer estimate of the future market chances of different NF techniques is provided below.

11.6 Excursus: Wood and Cotton

In the framework of the study, there has been an attempt to survey also the amounts of wood fibres and cotton that are used in German automotive production. Unfortunately, however, this was not possible in the framework of the investigation.

As important companies from the wood fibre and wood flour processing sector did not take part in the survey, merely about 16,000 t of wood fibres could be verified. In the framework of our market study of 2004, we had assumed about 25,000 t of wood fibres and about 36,000 t of wood fibre composites for 2003. Because growth was generally expected for this sector, we estimated the amount for 2005 at about 27,000 t of wood fibres and about 40,000 t of respective wood fibre composites.

The wood fibre composites used in the automotive industry have a large fibre content and an almost exclusively thermoset matrix. Wood plastic composite (WPC) granulates made from a thermoplast, wood flour and fibres, respectively, as well as additives,

constitute an exception. Their market share is still less than 1%, but will increase.

The data for cotton are even sparser. Only a few hundred tonnes could be verified, although our previous study (2004) had stated about 45,000 t of cotton and about 79,000 t of respective composites for 2003.

This discrepancy was due to the fact that the survey was conducted primarily among passenger car subsuppliers, while thermoset cotton composites today are used almost exclusively in lorry driver cabs.

11.7 Natural Fibre Shares for Different Production Techniques

Figure 11.4 shows the NF shares for different production techniques. This year, as mentioned in the introduction, these data were surveyed for the first time. As expected, the fibre shares for thermoset wood composites were the highest with just under 85%. When natural fibres (not including wood and cotton) are processed on a thermoset basis, the fibre share amounts to approximately 55%. The large share of NF in thermoplastic composites came as a surprise. While in the past, in accordance with the producers interviewed, we had assumed a fibre share of 30–40%, the current survey

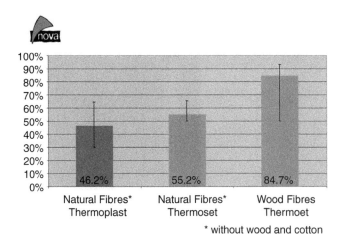

Fig. 11.4. Natural fibre reinforced composites (not including wood or cotton) in different processing technologies of composites in the German automotive industry in 2005 (nova-Institut, November 2006)

resulted in an average share of 46%. Averaged over all techniques, the average NF share amounts to 51.5%.

In Fig. 11.4, the fibre share ranges are shown. Thermoplastic composites range from 30 to 65%.

11.8 Natural Fibre Composites in Motor Cars

Based on these now confirmed NF shares and an assumed average edge trim of 20% (with compression moulding), the following amounts of NF composites arise. As discussed in the introduction, the data of 1999–2003 had to be corrected accordingly, making inconsistencies with previous nova-Institut publications inevitable here.

In addition to the increasing total amounts, Fig. 11.5 also shows the changing shares of thermoset and thermoplastic techniques. Since 1999, the share of thermoplastic composites has increased considerably; in the past 3 years, however, there have been no further shifts.

11.9 Natural Fibres per Passenger Car

According to the Association of Automotive Industry (*Verband der Automobilindustrie*, www.vda.de), 5.2 million (2004) and 5.4 million (2005) passenger cars, respectively, were produced in Germany. Based on these figures, together with the data from Fig. 11.1, the average NF content per passenger car can be calculated easily. For 2004 and 2005, this results in 3.6 kg/passenger car, a value only slightly higher than in 2003 (3.5 kg/passenger car).

11.10 Natural Fibre Composites Beyond the Automotive Industry

The automotive suppliers surveyed were asked whether they were also producing composites for other branches. Several small and medium suppliers affirmed this, which together amounted to approximately 150 t that were processed mainly with PP-NF granulates using the injection moulding

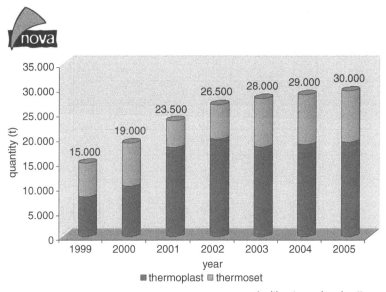

Fig. 11.5. Natural fibre reinforced composites (not including wood and cotton) in the German automotive industry 1999–2005 (nova-Institut, November 2006).

technique. The carrier material of grinding wheels is an example of this, with the aforesaid material successfully replacing polypropylene (PP) glass fibre injection moulding material.

11.11 Future Developments

Regarding the future market development of NF reinforced composites, there currently is no clear trend noticeable. Estimates within the automotive branch vary greatly. There is belief on the one hand that natural fibres have already passed their peak and their applications will decrease and, on the other hand, that there is stability with a (slight) market growth and interesting potential in the medium term. 'No clear direction for NF materials: successes in the past, weakening at the moment, and an interesting future' – this is how an insider summarized the situation in the summer of 2006.

Also, the material choice of original equipment manufacturers (OEMs) and tier-one suppliers is hard to assess, depending on the series of models and that decisions regarding the pros and cons of NF reinforced composites are made at the same time. At the moment, NF compression moulding is in a phase of stagnation, while NF press flow moulding and PP-NF injection moulding are increasing, however, but at a very low level.

It is clearly noticeable that the setting for new materials has changed substantially in recent years. Under heavily increased cost pressure, for which quality is also partly sacrificed, new materials have, since 2004, had considerably more difficulties. Suppliers want to use existing processing lines to capacity and not invest in new machines. New materials must be better *and* cheaper, which can hardly be achieved.

From an economic point of view, NF and wood materials exhibit decent price stability, being less dependent on oil prices than other materials, particularly if large NF and wood shares, respectively, can be realized. Should CO_2 emissions be financially punished more severely in the future, further economic benefits would come about.

11.11.1 Compression Moulding

NF compression moulding is an established and proven technique for the production of extensive lightweight and high-class interior parts in medium- and luxury-class cars. The advantages (lightweight construction, crash behaviour, deformation resistance, lamination ability, depending on the overall concept and also price) and disadvantages (limited shape and design forming, scraps, cost disadvantages in the case of high part integration in construction parts) are well known. Process optimizations are in progress, in order to reduce certain problem areas such as scraps and to recycle wastage. By means of new one-shot compression moulding presses, soft surfaces can also be integrated directly, which so far has not been possible with injection moulding.

As far as preferably inexpensive door concepts with a high part integration are concerned (up to the point of doing without lamination), NF compression moulding does not have a good chance against injection moulding. But, as far as high-class door concepts are concerned, NF compression moulding remains the first-choice technique. Against this background, it is not surprising that NF compression moulding is stagnating right now in the German automotive industry. The decreases among small and medium suppliers are fully offset currently by increases among large tier-one suppliers. One issue is that there are only a few compression moulding machine manufacturers and mat producers. Compression moulding is a specialized technique. Suppliers would prefer to use their existing (injection moulding) lines to capacity. Under heavy price pressure, this could become a disadvantage for NF compression moulding presses.

The future of NF compression moulding depends on numerous factors (price pressure, strategies and interior concepts of the OEMs and suppliers, oil, plastics and glass fibre prices, advancement of compression moulding, as well as correction concepts and materials). We assume that this technique will find a market in the future. This is also indicated by the fact that, currently, more NF compression moulding lines have been installed worldwide than ever before – not in Europe, but in China, India and Iran. There, in view of the current world market prices for natural fibres, NF compression moulding seems to be regarded as an economically interesting and seminal technique.

11.11.2   Press flow moulding

So far, only a few companies have dealt with the topic of NF press flow moulding. Those that have are seeing interesting technical and economic possibilities for this new technique.

11.11.3   PP-NF injection moulding

There are big controversies in the estimation of market developments in the field of PP-NF injection moulding. Some people do not see any relevant application of PP-NF in motor cars, neither their technical data nor prices are attractive, others attest to PP-NF' large growth rates and a huge potential. This applies especially to WPC material with wood flour and fibres.

NF granulates and their processing are often regarded as not yet sophisticated and too expensive, in addition, there are complaints that, as yet, there are no established larger suppliers with respective support. Should these problems be solved, there seems to be real interest among OEMs and suppliers. Furthermore, PP-NF granulates are becoming more and more interesting in view of increasing mineral oil and glass fibre prices.

These different estimates are due partly to the fact that the current market of PP-NF granulates is very confusing and the various granulates can differ in their mechanical properties and prices at a factor of up to 2. Many potential customers do not know the current biggest and best suppliers.

11.12   Political Framework

A favourable political framework could help biomaterials experience considerable growth. For example, forced measures for the reduction of CO_2 emissions should be mentioned here. In this sector particularly, natural fibres

can score well, the production of which is ten times less energy-intensive than that of glass fibres.

A new EU End-of-Life Vehicle Directive, which is under revision at the moment, could also have a substantial influence. If attempts were successful in achieving a renewable resources deduction at source, like, for example, the steel quota, as representatives of the NF branch have been claiming for years, there would be considerable advantages for NF reinforced composites. A practical solution could be that the actual share of renewable resources be credited to each vehicle as material recycling – regardless of whether the part is used energetically or materially. This approach would be justified by the fact that even in the case of burning the renewable resources shares, the CO_2 balance would be almost neutral. Right now, this would result in a mere 3.6 kg of natural fibres on average per vehicle; but vehicles with considerably larger amounts of 20 or 30 kg have been produced successfully in series for years and could credit these amounts in the future, according to the model mentioned above. A respective revision of the End-of-Life Vehicle Directive would cost Brussels nothing and would have crucial steering effects.

11.13   Sources of Previous Data

Comparable nova-Institut market studies from previous years:

* *Use of natural fibres in composites in the German automotive production 1996 to 2003.* The use of natural fibres has further increased despite recession and pricing pressure, PP natural fibre injection moulding with first series applications. Publisher: nova-Institut, Hürth, Germany, September 2004, 16 pages (long version); authors: Karus, M., Ortmann, S. and Vogt, D. (also a free short version is available (4 pages)).

* *Use of natural fibres in composites in the German and Austrian automotive industry – market survey 2002: status, analysis and trends.* Publisher: nova-Institut, Hürth, Germany, February 2003, 5 pages; authors: Karus, M., Kaup, M. and Ortmann, S.

Acknowledgement

The present study was financially supported by the Agency of Renewable Resources (*Fachagentur Nachwachsende Rohstoffe e. V., Gülzow*), in the framework of the project 'Study on the market and competition situation of natural fibres and natural fibre materials (Germany and EU)', and the Working Group Natural Fibre Reinforced Plastics, Federation of Reinforced Plastics, Frankfurt, in the framework of the project 'Status and future of natural fibre reinforced composites, wood plastic composites and biopolymers in the German automotive industry'.

Thanks are due for this help, for the study could not have been realized without this financial support.

12 Increasing Demand for European Hemp Fibres

Michael Carus

European Industrial Hemp Association (EIHA), c/o nova-Institut GmbH, Hürth, Germany

12.1 Introduction – A Sustainable Raw Material for Bio-based Composites

From an historic point of view, hemp has been an important raw material for industry for more than 2000 years. Hemp fibres were used for technical textiles such as ropes, hawsers and boat canvas, as well as for clothing textiles and paper. In the 1990s, hemp was rediscovered throughout the world as an important raw material for bio-based products in a sustainable bioeconomy, and has been in high demand ever since. The most important cultivation and manufacturing regions are Europe and China, and the most important applications are bio-based composites (natural fibre reinforced plastics), as well as construction and insulation materials. The bio-based materials sector, in particular, still has large, untapped market potential for both reinforcing oil-based plastics and, to an increasing degree, for bio-based plastics.

12.2 Success Story Automotive Industry: Current Trends and New Applications

In 2005 – more recent data are not available – 30,000 t of natural fibre composites (EU: 40,000–50,000 t), wood not included, were used in the automotive industry, requiring 19,000 t of natural fibres (EU: 30,000 t). European flax (about 65%) and hemp (about 10%) were used, with the remaining 25% covered by imports from Asia (jute, kenaf, coir, abaca). Natural fibre compression moulding is the dominant processing technique (share of >95%): it is an established and proven technique for the production of extensive lightweight and high-class interior parts in medium- and luxury-class cars. Its advantages are lightweight construction, crash behaviour, deformation resistance, lamination ability and, depending on the overall concept, price. The disadvantages are limited shape and design forming, offcuts and cost disadvantages in the case of high part integration in construction parts. These advantages and disadvantages are well known. Process optimization is in progress, in order to reduce certain problem areas such as offcuts and waste recycling. By means of new one-shot compression moulding presses, soft surfaces can also be integrated directly, something that so far has not been possible with injection moulding.

Between 2005 and 2009, the use of natural fibres in the European automotive industry did not expand, and even decreased slightly in Germany, after it had grown by double-digit figures each year between 2000 and 2005. Since 2009, however, there has again been an increasing demand: new models from almost all automotive companies that will be released on the market this year or next year do have considerably more interior parts once again made with natural fibre reinforcing. On the one hand, this is due to the high development of the materials and the fact they have proven themselves in practice, but on the other hand, it is due also to the increasing interest of the automotive industry in bio-based materials and lightweight construction – natural fibre construction parts can score in both fields. In addition, further cost and weight reductions have been achieved in recent years, especially with regard to compression moulding.

Furthermore, new trends are becoming apparent: automotive manufacturers do not want only to use bio-based materials but also to show them to their customers. While, up to now, natural fibre construction parts have disappeared under lamination, thus becoming invisible to the customer, in the near future vehicles will be released on to the market that will exhibit natural fibres under transparent films or lacquers, showing completely new surface effects. Another trend is seen in the development of making the plastics matrix bio-based as well; that is, producing interior parts from polylactic acid (PLA) or bio-based polypropylene (PP) and natural fibres. While such 100% bio-based composites will soon be found in Japanese cars, this will still take a while in Europe.

With demand again increasing and new concepts and the support of politicians for bio-based products, sales of 40,000–50,000 t of natural fibres could be achieved in Europe by 2015, at least 10–20% of which could be supplied by European hemp.

12.3 Insulation Materials and Many Other Applications

The second road to success is hemp insulation materials, of which about 3000–4000 t are produced annually and put on the market in the EU. The most important manufacturers and users are Germany, France and the UK, respectively. The properties of hemp insulation materials are very good and are appreciated by customers. Achieving bigger sales markets is

hampered solely by the relatively high price compared to mineral fibre insulation materials. Here, only suitable economic–political framework conditions can be of help.

Apart from the automotive and construction industry, there are numerous applications with a smaller volume, such as briefcases, other cases, various consumer goods (e.g. letter scales, battery chargers, toys) or trays of grinding/sanding discs and urns. The latter are a good example of a 100% bio-based product: the urns are produced from PLA, reinforced by hemp fibres, and are fully biodegradable.

As well as the examples mentioned, injection moulding plays an important role in addition to the aforementioned compression moulding. The increasing availability of high-grade natural fibre injection moulding granulates will help to develop new applications quickly here.

12.4 New Processing Techniques

For some decades in the EU and North America, there has been intense research going on in new processing techniques for flax and hemp fibres in order to make the development of new high-price fields of application possible for natural fibres. Two outstanding processing techniques that are close to commercial implementation are already today producing modified hemp fibres amounting to several hundred tonnes per year: first, the Crailar process from Canada, which focuses on the use of hemp fibres in the textile industry, and second, the ultrasonic processing technique of the ECCO Group from Germany, which focuses on high-grade technical fibres.

12.5 Availability and Price Development of Natural Fibres – A Chance for European Hemp?

While the technical natural fibre market is increasing worldwide, the question of prices and security of supply arises. In important cultivating countries in Asia, the cropping areas for jute and kenaf cannot be extended because there is considerable competition for land for

Price indices for natural fibres, crude oil and polypropylene (per Euro basia)

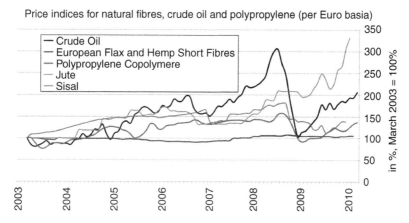

Fig. 12.1. Price indices for natural fibres, crude oil and polypropylene (on a per euro basis).

other purposes. The situation is better regarding sisal: here, an extension of cropping areas is possible in the dry regions of Africa and South America – places where hardly any other crop can be cultivated. But European production is also under pressure: the cropping areas of flax are decreasing due to strong competition from areas with subsidized bioenergy crops such as ethanol, as well as the dependency on exports to China, which is buying less textile-grade long flax fibres. As for hemp, an extension of cropping area is possible if rates of return similar to those of the food and feed sector and energy crops can be achieved. Areas under hemp cultivation are also on the rise in China, with hemp expected to replace cotton in the clothing textile sector.

In December 2009, Bangladesh imposed a ban on jute fibre exports for the first time and it was not before February 2010 that it was partly suspended for certain grades. The reasons for the embargo were to be found in 3 years of poor harvests and increasing demand, particularly from India (packaging) and China (composites), threatening a shortage of the necessary raw material from the Bangladesh jute industry. Due to the embargo, jute prices rose by 50–100%. At the same time, sisal prices were increasing too, due to a severe drought in East Africa.

Asian packaging (bags) accounts for 80% of the jute and kenaf used, sisal particularly in the form of tows and harvest belts. In contrast to these, natural fibre composites still constitute small markets that can be supplied quite easily.

As a result of farmers reacting more quickly to changes in demand, rates of return and a local shortage of area, there has been a general trend towards a more dynamic agricultural market with more volatile prices, and this, fuelled by speculators, is now affecting the world of natural fibres. For a long time, prices have been quite stable compared to other agricultural products or oil. But it is expected in the future that natural fibre prices will definitely stay below €1/kg so that they remain attractive for composites.

Figure 12.1 shows the price developments of important natural fibres and, as a comparison, the price development of oil and polypropylene. European flax and hemp short fibres have only recently shown moderate price increases after a long period of price stability and are currently showing particularly good price stability – a price rise of less than 10% in over 7 years.

To sum up: exciting times for European hemp, which, with adequate framework conditions, has considerable growth potential.

13 Hemp in Papermaking

Bernard Brochier
Formerly Centre Technique du Papier, France

13.1 Introduction

Although there are thousands of non-wood paper mills in the world, only a few of them use hemp as a fibre source (Fig. 13.1). About ten mills located in the western world (USA, UK, France, Spain, Eastern Europe and Turkey) use hemp to produce speciality papers.

The following papers can generally be produced by especially long fibres such as hemp, flax and cotton:

- Cigarette paper
- Filter paper (technical and scientific)
- Coffee filter
- Tea bags
- Speciality non-wovens
- Insulating papers (condensators)
- Greaseproof papers
- Security papers
- Various speciality art papers

The use of hemp is concentrated mainly on bast fibres, while the woody core of the plant is generally considered as waste. In the production of cigarette paper, however, all the fibre in the hemp stem is used.

13.2 History of Hemp in Papermaking

The use of hemp fibre and paper dates back more than 2000 years. In fact, it is believed the first paper sheets (reported in China in 105 AD) were made of hemp fibre. Although no one knows what these sheets look like, this does indicate that hemp is a suitable papermaking material. Paper historians do agree that the earlier Egyptian papyrus sheet should not be referred to as paper, because the fibre strands are woven and not 'wet laid'. Chinese papermaking craftsmanship was transferred to Arabic and North African countries, and from there to Europe.

Until the early 19th century, the raw material for papermaking was rags, or worn-out clothes. Since at that time clothing was made solely of hemp and flax (sometimes cotton or wool), almost all paper in history was thus made of hemp and flax fibres.

The first Bible was printed on hemp paper, and so were the first drafts of the Declaration of Independence.

With the Industrial Revolution, the need for paper started to exceed the available rag supply. Even though hemp was the most traded commodity in the world up to the 1830s, shortage threatened the monopoly for hemp and flax as papermaking fibres. This was the eventual push for investors and industries to develop new processes to be able to use the world's most abundant (and free) source of natural fibres, our forests.

Actually, only 5% of the world's paper is made of annual plants such as hemp, flax,

Fig. 13.1. Hemp fibre before treatment.

cotton, bagasse, wheat straw, sisal, abaca, reed and other exotic species.

The world hemp paper production was around 120,000 t/year, which was about 0.05% of the world paper production volume.

Hemp pulps are generally blended with other (wood) pulps for paper production.

13.3   Hemp Composition

The stem of the plant is composed of two fractions:

- 30% bast fibres
- 70% woody core.

The chemical composition of the different fractions of the plant (Table 13.1) indicates that the entire hemp stalk can be used for the production of cellulose fibres. The comparison with hardwood and softwood shows similar composition, but the bast fraction has the higher A-cellulose content available for papermaking. The low lignin content of the bast fraction is interesting because of the lower chemical required in the cooking processes.

The composition of the woody core is similar to a hardwood but is generally considered as waste and is used in other applications.

Compared to all annual plants used in papermaking, hemp bast fibre is one of the longest. The woody core fraction generates one of the shortest fibres (Fig. 13.2).

13.3.1   Raw material structure

Hemp is composed mainly of lignin, hemicelluloses and cellulose.

The separation of the fibres is carried out following the paper pulping processes, consisting of dissolution of the lignin and part of the hemicelluloses responsible for the linkages between fibres (which give the stem its rigid structure).

Lignin

Lignin is a polymer of complex and varied propylphenol units. Figure 13.3 illustrates the structure of lignin. Alkali can partially depolymerize this molecule, leading to soluble compounds.

Table 13.1. Chemical composition of hemp.

	Whole plant (%)	Bast (%)	Softwood (%)	Hardwood (%)
Holocellulose	80–83	81–86	75–85	60–70
Alpha-cellulose	50–55	65–67	58–61	45–55
Lignin	17–20	8–10	23–26	20–25
Ash	2–4	3–5	0.3–0.4	0.2–0.1
Solvent extract	2–3	1–2	0.1–0.5	2–13
Water soluble extract	5–8	9–11	2–3	3–13

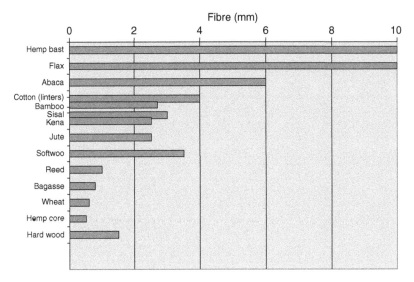

Fig. 13.2. Comparison of fibre lengths of annual plants.

Cellulose

Cellulose is a carbohydrate polymer built of glucose units (Fig. 13.4). It is less sensitive to cooking conditions compared to lignin.

Cellulose is the most important component of the fibre for papermaking. The fibre structure is built of cellulose chains, constituting protofibrils made of microfibrils (themselves made of macrofibrils). The fibre structure is illustrated in Fig. 13.5.

Hemicelluloses

These are sugar chains of low polymerization degree and different compositions, such as different hexoses and pentoses.

These compounds are easily hydrolysable by alkali or acid solutions, generating soluble elements. They are linked to the lignin and play an important role in the elimination of lignin.

13.3.2 Structure of a fibre

A fibre is composed of several cylindrical walls, as described in Fig. 13.6. In the figure: M represents the middle lamella, composed mainly of lignin (70%) and hemicelluloses; P represents the primary wall, which is very thin (0.1 to 0.5 μm) and composed of lignin (50%), pectins and hemicelluloses; and S represents the secondary walls, which are mainly cellulose containing and the thickness of the walls are 0.1–0.2 μm, 0.5–8 μm and 0.07–0.1 μm for S1, S2 and S3, respectively.

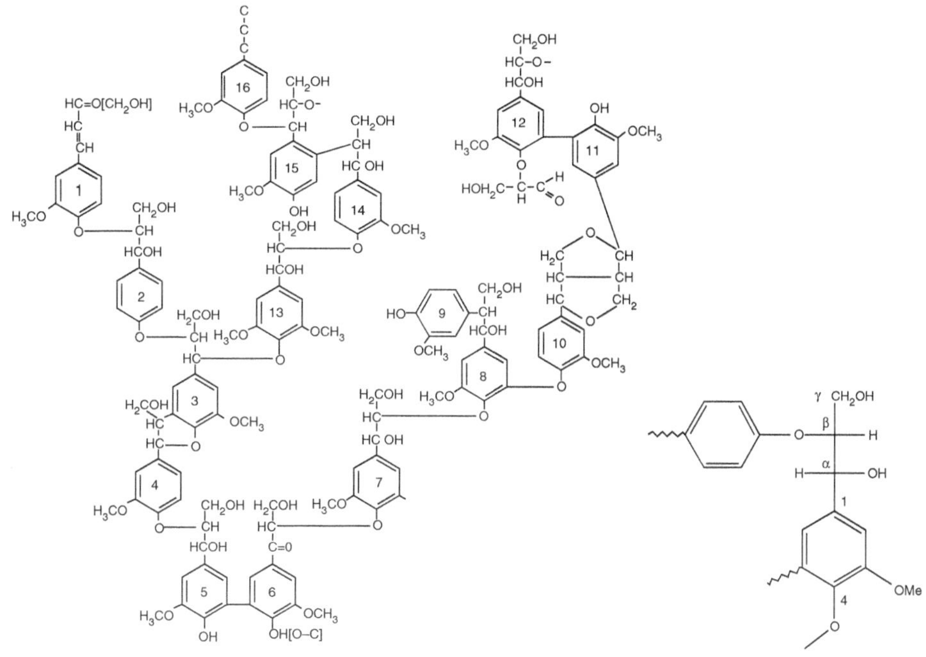

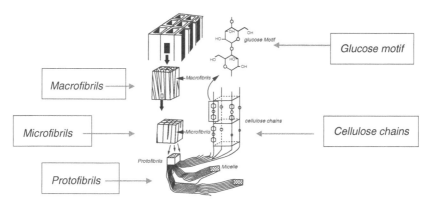

Fig. 13.3. Chemical structure of lignin.

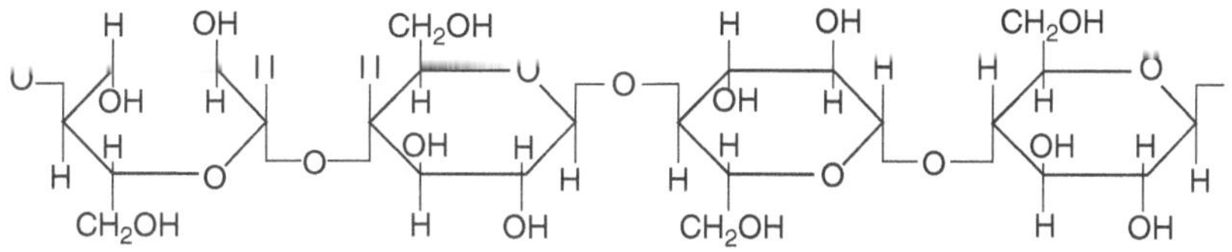

Fig. 13.4. Chemical structure of cellulose.

Fig. 13.5. Fibre cellulosic microstructure.

M : middle lamella   S2 : central secondary wall
P : primary wall   S3 : internal secondary wall
SI : external secondary wall   L : Lumen

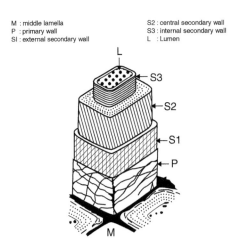

Fig. 13.6. Fibre macrostructure.

13.4   Pulping Processes

The pulping processes used with hemp are based on the application of chemicals. Four main processes can be employed:

- Kraft process
- Soda or soda anthraquinone process
- Neutral sulfite process
- Acidic sulfite process.

13.4.1   Kraft process

This is the worldwide chemical pulp manufacturing process most used for lignocellulosic materials. The cooking action is done at high temperature (150–170°C) using a mixture of caustic soda (NaOH) and sodium sulfide (Na_2S). The sodium sulfide is necessary as a buffer and to prevent cellulose oxidation, because of its reductive properties.

One advantage of the kraft process is the possibility of recovering all the chemicals by evaporation and combustion of the residual liquors, leading to self-sufficiency of the mill in terms of energy.

The cooking conditions are usually:

1. 15–18% NaOH and 4–6% Na_2S based on the dry raw material weight.
2. The chemicals are dissolved in water; the liquor is 4 or 5 times the raw material weight.
3. A temperature of 170–175°C is maintained for 1–3 h.

Cooking chemistry: action of sodium hydroxide on lignin

When the temperature reaches 140°C in strong alkaline conditions, the OH groups located in α position of the aromatic cores are partially ionized. A nucleophilic attack of the β carbons generates phenolic fragments. The lignin is then divided into small soluble units of phenolic type (Fig. 13.7).

The action of the sodium sulfide as a buffer helps to maintain a high alkaline medium, but secondary reactions of degradation are induced by HS⁻ (generated from Na_2S) reducing the size of lignin fragments. HS⁻ is responsible for the demethylation of aromatic groups generating methylmercaptan, MeSH, with a characteristic smell.

The demethyled groups are then transformed into quinines, leading to the brown coloration of the kraft pulp (Fig. 13.8).

Hemp bast fibre cooking yields are around 65–75%, depending on the quality of the decorticated hemp. The resulting pulp is brown and must be bleached for paper use.

13.4.2   Soda or soda anthraquinone process

This process is similar to the kraft process, using strong alkali charges, leading to the same reactions as described previously. Anthraquinone is used in place of sodium sulfide, in this case to prevent the peeling reaction responsible for the depolymerization of the cellulose. During cooking, some hemicelluloses are also protected and kept in the pulp.

Anthraquinone is used in small charges as a catalyst (0.05–0.1%).

This type of cooking leads to slightly higher yields (1–3%). The unbleached pulp is quite similar to the kraft pulp.

13.4.3   Neutral sulfite process

The neutral sulfite process uses the principle of sulfonation of the lignin using sodium sulfite

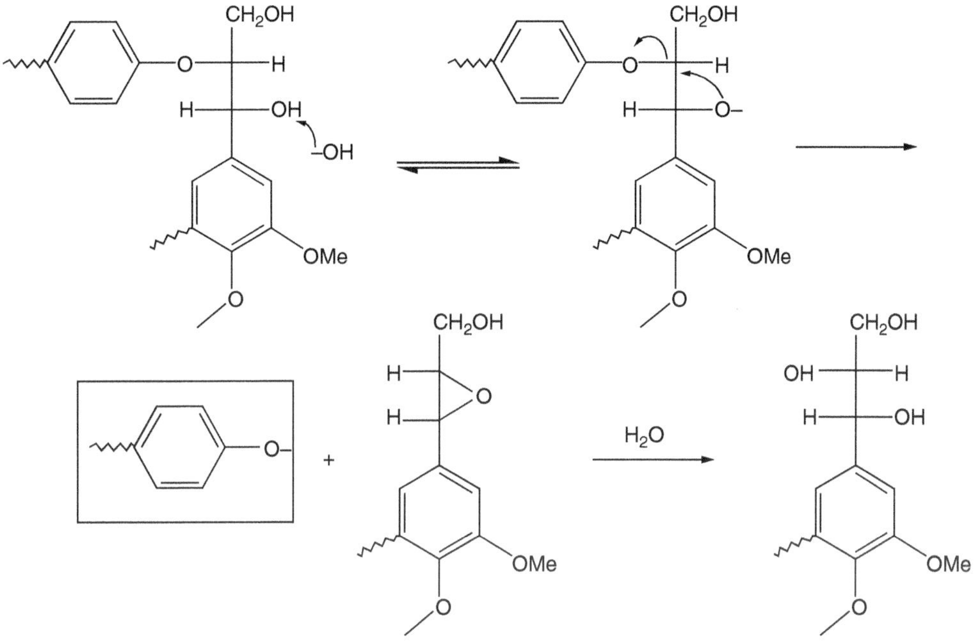

Fig. 13.7. Hydrolysis of the lignin structure by sodium hydroxide.

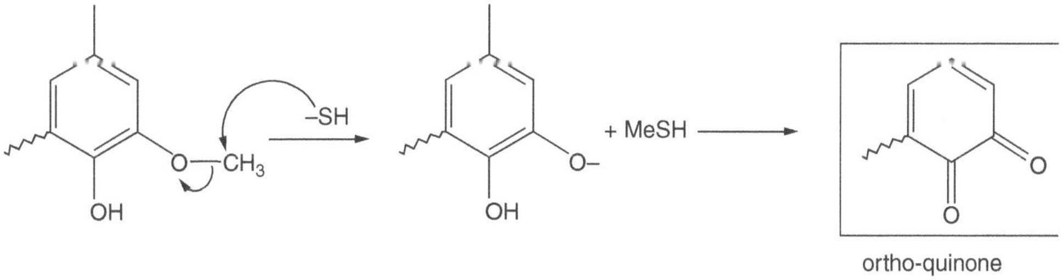

ortho-quinone

Fig. 13.8. Generation of chromophoric compounds responsible for pulp colour.

(Na_2SO_3) as a reagent. The chemical reactions are intended to achieve limited delignification through the combined effects of sulfonation and hydrolysis (sulfitolysis). High reaction temperatures (160–190°C) are used to accelerate sulfonation.

A near neutral pH is used in cooking to minimize carbohydrate loss, and the cooking liquor has a high buffering capacity (bicarbonate–carbonate) to compensate for pH drops caused by the formation of free acids through the decomposition of hemicelluloses.

13.4.4 Acidic sulfite process

Sulfite pulping derives its name from the use of a bisulfite solution as the delignifying medium.

The cations used are generally calcium, magnesium, sodium or ammonium. The bisulfite process operates at pH 2–3. The bisulfite solution is strengthened by the addition of SO_2 gas.

The cooking liquor is characterized as follows:

- Cooking liquor components (total SO_2): M_2SO_3 + H_2SO_3 + SO_2, where M is the cation

- Combined SO_2: M_2SO_3
- Free SO_2: ½ H_2SO_3 + SO_2.

Lignin reactions in sulfite pulping can be decomposed into three distinct phases: sulfonation, hydrolysis and condensation.

Both sulfonation and hydrolysis are important for lignin dissolution. Sulfonation makes the lignin more hydrophilic, and hydrolysis breaks lignin bonds so that new and smaller dissolvable lignin fragments are formed.

The hydrolytic carbohydrate reactions are also important in sulfite cooking. Both cellulose and hemicelluloses participate in these reactions. But in acid sulfite cooking, the attack on cellulose is minor, because the accessibility of cellulose is low. Dissolution of hemicelluloses through hydrolysis is substantially lower than in kraft pulping, especially when cooking is interrupted at high enough residual lignin content.

The chance of lignin condensation increases as the sulfite concentration decreases. Lignin condensation darkens the pulp and can be avoided by keeping the bound sulfite concentration high enough.

13.5   Other Processes

13.5.1   Chemimechanical hemp pulping

New developments and recent interest in paper grades for writing and printing require different kinds of pulps from those currently produced by the classic pulping mills. The development of recycled fibres leads to more and more degradation of the paper properties, due to the hornification of the fibres, and increases the need for reinforcement fibres. These processes can be developed either on high brightness level in printing and writing grades using the bast fibre fraction, or in semichemical pulping using the whole plant, as in the case of the production of reinforcement pulps for liner board.

13.5.2   Pulping equipment

Due to hemp characteristics such as fibre length (leading to the production of agglomerates like ropes), the design of the digesters is different when compared to the one used in the continuous cooking of wood chips.

Digesters

BATCH COOKING. Batch cooking is usually carried out in rotating spherical digesters (Fig. 13.9).

The raw material and the cooking liquor are introduced in the upper part of the digester. The temperature is regulated by steam injection. After cooking, the liquor is drained and the pressure released in order to open the cover. The digester is then reversed to discharge into a vat.

CONTINUOUS COOKING.   The continuous cooking of hemp is generally carried out in a screw

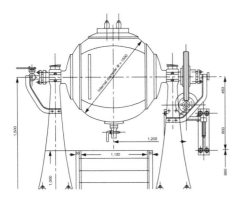

Fig. 13.9. Rotating spherical digester.

system maintained under pressure. The raw material is impregnated with the cooking liquor before feeding the digester. The process is illustrated in Fig. 13.10.

After cooking, the hemp pulp needs to be washed, shortened and bleached. Washing is an important operation to eliminate residual dissolved material that could consume the chemicals in the bleaching stages.

Shortening the fibre is necessary to make it usable in the next part of the paper process.

Bleaching can be carried out when the previous steps are completed.

Washing and shortening are usually carried out simultaneously in a Hollander beater (Fig. 13.11). The Hollander beater is a chamber into which the pulp is introduced. The pulp is moved in this equipment by the rotation of a cylinder equipped with special blades. The cutting action is regulated by the pressure applied on this cylinder. The Hollander beater is usually fitted with a drum filter. Washing is performed during the cutting time.

The time spent for cutting can be 6–8 h, allowing for the bleaching stage using sodium

hypochlorite in some applications. This process is described in Fig. 13.12.

The pulp preparation is made batch by batch.

13.5.3 Bleaching

After cooking and pulp preparation, bleaching stages are necessary to obtain pulps with a high level of brightness.

The brown pulp obtained after cooking contains residual lignin and diverse chromophoric compounds. The aim of bleaching is to make these compounds soluble.

The principal chemicals used are:

*   Chlorine
*   Sodium hypochlorite
*   Caustic soda (itself for alkaline extraction or in the presence of oxygen and/or hydrogen peroxide)
*   Chlorine dioxide
*   Oxygen
*   Ozone
*   Hydrogen peroxide
*   Peracetic acid
*   Enzymes

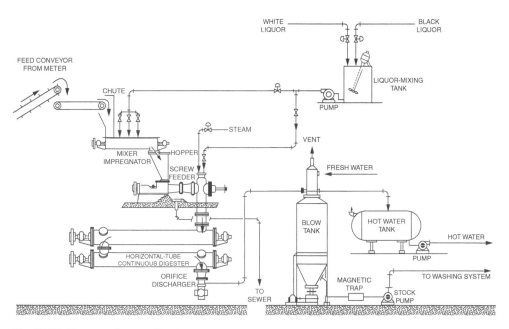

Fig. 13.10. Hemp continuous digester process.

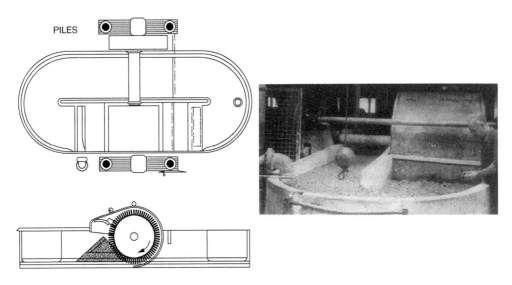

Fig. 13.11. Hollander beater (fitted with a drum washer).

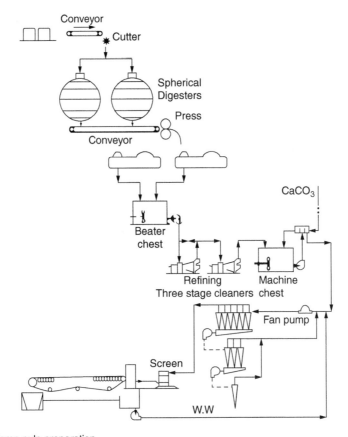

Fig. 13.12. Hemp pulp preparation.

In past decades, chlorine and sodium hypochlorite were used as bleaching agents. These chemicals, although very efficient and low cost, are known to generate chlorinated organic compounds toxic to the environment. Due to environmental concern, industrial countries are thus limiting the use of elemental chlorine in bleaching stages.

Actually, the bleaching of hemp pulps is carried out mainly using chlorine dioxide (reacting as an oxidative agent, being more environmentally friendly than chlorine) or hydrogen peroxide.

Another way to prepare the pulp is to perform cutting and washing simultaneously and to mix the bleaching chemical in a double-screw extruder (Bivis). This process can be associated with a continuous digester to organize a perfectly continuous plant (Fig. 13.13).

After this mechanical treatment, the pulp needs to be refined in order to homogenize fibre length.

13.5.4 Refining

Refining is an important step in the preparation of pulp for papermaking. The refining action consists of the development of the fibre surfaces. A mechanical treatment of the fibre opens the S layers, generating fibrils and, as a result, increasing the linkages between fibres.

The consequence of this is an improvement of the mechanical properties. In reverse, the increase of linkages between fibres has an effect on paper porosity. This property must be considered in the use of cigarette paper.

13.6 Cigarette Paper Manufacturing

The characteristics of cigarette paper are:

1. Porosity, leading to the manufacturing of 20–25 g/m² paper.
2. Wet strength resistance and tear index, especially for paper manufacturing (the thinner the paper, the weaker the strength). Hemp fibre resistance and fibre length are adapted properly for this application.
3. Opacity, which is essential for cigarette paper to avoid transparency showing the brown hue of the tobacco inside. Due to the thinness of the paper, a high mineral filler charge (up to 30%) has to be blended with the fibres to give enough opacity of the paper. The filler (generally calcium carbonate) will affect the mechanical properties of the paper but will have a positive effect on porosity.

Paper manufacturing is a continuous process. The pulp preparation is a mixture of fibres, mineral fillers and diverse retention agents (starch for about 1% of the dry matter). This

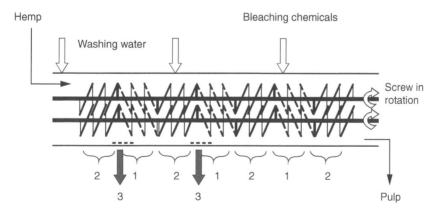

1 Reverse screw : Cutting and mixing zones
2 Convoying zones
3 Washing zones

Fig. 13.13. Bivis extruder principal schema for cutting, washing and bleaching.

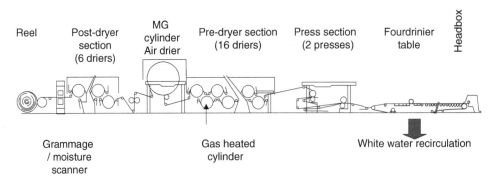

Fig. 13.14. Diagram of a paper machine.

Fig. 13.15. An industrial paper machine. Foreground: the headbox feeding the fourdrinier table; middle distance: a vertical press section; background: the drying section.

preparation is spread on a wire on which the pulp suspension is dewatered from 3 to 5 g/l concentration to a wet sheet of 20–25% of dry matter. This sheet is then transferred on a felt to the press section to improve dewatering before moving to the drying section, in which the water is evaporated on rolls heated to 100–110°C. The paper is rolled at the end of the machine and stored before use.

Figure 13.14 shows a diagram of a paper machine.

An industrial paper machine can produce 10–30 t/day. The wire width is 2–3 m and the machine speed is 200–300 m/min. Figure 13.15 shows an example of an industrial paper machine.

13.7 Conclusion

The use of hemp in papermaking is focused mainly on cigarette paper production, because its long and resistant fibres allow the manufacture of thin paper. The whole hemp production is devoted to this use, as the whole hemp plant can be considered for pulp production, generating long and short fibres, which can be used mixed or separate. The application of long hemp fibres in tissue paper increases wet resistance, improving the quality of such products. Development of the uses of hemp can also be orientated to non-woven production.

14 Hemp and Plastics

Gérard Mougin
AFT Plasturgie – Agro Fibres Technologies Plasturgie,
Fontaine les Dijon, France

14.1 Introduction

For over a decade, a number of laboratories and technical centres across the world have been undertaking research into the incorporation of natural materials into plastics. This work is motivated largely by the need to protect the environment and reduce our dependency on non-renewable resources.

The encouraging results from these programmes have led to a number of attempts to seek industrial applications; these have often met with failure, for a range of reasons.

Developments seeking to enhance the value of an agricultural by-product (e.g. ground straw or wheat, maize cobs and other plant crop by-products) have found that the processing costs are not covered by the value of the product, thus assuring their failure.

The production of compounds containing plant cellulose materials, using traditional production methods, necessitate the use of costly products in order to make the cellulose compatible with the polymers. This is not economically viable.

The creation of composite materials containing natural fibres can only be achieved industrially by creating a complete, structured production sequence from the farmer to the plastics manufacturer. The variability of the raw materials necessitates precise control of their production and processing, without which the minimum quality for the product cannot be attained.

In view of these difficulties, hemp producers created, in 2001, the company *AFT Plasturgie*, with the specific objective of creating and developing a procedure that would allow an economically viable production of a modified polymer containing natural fibres. This polymer also needed to demonstrate improved technical properties.

14.2 Plasturgy

14.2.1 Polymers

The worldwide market for plastics is larger than that for steel, with over 80 million tonnes (Mt) per annum. Its potential for growth is increasing. Among the countries consuming plastics, France lies fifth behind the USA, Japan, Germany and Italy. While similar amounts are consumed in Europe and the USA, there are variations in the amounts of different plastics used.

Among the commonly used polymers, polyethylene (PE) and polyvinyl chloride (PVC) are more popular in the USA than in Europe. This is due to higher individual consumption in the USA, where the plastics are much used in packaging (Fig. 14.1).

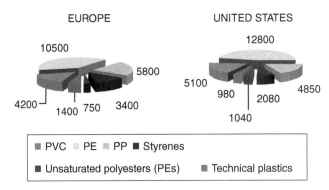

Fig. 14.1. Consumption in kt.

The second notable difference is the lower consumption of polypropylene (PP) and the greater consumption of unsaturated polyesters in the USA. This is due to the intensive use of temperature-hardened plastics in the American automobile industry.

Excluding technical plastics, polymers that are reinforced easily represent a demand of 6 Mt, both in the USA and the EU.

14.2.2   Reinforcing materials

The plasturgy industry has at its disposal a large number of polymers suitable for transformation. These are produced by the chemical industry, but as raw materials, do not possess the technical properties required for the end product to be constructed. In order to attain these properties, the plastics manufacturer must modify the polymers by incorporating various additives, ingredients, adjuvants (stabilizers, pigments and lubricants), reinforcing fibres, etc.

The rapid growth in demand for plastic materials has led, in the space of a few decades, to polymer production that accounts for close to 100 Mt of plastic per annum.

Concurrent with the growth in applications, the properties of the materials produced have continued to improve. This has led to them replacing traditional materials such as metal, wood and glass.

One of the most significant developments contributing to the increased use of plastic materials has been, without doubt, the improvements made to their mechanical properties

and, in particular, their rigidity and traction. Improvements to these parameters have been achieved through the incorporation of reinforcing materials.

Reinforcing materials are made of unwoven fibres or from fabric. Single fibres can be made from glass, carbon or from polymers (aramid, PE, liquid crystal polymer (LCP)). Since the beginning, glass fibre has been the predominant fibre, with the other fibres making their appearance only recently. The quantity of fibres used across the world amounts to some 2 Mt per annum; the quantity of natural fibres used in the same way has not, however, been quantified.

14.2.3   Composites and compounds

Historically, reinforcing fibres started being used in the industrial manufacture of temperature-hardened plastics. These are polymers with a three-dimensional structure produced by a chemical reaction between two or more mostly liquid components. The structure of these materials produced by this chemical reaction cannot subsequently be changed by heating. The fibre-reinforced products obtained in this way are generally described as 'composites'. Their most important applications are in the manufacture of automobile parts (body panels, bonnet, tailgate, wing...), in the leisure industry (boat hulls) and in electrical engineering (fuse boxes), as well as in the manufacture of body panels for agricultural and specialist machinery that is produced in relatively moderate quantities.

Europe produces approximately 1 Mt of temperature-hardened composite plastics. Germany is the main producer and the automobile and electrical markets are the main consumers of these materials, except in the UK, where the construction industry is the main consumer.

Thermoplastic polymers, whose shape could be changed simply by increasing the temperature and thereby softening the plastic, were soon reinforced with fibres (first glass, then carbon and polymers). At the present time, nearly 20% of the 35 Mt of thermoplastics transformed in Europe are reinforced with fibres. There are few reinforced polymers and, currently, polyamides are the main source. The principal applications for these products are in the automobile, electricity, sport and leisure industries. Recently, the use of long fibres with the same polymers has further improved the mechanical properties of these products. The volume of thermoplastics produced per year is increasing faster (5% per year) than the volume of temperature-hardened plastics (Fig. 14.2).

The situation in the USA is much the same as in Europe. North America has, however, shown a greater tendency to develop composite parts for the automobile industry (130 kg/vehicle in the USA, compared to 30 kg/car in France). The growth in this market is very significant, especially for the car manufacturer, SMC, who has over 100 applications for composites. Ford makes the greatest use of composites: using them for almost all the bonnets and radiator grills in its vehicles. The latter was, in fact, the first to introduce composite plastic reinforced

with natural fibres, in particular for the back of its pick-up trucks. The consumption of reinforced polymers in the USA is of the order of 1800 kilotonnes (kt), of which 650 kt are thermoplastics. The relative importance of each market is comparable to that in Europe (Fig. 14.3).

14.3 Natural Fibre Reinforced Polymers

14.3.1 Natural fibres

Natural fibres can be of animal origin (wool, silk…) or of plant origin (cotton, flax, hemp, jute, sisal, kenaf, cocoa, abaca and wood). Plant-derived fibres are the most common natural fibres used to reinforce plastics. In Europe, wood, flax and hemp are easy to produce and are consequently the three fibres most commonly used. In terms of the management of resources, it should be noted that both hemp and flax are annual plants. Their production can therefore be increased significantly at short notice. Wood, by comparison, requires many years of growth before it can be exploited.

Vegetable fibres are all made of four constituent principal components. The respective proportions of these components vary between plants (Table 14.1).

The main constituent of plant fibres is cellulose, a polymer with entire blocks in crystalline form. The different constituents are not distributed homogeneously within the complex microstructure of the fibre. The elementary

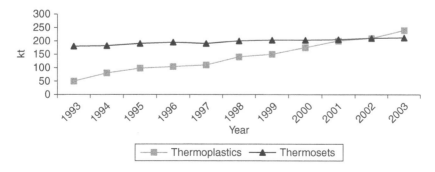

Fig. 14.2. Consumption of reinforced fibres.

fibres of hemp and cellulose have a length varying between approximately 5 and 60 mm. They have a density of 1.4 g/cc.

14.3.2   Hemp

Hemp provides fibres with properties comparable to those of other technical fibres (Table 14.2).

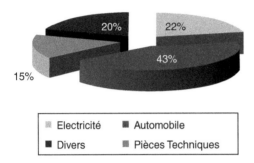

Electricité   ■ Automobile
■ Divers   ■ Pièces Techniques

Fig. 14.3. Market for reinforced plastics.

It also presents a number of specific advantages that suggest it will be developed in preference to other plants:

- Hemp prices are stable and unlike other fibre-producing plants such as flax are not susceptible to changes in fashion.
- The hemp plant is not destroyed by parasites or disease and therefore does not require any chemical treatment (insecticides, pesticides) and its need for nitrogen fertilizers is low.
- Hemp is less fussy about the ground it is grown on, making it easy to expand its cultivation in response to increases in demand.
- Hemp is a beneficial plant for other crops as it smothers out all weeds and thereby avoids the need to use chemical weedkillers.
- In addition to the use of hemp fibre in the papermaking industry, hemp can be used to produce a number of co-products (hurds, hemp seed and powder) that therefore provide added value.
- The cultivation and transformation of hemp are well practised in the EU.

Table 14.1. Content of vegetable fibres, showing percentage of four principal components: cellulose, hemicelluloses, pectin and lignin.

Fibres	Cellulose (%)	Hemicelluloses (%)	Pectin (%)	Lignin (%)
Cotton	92	6.0		
Flax	71.2	18.6	2.3	2.2
Hemp	76	11.5	1.3	3.2
Jute	71	13.3	0.2	13.1
Sisal	75	21.8	0.9	11
Abaca	70	21.7	0.6	5.7
Kenaf	61	20.4		11
Wood	50	24.0		26

Table 14.2. Tensile properties of fibres.

Fibre properties under tension			
Nature	Density (g/cm³)	Resistance (gpa)	Lengthening (%)	Module (gpa)
---	---	---	---	---
Glass	2.54	1.4–2.6	2	75
Wood	1.54	0.14–1.5		20–40
Flax		0.3–0.96	1.5–4.0	27–80
Hemp	1.4–1.5	0.5–1.04	1.0–6.0	32–70
Jute		0.4–0.8	0.8–2.0	13–26.5
Viscose	1.5	0.31		11

Note: According to FINK, WEIGEL IAP Potsdam – GEIGER ICT Pfinztal – BUSCH IWM Halle.

In addition to hemp, all fibres of plant origin can be considered for use in the reinforcing of plastics, as the aim is to achieve certain technical characteristics using an appropriate mix of fibres with different characteristics. Additionally, natural fibres that are not abrasive do not require treatment of the surfaces of the transformed material. These treatments can be expensive and have a significant negative impact on the environment.

The use of hemp as a reinforcing material in plastics does, however, require the close control of the production and treatment of the plant: cultivar, the nature of the soil, the method of cultivation and of fibre separation can all impact on the properties of the reinforced products produced.

14.3.3 Hemp/polymer compounds

The amount of reinforcement that can be achieved using hemp exceeds 70% by weight. That said, the transformation of compounds can only proceed without problems where the amount of hemp is less than 40%.

Procedure

Cellulose is a hydrophilic compound that is incompatible with the majority of polymers, which are generally hydrophobic. In order to bind cellulose fibres with a polymer, it is necessary, generally speaking. to incorporate a third component. This additional component must be compatible with both the fibre and the polymer and is there as a binding agent.

The originality of the procedure developed by AFT Plasturgie (Brevet No W0 2004/071744) lies in the use of a thermomechanical treatment that creates a fibrillation on the surface of the fibre (Fig. 14.4b). The single fibres thus freed up can then anchor themselves to the parent fibre within the polymer matrix.

The fibres incorporated after treatment have a final average length of 4 mm.

Properties

MECHANICAL PROPERTIES. The procedure introduced by AFT Plasturgie allows a reinforced composite plastic to be obtained using hemp that compares well with that obtained using the same amount of glass fibre (Fig. 14.5 and Fig. 14.6).

With a content of 30% hemp, a PP compound becomes extremely rigid, but also increasingly easy to break. The ability to withstand shock loading can be improved significantly following treatment with a compatibility agent. The pretreatment of fibres with a lubricant does not improve performance. By contrast, the use of both treatments produces a synergistic effect that allows the initial resistance to shock loading to be doubled (Fig. 14.7).

THERMAL PROPERTIES. The thermal performance of polymers is improved significantly by the incorporation of hemp fibres. A piece of PP composite containing 30% hemp exposed to 150°C for 400 h showed virtually no deformation. By contrast, the same piece made from ABS[1]/polycarbonate (PC) was

Untreated fibre   Treated fibre

Fig. 14.4. Untreated (a) and thermomechanically treated (b) hemp fibres.

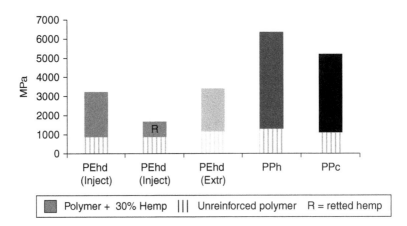

Fig. 14.5. Tensile modulus of reinforced and non-reinforced polymers.

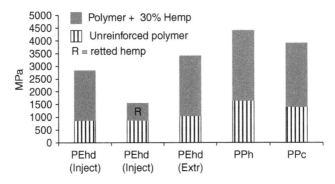

Fig. 14.6. Flexion modulus of reinforced and non-reinforced polymers.

completely deformed under its own weight. PP's thermal performance increases from 90°C to 145°C when it is mixed with 30% hemp. The Vicat[2] point of PVC similarly is increased by 5°C under the same conditions.

For a copolymer of PP, we arrive at the properties shown in Fig. 14.8.

WATER ABSORPTION. A well-known property of natural fibres is their ability to absorb water. Fibres possess a porous structure and can, when incorporated within a polymer, absorb liquids by capillary action. A piece of composite reinforced with 30% natural fibre can therefore absorb a significant amount of water (approximately 7%) when submerged. If, once saturated, the composite is left to air dry under normal temperature and humidity conditions,

the water is redistributed (Fig. 14.9). The cycle is reproduced without affecting the compound.

These properties can be used in the creation of active objects. We have demonstrated that a storage pallet made from high-density PE reinforced with 20% hemp fibre possesses, in addition to its interesting mechanical properties, the ability to store a bactericidal liquid. It can take up this liquid in the space of a few hours and then release it slowly over a period of 8 months, thus killing any bacteria coming into contact with the pallet. Such materials can therefore be reinforced mechanically and used in such a way that they control smells and humidity. They can also be used to produce particular surface properties (e.g. lubricated and antistatic surfaces).

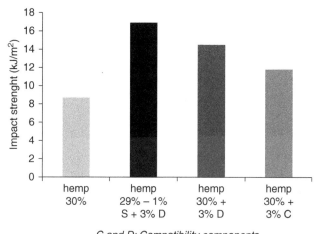

C and D: Compatibility components
S: Silane

Fig. 14.7. Charpy impact test (without notch).

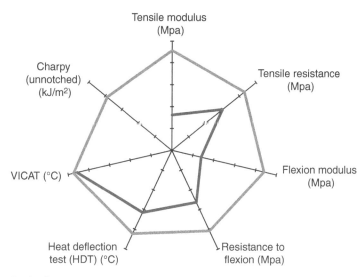

Fig. 14.8. Synthesis diagram.

ISOTROPIC PROPERTIES. Glass fibres are oriented with the flow of the material, according to whether it is laminar or turbulent, adopting an orientation that is parallel or perpendicular to the flow. Properties are therefore anisotropic according to the direction. In a complex piece exhibiting variation in thickness and flow direction when filled, the mechanical properties of different parts will vary. Thickenings and differential retractions will also be seen following moulding.

Hemp fibres are not orientated with the flow (Fig. 14.10). Hemp therefore produces moulded plastic that possesses isotropic mechanical properties. Additionally, the random arrangement of hemp fibres also blocks retractions in all directions, giving rise to the following advantages:

• Absence of thickenings, even in areas where the compound is thickest. Inflating agents become useless.

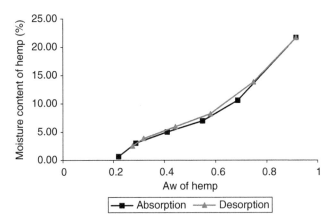

Fig. 14.9. Water absorption and desorption for hemp fibres at 25°C.

Fig. 14.10. Hemp fibres are not oriented in the direction of the flow of material when an object is moulded.

- Increased geometric stability even when hot. No post-moulding deformation.
- Unmoulding of pieces possible at very high temperatures. This economizes time in the production cycle (currently a saving of 30%).
- Very little retraction.

ENVIRONMENTAL IMPACT. The act of reinforcing a polymer such as PVC with 30% natural fibre produces a final product with a fossil fuel content that is only 70%. The ecological impact is therefore significant, especially in view of the improvements in the properties listed above.

These can be attributed to the inclusion of hemp, which itself is an important carbon sink.

If we compare a compound reinforced with natural fibres to a compound reinforced with glass fibre, setting aside the lack of toxicity, the low cost, low density, the absence of abrasion and the improved dimensional stability, we can identify enormous advantages in terms of sustainable development (see Tables 14.3 and 14.4).

A 1 kg reduction in the weight of a vehicle arising from the use of hemp fibres will result in the savings shown in Table 14.5.

Furthermore, hemp is able to store 0.79 kg of CO_2/kg of fibre and can release 10 MJ/kg of energy when incinerated at the end of its life.

It should also be remembered that hemp has the ability to smother undesirable weeds, thereby economizing on herbicide treatments and reducing the cost of cultivation for the subsequent crop.

Hemp fibres are also sufficiently supple to allow them to withstand mixing in a viscous environment in a vice/press for injection or extrusion many times. Their length, and therefore their L/D ratio, is preserved and their properties as reinforcing agents are not affected by multiple recycling.

Other properties

Resistance to ultraviolet (UV) light is unaffected by the presence of hemp, which itself is resistant to UV.

Table 14.3. Production of 1 kg of fibres.

	Hemp	Glass fibre
Energy consumption	3.4 MJ	48.3 MJ
CO_2 emissions	0.64 kg	20.4 kg
SO_2 emissions	1.2 g	8.8 g
NO_2 emissions	0.95 g	2.9 g
Biochemical oxygen demand	0.265 mg	1.75 mg
Chemical oxygen demand	3.23 g	0.02 g

Table 14.4. Substitution of 1 kg of natural fibres to 1 kg of polymer.

	Hemp	Polypropylene
Energy comsumption	3.4 MJ	101.1 MJ
CO_2 emissions	0.64 kg	3.11 kg
SO_2 emissions	1.2 g	22.2 g
NO_2 emissions	0.95 g	2.9 g
Biochemical oxygen demand	0.265 mg	38.37 mg
Chemical oxygen demand	3.23 g	1.14 g

Table 14.5. Savings over the lifespan of a vehicle consuming 7 l/100 km of petrol, assuming the combustion and production of the amounts shown.

Energy saved	273 MJ
Reduction in CO_2 emissions	17.76 kg
Reduction in SO_2 emissions	5.78 g
Reduction in NO_2 emissions	163 g

Fire resistance is reduced slightly. The addition of 15% hemp to PVS will reduce the composite's oxygen index from 43.8 to 36. Where necessary, the fibres can be pretreated with an intumescent product in order to restore the fire-resistant properties.

Polypropylene reinforced with 15% hemp, tested according to standards XP ENV 1186, CEE No 85/572, CEE 97/48 and CEE 2002/97, has been approved for contact with foodstuffs.

Hemp's properties as a sound and thermal insulator can be used to improve these characteristics in the plastic to which it is added.

Hemp compounds can be coloured using the same traditional methods in use in the plastics industry. Additionally, the fibres can be coloured before compounding in order to produce materials where the fibre's colouring is more obvious.

In view of the natural humidity content of reinforcing hemp fibres, the surface of compound plastics possesses a conductivity that favours the dispersal of electrostatic charge.

Fogging tests, using the Renault D45 1727/D method, on a 70:30 PP/hemp compound yielded a translucent homogeneous fog deposition with a fogging index of 90.2.

Suitable polymers and fibres

The principal component of natural fibres is cellulose, which degrades at temperatures in excess of 250°C. Consequently, the only polymers that are suitable for use with reinforcing natural fibres are those whose transformation temperatures are below 250°C. The most commonly used polymers are the polyolefins, the styrenes and PVC (Fig. 14.11).

Hemp provides the best mechanical properties among the fibres in common use (Fig. 14.12).

Implementation

The implementation parameters are virtually the same as for virgin polymers, although the fibres do possess a number of peculiarities that must be taken into account.

The compounds are normally conditioned with a residual humidity less than 0.5%. After a few months' storage, however, or after a few hours left uncovered, the compound will need to be desiccated before use.

The temperature at which these compound plastics are used is generally lower than that for pure polymers (e.g. PP/hemp is injected at approx 180°C). It is imperative that mass temperatures in excess of 220°C are avoided.

It is generally recommended that sudden increase in cutting force be avoided and that the threshold for injection be matched carefully to the viscosity of the compound. The behaviour of hemp-reinforced polymers is clearly pseudo-plastic.

Natural fibres contain a certain amount of pectin, which is degraded at temperatures in excess of 160°C. This can confer a light brownish colour to products without their properties being affected. That said, a dark brown coloration together with a strong-smelling product indicates that the transformation conditions are maladjusted.

When conducting initial tests, it is recommended that the compounds be diluted to obtain a fibre concentration of approximately 20%. This approach allows materials to be transformed without there being a risk of degradation and allows operators to familiarize themselves with the use of the compound.

14.4 Applications

14.4.1 Plastics

Plastic materials reinforced with hemp fibres are used in all the main plastic markets (Figs 14.13–14.17).

14.4.2 Unwoven felt

Natural fibres have been used by the automobile industry for more than a decade in the creation of car interior fittings. The products are made of felt with a density of approximately 500–2000 g/m². These felts are produced from a 50:50 mix of natural fibres (hemp, flax, sisal, jute, kenaf, abaca) and thermoplastic fibres (typically, PP). The mix is carded and needled to produce sheets with a thickness of 5–30 mm.

The felt is heated in flat presses to a temperature approaching 200°C before being placed in a cold mould. The end product is light and rigid and is not brittle. The manufacturing process only allows the production of pieces with a constant thickness and low stamping rate. The majority of German and US car manufacturers use these materials to produce door panels, parcel shelves and boot panels.

The production of felts with a density of 400–1000 g/m²without the addition of synthetic fibres allows sheets to be produced for use in mulching and soilless cultivation.

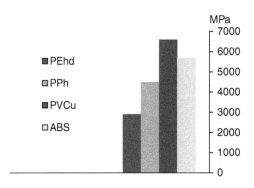

Fig. 14.11. Section modulus for different polymers reinforced with 30% hemp fibre.

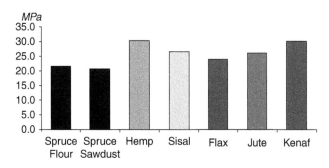

Fig. 14.12. Maximal tension resistance (compounds with 30% plant fibre reinforcement).

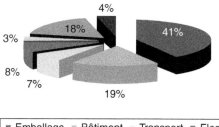

Emballage ■ Bâtiment ▨ Transport ■ Electr.
▨ Agriculture ▨ Electroménager ■ Divers

Fig. 14.13. European plastics markets.

Structural piece   Trim part

Fig. 14.14. In cars.

Fig. 14.15. In packaging and handling.

Finally, it is also worth noting the recent attempts to use natural fibre felt as a means of reinforcing temperature-hardened materials. These materials have attracted interest from the nautical, automobile and aeronautic industries because of the 15–20% weight improvement they offer when compared to glass fibre.

14.5 Conclusion

Finally, the essential properties of hemp-reinforced polymers can be summarized under the three thematic headings.

14.5.1 Economic

- Low cost.
- Significant reduction in the duration of the production cycle.
- Energy saving during transformation.
- Low density compared to other reinforcing materials.

14.5.2 Technical

- Mechanical properties identical to those of traditional reinforcing materials.

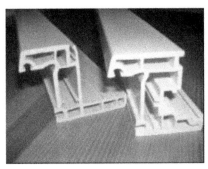 

Fig. 14.16. In building.

Fig. 14.17. In technical products.

- Products approved for contact with foodstuffs.
- No abrasion.
- Improved thermal resistance.
- No deformation or retraction.
- Significant geometric stability of items produced.
- Good sound and temperature insulation.

14.5.3 Ecological

- Completely recyclable.
- The reinforcing material is not toxic.
- Reduced content of fossil fuel derived materials.
- Use of a plant with good ecological credentials.
- Creation of pieces made from 'active plastic'.

- Manufacture of pieces with a natural side.
- No residues left following incineration at the end of the product's life.

These characteristics are leading more and more manufacturers to adopt these new materials. Among the significant potential markets, it is worth citing the following.

In the USA, some 500 kt/year of moulded plastic containing natural fibres are produced for the creation of flooring, terraces, swimming pool sides, etc… This market is starting to develop in Europe and should, over the next few years, become increasingly important, even if it does not achieve the importance of the US market.

The construction industry is a promising market. Of the 800 kt of polymer used in France per year, the natural fibre compounds have a potential of 60 kt, without counting the emerging market mentioned above.

The market for pallet boards and plastic boxes represents some 100 million units, with a lifespan/replacement rate of 10 years each. In Europe, the pallet market numbers some 32 million units/year. The weight of a plastic pallet is approximately 12 kg; the European market therefore accounts for an annual demand of 400 kt. The advantages of these pallets in terms of their low weight, rigidity, rapid construction and the possibility of making 'active pallets' mean that hemp-reinforced high-density polyethylene (HDPE) and PP have the potential for significant further development.

Technical items and automobiles are also areas of potential development for natural fibre reinforced polymers. The many specific properties of these compounds will be of interest to a large number of manufacturers. The most important characteristics are rigidity, low weight, shorter production cycles, geometric stability and the absence of clumping. The potential development in this market has been estimated at 10–20 kt for compounds of hemp and PP, PE or ABS.

Various other markets also have interesting potential: toys, sport and leisure, electrotechnical, etc... The value of these compounds lies in both their technical properties and aesthetic criteria. AFT Plasturgie has developed a collection of materials with coloured fibres that can be seen within the mass of the polymer.

Taken together, the plastics market for compounds reinforced with hemp represents a potential in excess of 100 kt. In order to realize this share of the market, the following will prove very important:

• A well-structured industry where all aspects from fibre production to the final formation of the plastic end product are controlled efficiently.
• A network of specialists capable of supporting the development of the industry.
• The capacity to grow rapidly in response to market demand.

Notes

[1] ABS = acrylonitrile, butadiene and styrene.
[2] The *Vicat* is the temperature at which a flat-ended needle penetrates a test sample to a depth of 1 mm under a particular load and uniform heating rate.

15 Medicinal Uses of Hemp

Denis Richard and Catherine Dejean

Centre hospitalier Henri Laborit, Poitiers, France, and Faculty of Medicine and Pharmacy, University of Poitiers, France

15.1 Introduction

The properties and virtues of hemp, or cannabis (*Cannabis sativa* L.), have been recorded for several thousand years, although the medical applications of its extracts and molecular derivatives are still not universally recognized. Studies undertaken since the start of the 1990s on cannabinoid receptors and their physiological role within living organisms allow us to anticipate developments on the therapeutic use of synthetic cannabinoids. For a review of this subject, see Hanus (2009) and the section on further reading at the end of the book.

15.2 Well-established and Recognized Medicinal Properties

The medicinal properties of cannabis were exploited as early as 5000 BC. It was used in China for a number of indications, including malarial fever, rheumatic pains, menstrual pains and constipation. In India, hashish was used in various forms (*ganjah*, *charras*, *bhang*), as an antipyretic but also as a hypnotic, for the treatment of migraines and as an appetite stimulant, applications that are still popular today among supporters of its use as a therapeutic agent. For historical reviews of the subject, see Russo (2007), Frankhauser in EMCDDA (2008, Vol 1) and Hanus (2009).

The start of the 19th century saw the golden age of cannabis use as a medicinal drug. It was included in the American pharmacopoeia in 1854 and could be purchased in pharmacies. It soon replaced opium as a drug, thanks to the relative lack of narcotic side effects. Interest in the plant declined towards the end of the 19th century – as with other medicinal plants – as doctors started to favour industrially produced medicines, the effects of which were faster in onset, more spectacular and more reproducible. In the USA, the prescription of cannabis was outlawed in 1937, when the drug itself became prohibited. It disappeared from the American pharmacopoeia in 1941 and from the French pharmacopoeia in 1953. Cannabis has, however, retained its listing in some pharmacopoeias around the world, including that of Italy (Caballero and Bisiou, 2000).

A number of patient groups, together with many doctors, are calling for cannabis to be licensed again and therefore available for medicinal use in cases where conventional treatments are ineffective and/or poorly tolerated (Michka *et al.*, 2009).

15.3 The Pharmacological Action of Cannabis

Cannabis contains several hundred components, of which approximately 60 are derived from the cannabinoids. Among this chemical family, tetrahydrocannabinol (THC) is the main psychoactive principal and is responsible for the popularity of the plant among people seeking to exploit it as a drug for hedonistic purposes. This family includes all the substances similar to THC, whether they are obtained from cannabis or synthesized chemically. Considerable progress has been made since the early 1990s in furthering our understanding of the various effects of cannabinoids on living organisms and the many, as well as fundamental, roles of the endocannabinoid system.

15.4 Cannabinoid Receptors

Both naturally occurring and synthetic cannabinoids exert their pharmacological effect by binding to specific cellular receptors. These receptors are termed cannabinoid receptors. Their existence was demonstrated in 1990 with the isolation of receptor type CB1 by the American pharmacologist, Lisa A. Matsuda (Richard, 2009). Two types of cannabinoid receptor have since been identified.

1. The CB1 receptor, isolated from a rat brain, is expressed in both the central and peripheral nervous systems, in testicular and uterine tissue, within the immune system, the intestine, bladder, retinal cells, endothelial cells (blood vessels) and in adipocytes. It is one of the most common neuronal receptors found in the brain. The location of these receptors correlates well with the behavioural effects (action on memory, action on sensory perception and on motor function) produced by natural and synthetic cannabinoid agonists such as levonantradol, nabilone, CP55940, etc. The CB1 receptor is not expressed within the brain stem, where the cardiovascular and respiratory control centres are located. This probably explains the fact that cannabinoid derivatives do not demonstrate any acute somatic toxicity.

2. The CB2 receptor, isolated in 1993, is expressed in peripheral tissues and, in particular, in cells of the immune system.

15.5 The Endocannabinoids

The existence of endogenous physiological ligands capable of binding to cannabinoid

receptors was established in the early 1990s (Richard, 2009). These substances are termed endocannabinoids (eCB) and are produced by the organism and can act like cannabinoids. The principal eCB are anandamide (AEA) and 2-arachidonoglycerol (2-AG). These are lipid in nature and are eliminated rapidly by the organism following their breakdown, a process catalysed by the enzyme, fatty acid aminohydrolase (FAAH). They are released in a range of tissues, in particular the central nervous system, where they modulate neurotransmission. Levels of anandamide in the brain are comparable to those of other neurotransmitters such as dopamine and serotonin. This substance is particularly produced in areas of the brain that show a strong expression of CB1 receptors (cortex, hippocampus, striatum and cerebellum) and shows a stronger affinity for this receptor than for the CB2 receptor.

15.6 Cannabis and Cannabinoids in Therapeutics

Many patients in the Anglo-Saxon countries, as well as in Holland and in Germany, provide testimony in favour of the use of cannabis to treat certain pathologies (medical problems). The claimed indications include diseases that are liable to benefit from other types of treatment and include nausea, vomiting, asthma, glaucoma, epilepsy, anorexia in AIDS patients and the treatment of certain painful conditions (phantom limb pain, rheumatic pain, etc.) (Grinspoon and Bakalar, 1993; BMA, 1997; Rosenthal et al., 1997; EMCDDA, 2008, Vol 1). That said, it is synthetic cannabinoids rather than the plant itself that have been the subject of clinical investigations and will continue to be researched (EMCDDA, 2008, Vol 1; Hosking and Zajicek, 2008; Hanus, 2009).

15.6.1 Cannabis and natural cannabinoids as a therapeutic agent

There are two forms of natural cannabinoids that can be used in therapeutics. In the first instance, the plant or its immediate derivative (hashish) can be used, or alternatively, titrated extracts of the plant can be used.

The plant

Those who extol the virtues of the plant and its derivatives (in particular hashish, as well as food containing cannabis, such as cakes, sweets and milk-based beverages) emphasize the following (Richard *et al.*, 2004):

- Cannabis is very well tolerated.
- The ecological/natural side of this treatment.
- The effect of the plant can be 'dosed' by adapting the rhythm of inhalation.
- The restrictive nature of legislation (Caballero and Bisiou, 2000).

Detractors of the use of cannabis in this manner emphasize:

- The lack of scientific studies to support the innocuous nature of cannabis and its derivatives relative to the number of studies that highlight its somatic toxicity (cancer) and psychic toxicity (apathetic syndrome, psychotic symptoms, etc. ...), as reviewed by Richard (2009).
- The availability of licensed industrially produced medicines for the various indications specified for cannabis.
- The existence of legal medicines based on THC (dronabinol – see below) or synthetic cannabinoids (dronabinol, nabilone, etc.).
- The impossibility of obtaining reproducible results from the use of a plant rich in pharmacologically active products, with variable levels and in varying proportions.
- The dangers presented by the concurrent use of tobacco and/or the inhalation of various combustion products (e.g. tars) from the plant when consumed in the form of cigarettes or smoked in a pipe.
- The potential for confusion, both within society and legally, that is likely to arise if a therapeutic form of cannabis is available at the same time as its drug form.

Many patients wishing to self-medicate with cannabis will choose to grow the plant themselves, either indoors, sometimes hydroponically, or outdoors. They can then smoke the plant directly or make their own hashish, using specially conceived systems commercially available in countries such as the Netherlands and Switzerland (e.g. the Pollinator® system).

Plant extracts

Besides using the plant itself, it is possible to use a prescription drug called Sativex® in those countries where its use is licensed. This product contains cannabis plant extracts that have been selected for their reliably high THC and cannabidiol (CBD) concentration (Hosking and Zajicek, 2008). CBD binds to CB1 receptors and limits the psychotropic activity of the drug. It is administered orally and is presented as a spray formulation. Each 100 µl spray delivers 2.7 mg of THC and 2.5 mg of CBD. It is commercialized jointly by a British and a German pharmaceutical company. It is available in Canada and is used for patients suffering from severe neurological problems.

15.6.2 Synthetic cannabinoids as therapeutic agents

The pharmaceutical industry synthesized and developed several cannabinoids during the 1970s.

Levonantradol is an isomer of nantradol. It is a synthetic cannabinoid that can be administered by intramuscular injection. It is stronger than THC and, given the incidence of side effects on the nervous system, it has not been commercialized.

Nabilone (Cesamet©) is a synthetic derivative of THC that can be administered orally. It is available as 1 mg capsules and can be used to treat vomiting and nausea. It is also effective against neurological pain (Davis, 2008).

Dronabinol (Marinol©) is none other than THC itself, obtained synthetically rather than being extracted from the plant. It is available in North America in 2.5 mg, 5 mg and 10 mg capsules that can be administered orally. It is used in the treatment of nausea and vomiting and can stimulate the appetite of cachectic AIDS patients.

Studies have suggested that CBD, the main non-psychoactive component of cannabis resin, has the potential to be used as a sedative, anticonvulsant, antipsychotic, anti-inflammatory and neuroprotective drug without provoking the side effects associated with THC (Scuderi *et al.*, 2008; Iuvone *et al.*, 2009). It is not available commercially, however.

Other synthetic cannabinoids are currently the subject of experimental studies. These include menabitan, pirnabine and naboctate. Rimonabant, a cannabinoid receptor antagonist, was commercialized as an appetite suppressant and for the treatment of obesity, but was withdrawn from the market in 2008 after it was found to have provoked serious psychological side effects, including depression and attempted suicide.

A review of 31 studies on cannabinoids published between 1966 and 2007 shows that their use in the short term has not resulted in any loss of life, serious problems, hospitalization or serious disability. In 96.6% of published cases, the iatrogenic effects were described as central or peripheral nervous system disturbances (loss of balance, dizziness and paraesthesia) (Wang *et al.*, 2008). The studies conducted do not, however, allow us to confirm the long-term innocuity of such treatments (Wang *et al.*, 2008). For further information, see Degenhardt and Hall (2008).

15.7 Various Controversial Medical Indications

The range of recommended indications for cannabinoids is very broad. There have, however, been very few rigorous scientific trials to evaluate these uses. Many publications deal only with a small cohort of patients or report on individual cases. The status and reputation of cannabis does not lend itself well to further studies, even if public opinion appears to the use of this plant and its derivatives in medicine.

15.7.1 Prevention and treatment of nausea and vomiting

Various cannabinoids exert a significant anti-emetic effect and anti-nausea effect in patients receiving chemotherapy or radiotherapy. As is often the case, the opinion of various authors on this subject varies:

- The synthetic cannabinoids that have been trialled clinically (nabilone, dronabinol and levonantradol) have not shown themselves

to be any more effective than standard treatments (antipsychotic drugs used as anti-emetics, the sulpiride for example) and have a poor therapeutic index according to a review of the literature up to 2000 (Tramer *et al.*, 2001).

- A recent literature review, however, suggests that the therapeutic index is, in fact, superior to that of antipsychotic drugs used for this indication (Machado Rocha *et al.*, 2008).

It is important that we put into context the potential value of cannabinoids in the treatment of chemotherapy-induced and radiotherapy-induced nausea and vomiting following the commercialization in the early 1990s of a group of highly effective anti-emetics (5HT3 serotonin antagonists called setrons) (Michalon, 2005). Cannabinoids can potentiate the effect of these anti-emetics and therefore increase their effectiveness (Michalon, 2005).

The use of cannabis can limit the nausea induced by certain antiretroviral treatments and can therefore improve treatment compliance (De Jong *et al.*, 2005).

15.7.2 Treatment of pain

Cannabinoids possess a well-recognized central analgesic effect (Manzanares *et al.*, 2006). This is probably linked to the inhibition of GABAergic conduction associated with their interaction with CB1 receptors (Hosking and Zajicek, 2008). This action is similar to that of codeine and therefore inferior to that of morphine (Campbell *et al.*, 2001). Cannabinoids can act to depress the central nervous system and this effect can limit their worth (Campbell *et al.*, 2001; Michalon, 2005).

It may, however, be worth investigating further the value of cannabinoids and cannabis in the treatment of certain pain and, in particular, pain associated with deafferentation (Manzanares *et al.*, 2006; Hosking and Zajicek, 2008; Martin Fontelles and Goicoechea Garcia, 2008). If trials on patients suffering from cancers or multiple sclerosis have demonstrated a reduction in pain associated with the disease following the administration of cannabis extracts (Sativex©), it is possible that synthetic cannabinoids may show themselves to be more effective (Radbruch and Elsner, 2005). Nabilone has been shown to be effective in the treatment of neuropathic pain, despite showing little activity against acute pain (Davis, 2008).

15.7.3   Treatment of neurological pathology

It is probably in neurology that cannabinoids are likely to prove most useful, for the endo-cannabinoid system is implicated heavily in the functioning of the brain, as well as in the inter-connections between neuronal pathways (Ramos *et al.*, 2005). Only a small number of synthetic cannabinoids have been subjected to extensive trials as treatments for Parkinson's disease, amyotrophic lateral sclerosis, trigeminal neuropathies (Liang *et al.*, 2004) and Alzheimer's disease (Ramirez *et al.*, 2005). Others are licensed already for certain limited indications.

Multiple sclerosis

A number of clinical publications emphasize the interest of cannabinoids in the treatment of multiple sclerosis (Brady *et al.*, 2004; Croxford and Miller, 2004; Vaney *et al.*, 2004; Correa *et al.*, 2005; Aragona *et al.*, 2008). They point to their efficacy in the treatment of spasticity, muscular rigidity, trembling and pain (associated with this condition). Other reports are less favourable, and some are even unfavourable (Killestein *et al.*, 2002; Zajicek *et al.*, 2003; Killestein and Polman, 2004; Zajicek, 2004) and do not underestimate the undesirable psychological side effects (Aragona *et al.*, 2008).

Sativex© is indicated officially in the treatment of spasticity problems in patients suffering from multiple sclerosis (Smith, 2004).

Parkinson's disease and other neurodegenerative diseases of the central nervous system

Cannabinoids exert a neuroprotective effect that may be of use in the treatment of neuro-degenerative diseases such as Alzheimer's disease, Huntington's chorea and Parkinson's disease (Michalon, 2005). CBD does not have psychoactive effects and may therefore be worth investigating further as a potential treatment (Scuderi *et al.*, 2008; Iuvone *et al.*, 2009).

The value of administering cannabis as a treatment for dyskinesia in patients suffering from Parkinson's remains controversial. It may be indicated where the dyskinesia is iatrogenic, arising from the use of levodopa (Sieradzan *et al.*, 2001), but it is also sometimes unhelpful where the motor problems are a product of the condition's evolution (Carroll *et al.*, 2004).

Epilepsy

The theoretical value of cannabinoids in the treatment and prevention of epileptic crises has given rise to few in-depth studies (BMA, 1997; Michalon, 2005), despite patients advocating the use of cannabis in this context (Gross *et al.*, 2004).

Treatment of anorexia

THC exerts a powerful appetite stimulation effect and an increase in appetite has been reported in cannabis smokers since the 1960/70s (BMA, 1997). This effect favours sugary foods. Tolerance is seen and the effect appears to disappear or even reverse itself where cannabis use is sustained (BMA, 1997). While cannabinoids are ineffective in the treatment of eating disorders (BMA, 1997), they may be of use in patients whose anorexia is due to cancer or advanced-stage infection with HIV. The benefits in such cases arise through the drug's action on gut motility, its use as an anti-emetic and its anxiolytic potential at low doses (Martin and Wiley, 2004).

15.7.4   Other potential therapeutic uses

The many recommendations and indications for cannabis use (made by supporters of its use) are often based more on anecdote than on validated medical research. The following examples illustrate some of the claims put forward by advocates of its use.

Glaucoma

Proposed by advocates of the therapeutic use of cannabis since the early 1970s (BMA, 1997), the value of this plant and of synthetic cannabinoids in ophthalmology remains very limited (Watson *et al.*, 2000). Only THC and 11-hydroxy-THC appear capable of lowering intraocular pressure (BMA, 1997). At present, there is no evidence to support an indication for cannabis use in the treatment of this ocular disorder.

By contrast, the neuroprotective action of cannabinoids, together with their pharmacological action, makes them good candidates for the treatment of retinal pathologies (Yazulla, 2008).

Asthma

THC is a powerful bronchodilator and does not produce its action by the same mechanisms used by other commercially available anti-asthmatic drugs (BMA, 1997). That said, despite the many asthmatic patients claiming that cannabis consumption has a favourable effect on their asthma (using vaporizers), few studies have been published to support this (BMA, 1997).

Inflammatory syndromes

Cannabinoids exert protective effects vis-à-vis cellular inflammation, particularly within the central nervous system (Correa *et al.*, 2005; Klein, 2005). CBD modulates the balance between pro- and anti- inflammatory prostaglandins and can exert an anti-inflammatory effect (Sacerdote *et al.*, 2005). These potential benefits warrant further study, even though the therapeutic arsenal in this area is extensive.

Cardiology

The cannabinoid system plays an important role in the development of atherosclerosis and its clinical signs. Low doses of THC (1 mg/kg/day – lower to those inhaled by a cannabis user) can reduce the progression of atherosclerosis in mice. It is proposed that this is mediated by an anti-inflammatory action arising from interaction with CB2 receptors at the time of plaque formation. The complex and often contradictory interactions between CB1 and CB2 receptors, platelet function, arterial pressure and heart rate suggest that pharmacological targeting of the endocannabinoid system may be of some therapeutic value in the treatment of atherosclerosis (Mach *et al.*, 2008). The development of synthetic cannabinoids that do not cross the blood brain barrier, and do not therefore act on the central nervous system, may lead to their use in cardiology (Ribuot *et al.*, 2005; Mendizabal and Adler-Graschinsky, 2007).

Gut motility problems

A small number of studies suggest that cannabinoids may have beneficial effects on various gut motility problems, including functional colopathies, diarrhoeas, etc. (Aviello *et al.*, 2008).

Hepatic fibrosis

Preliminary observations suggest that modulation of the endocannabinoid system may prove to be a suitable target for drug therapy in the treatment of hepatic fibrosis (Teixeira-Clerc *et al.*, 2008).

Cancerology

Certain publications outline the value of cannabinoid treatments in cancerology (e.g. Sarfaraz *et al.*, 2005; Widmer *et al.*, 2008).

Cannabis dependency

Synthetic cannabinoids exerting an action antagonistic to that of THC or cannabinoids permitting a therapeutic substitution may be used to treat cannabis dependency (Clapper *et al.*, 2009).

15.8 The Legal Situation: Where Confusion Reigns

The status of cannabis and its direct derivatives, as well as the status of synthetic cannabinoids, remains both confusing and unequal: certain national governments have chosen to prohibit clinical trials of cannabis and the therapeutic use of synthetic cannabinoids. Other national governments facilitate, or hope to facilitate, and promote patient access to cannabis for medical purposes. Such access may be through its sale by chemists, but may also be through providing patients with the possibility of cultivating the plant themselves. This last position is in direct opposition to international law.

15.8.1 France

Cesamet© and Marinol© may be prescribed provided the onerous and dissuasive administrative formalities are respected. A lessening of these procedures seeking to promote the compassionate use of cannabis and its derivatives together with the prescription of synthetic cannabinoids, as promised in 2001 by the then Health Minister, Bernard Kouchner, has not been enacted. The prescription or use of cannabis for therapeutic purposes remains illegal.

15.8.2 Other countries

The situation is very different in many other countries. In Canada, the UK, Australia, Holland, Belgium and Switzerland, as well as in 35 US states, the prescription of cannabis is possible, although strictly controlled. The following serve as examples.

Since 2003, certain chemist shops in Holland have been authorized to supply cannabis when presented with a doctor's prescription (Engels *et al.*, 2007). In 2008, the Dutch Supreme Court went further when ruling on the case of a patient suffering from multiple sclerosis who was unable to find the exact form of cannabis needed to control his pain. Their ruling now makes it possible for individuals, in Holland, to cultivate cannabis for medical purposes.

In the USA, only the University of Mississippi is authorized by the federal government to produce cannabis for the purposes of clinical research. Legal proposals aimed at clarifying policy to protect from prosecution individuals who use cannabis therapy outside official trials (on prescription by their doctor,

for example) are repeatedly presented to the US Senate. US federal law forbids the cultivation and sale of cannabis, considering it to be a narcotic. However, 13 states have adopted laws that legalize the use of cannabis for therapeutic purposes. The compassionate use of the plant was authorized in California in 1996. Other states, including Colorado and Alaska, have decriminalized in part the possession of small quantities of this drug. Many associations (Cannabis Buyers' Clubs) campaign for the drug to be made available to patients and, in particular, those suffering from AIDS, for whom the drug represents an important line of therapy. Three dispensing machines installed in Los Angeles, where they are used by holders of a specific card that allows the collection of up to 1 oz of 'grass'/patient/week, were adjudged by the International Narcotics Control Board (INCB), in early 2008, to contravene the international treaties controlling access to narcotics.

In 2001, Canada was the first country to codify the medical use of cannabis (Elkhasef *et al.*, 2008). It is hardly surprising, therefore, that Canada boasts more cannabinoid-based medicines than any other country. Other than the plant, which is itself licensed, Sativex©, Marinol© and Cesamet© are all sold in pharmacies. That said, in 2008, the founder of the Grant W. Krieger Cannabis Research Foundation, an advocate of the therapeutic use of cannabis and victim of multiple sclerosis, was convicted of trafficking drugs to a number of clients suffering from chronic pain or in the terminal stages of their condition and keen to improve their quality of life.

A consequence of the prohibition of medical cannabis in a number of countries has been the rise in the number of cases of employees using cannabis for medical purposes being detected during the course of drug-testing programmes. Public opinion, however, would tend to support the use of cannabis for medical purposes. Thus, 72% of American adults would support the legalization of cannabis therapy, while 70% of the population in France surveyed a few years earlier, thought the use of cannabis for medical purposes was justified.

16 Hemp Seeds for Nutrition

Gero Leson

Leson and Associates, Berkeley, California, USA

16.1 Introduction

In several Western countries, hemp seeds and oil are gradually making a comeback as ingredients in food and cosmetics products. The best example is North America, where the recent steady increase in Canadian hemp acreage is driven almost exclusively by demand from the US market for 'natural foods'. But also in the UK and Germany, hemp foods, that is, any food products containing hemp seeds or oil, are beginning to appear in stores and receive press coverage. In the 1990s, much of the coverage of hemp foods was driven by hype or, in the USA, the issue of contamination by trace amounts of THC, the major psychoactive ingredient of marijuana. Nowadays, the potential health benefits and taste of hemp foods have become an important buying consideration. This chapter reviews the drivers for the recent expansion of the hemp food market and discusses the opportunities and challenges it offers to the global hemp industry.

While the use of hemp as a fibre source in, for example, ancient China is well documented, less is known about the use of hemp seeds as a food source. However, European sources and archaeological finds suggest that hemp seeds were used for both food and medicinal applications by the first centuries AD. Despite the lack of written records, it is most likely that in European and Asian countries, wherever hemp

was grown for fibre, the by-product seeds were also always used as a food ingredient. An indication is the still common traditional use of roasted hemp seeds as snacks in China and Turkey. In recent centuries, hemp seeds were an ingredient in traditional recipes in rural Central and Eastern Europe. Examples include a 'hemp soup' described in a German recipe and a hemp butter used traditionally in Baltic countries. Historic sources from all over Europe describe medicinal applications of hemp seeds to treat symptoms such as stomach or ear pain, coughing and incontinence. However, it is not clear whether these benefits are derived from the nutrients provided by the seeds or from the cannabinoids, including THC, which may have been present and incorporated into the formulation. Since fibre quality was usually best for textiles before the seeds were fully mature, hemp seeds were often not harvested and probably never a major food source, only a welcome addition to the often monotonous make-up of rural cooking. Hemp seeds were used extensively as bird food and the oil as a drying oil, with properties similar to those of linseed oil.

As the farming and use of hemp for fibre declined in Europe even long before World War II, the use of hemp seeds for food also disappeared gradually. Figure 16.1 shows that, in recent decades, China has been producing the vast majority of the world's hemp seed

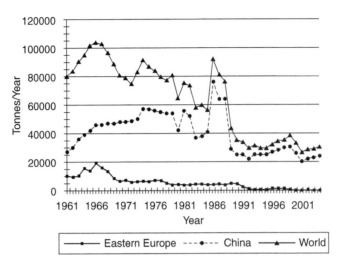

Fig. 16.1. World production of hemp seeds.
Source: http://faostat.fao.org.

crop, mostly for bird food. Much smaller quantities were produced in France, for example, as a by-product of hemp fibre for pulp and used for fish and bird food.

16.2   Revival of Hemp Foods

This situation changed as Europeans and North Americans rediscovered hemp stalks in the early 1990s as a versatile source of fibre and hurds. Subsequently, more attention was also paid to the nutritional composition of hemp seeds, particularly of their fatty acid spectrum, which was found to have a unique and possibly health-promoting balance of omega-3 and omega-6 fatty acids. Because of the 'hemp hype' in the mid-1990s and these early findings on potential health benefits, whole hemp seeds and cold-pressed hemp oil items became available in North America and in Germany and the UK. Initially, seeds and oil came from China and Eastern Europe. Because of their high content of triple-unsaturated fatty acids, hemp seeds and oil are very sensitive to oxidation, that is, their fat becomes rancid if the seeds are not stored and processed properly. Particularly in China, quality control for birdseed was not a high priority. Consequently, seeds and oil were often of poor quality, had a

short shelf life, unpleasant taste and were not well accepted by consumers. In addition, many of the varieties grown in China and even in the EU contained sufficiently high levels of THC to produce measurable amounts of the drug in the seeds and oil. This created concern with the authorities about inadvertent exposure of consumers to THC, the resulting health impacts and the potential for producing false positive results in workplace drug tests for marijuana, which are common in the USA.

    This situation has changed since commercial hemp farming became legal again in Canada and as European hemp producers increasingly began harvesting and utilizing hemp seeds as a food source and for use in cosmetics. The following gives a brief overview of developments in North America and Europe, the two major emerging markets for hemp foods and cosmetics.

16.2.1   North America

The 1998 legalization of commercial hemp farming in Canada has affected global trends strongly in the production and use of hemp. While Canadian hemp acreage in 1998–2000 was driven largely by speculation, since 2001, the total hemp farming area has increased consistently from 1300 ha to about 6000 ha in 2005. Figure 16.2 shows the Canadian hemp acreage since 1998. Unlike in the EU, where hemp acreage is driven largely by fibre production, the vast majority of Canadian hemp is grown to supply the increasing demand for hemp foods in the US market for natural foods and body care products. A recent survey indicated that retail sales of hemp foods in the USA grew by over 60% compared to 2003. Total US retail sales are now estimated at US$12 million/year for foods and US$50 million for cosmetics. This volume is still small, but its consistent upward trend suggests that demand for hemp seeds ultimately may become as important a driver for global hemp demand as the fibre. The growth in North American demand for hemp foods is driven by several factors.

Product diversity

Since 1997, hemp processors and food manufacturers in Canada and the USA have developed a wide range of hemp seed derivatives, finished food and cosmetics products containing hemp seeds and oil. In particular, the tasty and versatile hulled hemp seeds – also called hemp nuts – are increasingly sold in bulk to bakeries and manufacturers of nutrition bars, and packaged directly to consumers. The defatted and milled seedcake is now sold as protein-rich flour – at a good profit. Table 16.1 lists the most common hemp food materials and products available in North America.

Improved product quality

The Canadian Prairies, where large areas of flax and rapeseed are grown, offer a good infrastructure for the cleaning and processing of oilseeds. Thus, Canada now produces hemp seeds of high quality and low THC content necessary to meet the demands of US consumers. In particular, the shelf life and the flavour of these products are much improved over those offered a decade ago.

Increased visibility of hemp foods through court battle

Irritated by the growing popularity of seeds from a plant which looks like marijuana and still cannot be grown legally in the USA, the

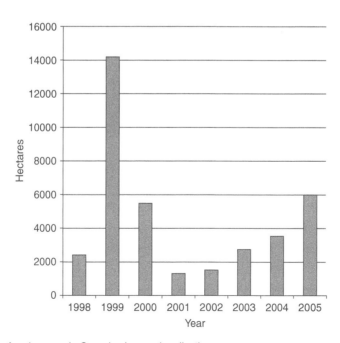

Fig. 16.2. Hemp farming area in Canada since re-legalization.

Table 16.1. Hemp food raw materials and products.

Raw materials:
   Whole seeds (raw, toasted)
   Cold-pressed hemp oil
   Hulled seeds (nuts)
   Defatted seedcake – protein flour
Food products:
   Cereals (muesli)
   Snack/nutrition bars
   Bread, cookies, pretzels, chips
   Nut butter, pasta
   Salad dressings
   Ice cream
Body care products:
   Soap, cream, lip balm, shampoo

federal Drug Enforcement Administration (DEA) passed a rule in 2001 which would have banned the sale of all food items containing 'any amount of THC'. Because all hemp seeds contain measurable trace amounts of THC, this would have outlawed all hemp foods and destroyed the fledgling North American hemp industry. However, the industry successfully challenged the DEA in federal court and, based on legal and technical grounds, ultimately won in 2004. This court battle generated much media attention and awareness of hemp foods by US consumers.

Health attributes

This publicity also drew attention to the nutritional composition of hemp seeds and oil, which appears to offer potential health benefits and is in line with several emerging trends in nutritional science. The strong US retail network for 'natural' foods, including several chains of supermarkets, has been a natural ally in bringing hemp foods to retail customers.

By developing the domestic food market, the North American hemp industry followed a uniquely different route, from which Europeans may benefit. Conversely, there are signs that the hemp fibre grown in the Canadian Prairies, and which is now either ploughed under or burned, will increasingly find technical uses. In its growth, the North American hemp foods industry now faces two main challenges: (i) supplying the fast-growing demand for seeds while maintaining quality; and (ii) making sure

that demand continues to expand by educating customers about the culinary and potential health benefits of hemp foods.

16.2.2 EU

In the EU, the demand for hemp foods has grown much more slowly than in North America. This is despite the fact that whole seeds and oil have been available to consumers, particularly in Germany, since 1995 and that hemp has been legal to grow in most member countries since 1997. As in North America, the inconsistent and often poor quality of early hemp products, combined with the hype in the 1990s, likely made many consumers sceptical. Furthermore, many of the national European food markets are more traditional and thus slower to respond to novelty. Unlike in the USA and Canada, distribution to supermarket customers has started only recently in the UK. In addition, while European governments have generally been reasonable about regulating THC levels in hemp foods, the lack of a high-profile court battle, as in the USA, has, ironically, also created less visibility for hemp foods.

However, since 2003, hemp food companies in Germany and the UK have experienced growing demand for hemp seed oil and snacks containing hemp seeds or nuts. Moreover, as in North America, the drivers increasingly are the proclaimed health benefits and taste of hemp foods.

16.3 Culinary Uses of Hemp Seeds and Oil

Even in North America, hemp foods are still in their infancy and the range of culinary uses has not been fully exploited. The basic raw material is, of course, *whole hemp seed*. Because of their crunchy shell, seeds have never become very popular with food manufacturers and consumers. Some roasted whole hemp seeds are used as snacks, salted, with a caramel coat, or as a minor ingredient in energy bars. Several US firms also sell coffee blends containing a portion of roasted seeds. The main use of whole seeds is as the raw material for cold-pressed hemp oil, hulled seeds (often called hemp nuts) and the defatted seed cake or meal.

16.3.1   Hemp oil

Hemp oil, when pressed from mature, undamaged seeds, has a delicious nutty flavour. Depending on the variety, it may have a slight grassy or bitter component to it. Hemp oil can be used for almost anything for which olive oil is used. Because of its high content of triple unsaturated fatty acids, it does not tolerate longer exposure to temperatures above 160°C and develops off-flavours and possibly toxic by-products. Thus, hemp oil should not be used for frying, while light sautéing of vegetables and other moisture-containing foods is acceptable. Currently, the main *finished* product made with hemp oil is salad dressing. The oil is also by far the main hemp ingredient in a wide range of cosmetics products: liquid and bar soaps, creams, lotions, lip balm and others.

16.3.2   Hemp nuts

Hemp nuts are emerging as the dominant hemp seed derivative in North America. They are slightly larger than sesame seeds and, if hulled properly, contain little shell. If eaten raw, their taste resembles that of sunflower seeds, but a *very* light roast in a pan brings out their full flavour potential. Hemp nuts can replace other nuts in any recipe, except where larger chunks are needed. They go well in soups and salads, in the sauces for meat and tofu dishes, over vegetables, in desserts, baked into bread and pastries. Major finished products made with hemp nuts are bread and pastries, breakfast cereals, a large number of snack or energy bars and hemp-nut butter.

The defatted seedcake was considered initially a by-product and used for animal feed and as flour in speciality breads and pastries. In 2003, several Canadian manufacturers started selling the flour as 'protein powder' or 'protein flour'. Depending on how well the larger hull pieces are screened out after crushing, the flour may contain up to 50% protein

by weight. Since protein powders are widely used in the USA by athletes and the general population, hemp flour did not have to create a new product category. Even though the protein content is low compared to whey or soy protein concentrates or isolates, the dietary fibre and small amounts of hemp oil make this product an attractive nutritional package – and one of the best-selling hemp foods in North America.

16.4   Composition and Nutritional Attributes of Hemp Seeds

Without doubt, much less research has been conducted on the nutrient composition of hemp seeds and oil compared to other common oilseeds. However, the limited data now available indicate that hemp seeds in fact have a rather unique composition and their consumption may offer considerable health benefits.

16.4.1   Fatty acid spectrum

Best researched of all nutritional properties of hemp seed is the fatty acid composition. Relative dominance of fatty acids will vary with hemp variety and growing conditions, yet the comparison with other nuts and oilseeds shown in Fig. 16.3 is typical. Hemp oil contains the two *essential fatty acids* (EFAs), linoleic acid and alpha linolenic acid, the origins of the omega-3 and omega-6 fatty acid families, in a desirable ratio of about 3:1. EFAs cannot be produced by our body and, like vitamins, they or their metabolites must be consumed in the diet.

In addition to these EFAs, hemp oil contains gamma linolenic acid (GLA) and stearidonic acid (SDA), two nutritionally relevant 'higher' omega-6 and omega-3 fatty acids, respectively. What is so significant about the omega-3 to omega-6 ratio? Mounting scientific evidence links many common ailments in Western societies to an imbalance in omega-3 versus omega-6 fatty acids in the typical Western diet. Put simply, we eat *too much omega-6* and *not enough omega-3*. The typical North American diet contains 10–30 times

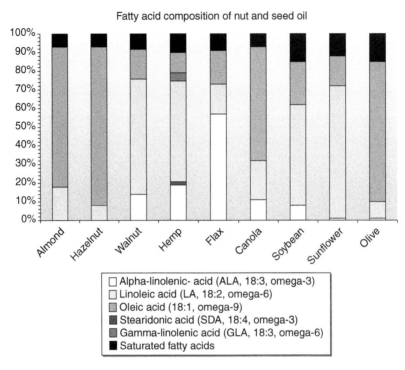

Fig. 16.3. Fatty acid composition of nut and seed oils.
Note: Fatty acid spectrum of hemp is for the Finola variety.
Source: Leson and Associates, 2005.

more omega-6 than omega-3 fatty acids, yet nutritional scientists recommend maintaining a ratio of between 2:1 and 4:1; that is, a much higher *relative* omega-3 consumption.

Many clinical studies implicate this imbalance as a key factor in the rising rate of inflammatory disorders and have demonstrated the benefits of a balanced dietary omega-3/omega-6 intake. Such benefits include a reduced risk of atherosclerosis, sudden cardiac death and some forms of cancer, alleviation of the symptoms of rheumatoid arthritis, mood improvement in bipolar disorders and optimized development in infants. It is this proven need to increase relative omega-3 intake that ultimately drives public attention and growing demand for plant- and animal-based sources of omega-3 fatty acids.

As Fig. 16.3 shows, most common cooking oils contain significant amounts of linoleic acid, the basic omega-6 fatty acid, but little if any omega-3s. The reason is not malignance or ignorance on the part of food companies,

rather convenience. As a triple-unsaturated fatty acid, the main plant-derived omega-3 fatty acid, ALA (alpha-linolenic acid), oxidizes, or turns rancid, rapidly, something neither producers nor consumers like. A high ALA content also limits the versatility of food oil, as it cannot be used for frying. To improve the stability of cooking oils, industry has, over the last decades, moved to the use of oils with lower omega-3 content. This was done through plant breeding or hydrogenation (hardening) of the omega-3 present, for example, in soybean oil. That process is still the main source of trans fatty acids in our diet, increasingly a suspect contributor to cardiovascular disease.

Figure 16.4 visualizes the metabolic pathways for the omega-3 and omega-6 families. Part of the EFAs are converted to longer and more unsaturated fatty acids and ultimately to eicosanoids, a series of potent hormone-like substances, such as the well-known prostaglandins, which control the process of inflammation, fever and pain, reduce blood pressure and

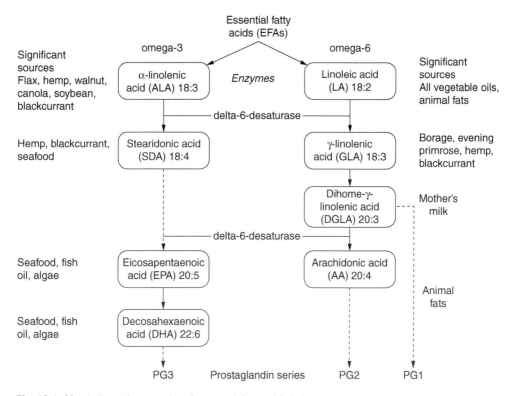

Fig. 16.4. Metabolism of omega-3 and omega-6 fatty acids in humans.

affect the coagulation of thrombocytes. A severe shortage in omega-3 will shift the balance in some functions controlled by these prostaglandins – with the above listed health consequences. In addition, the omega-3 fatty acid, DHA, is abundant in brain tissue and required for brain development and special neural functions.

Not all omega-3s are equally 'potent'. Clinical studies show that even healthy people convert only a fraction of ingested ALA to the ultimately needed omega-3s, EPA and DHA, and their respective prostaglandins. The 'metabolic bottleneck' is the conversion of ALA to SDA by the delta-6-desaturase enzyme, with an efficiency of only 20%. This enzyme's effectiveness in producing 'higher omega-3s' is inhibited further in older people and by diabetes, obesity, excessive omega-6 intake and elevated levels of insulin, coffee, trans fatty acids and alcohol. These common conditions aggravate omega-3 deficiency symptoms further, while direct intake of SDA, DHA and EPA

would alleviate them. Unfortunately, only marine organisms contain DHA and EPA. These fatty acids are commonly consumed with fish or fish oil supplements. The predominant omega-3 fatty acid supplied by plants is ALA, while only a few seed oils, among them hemp as the only food oil, also offer SDA, an omega-3 fatty acid with a 'potency' about 5 times that of ALA.

To summarize the nutritional attributes of hemp oil: hemp oil offers the two EFAs in a very desirable LA to ALA ratio of about 3:1 and contains both SDA and GLA in nutritionally relevant amounts, thus making up for a potentially impaired fatty acid metabolism. According to the current scientific thinking about fat nutrition, no other plant-based oil offers such desirable composition. In addition, cold-pressed unrefined hemp oil adds flavour to many dishes – and some people even enjoy taking an occasional tablespoon of the oil, both for health and for taste. Cold-pressed oil also contains significant amounts of gamma-tocopherol, a member of the vitamin E group, whose importance compared to the long preferred alpha-tocopherol has only recently been recognized. Finally, hemp oil contains relatively high concentrations of phytosterols, compounds known to reduce total and LDL cholesterol.

16.4.2 Protein, vitamins, minerals and other micronutrients

We know much less about the composition and content of other nutrients in hemp seed and oil, particularly vitamins, minerals and other micronutrients. The scattered data do not yet provide a comprehensive picture of the level of some of these nutrients and their variability as a function of variety and growing conditions. The following sections summarize what we know today.

Macronutrients

Figures 16.5 and 16.6 show the typical composition of whole and hulled hemp seeds. Most carbohydrates are present in the hull, that is, non-digestible dietary fibre, rather than sugar or starch. The fat or oil content in whole seeds may vary between 30 and 34%.

Protein

Hemp protein contains all nine essential amino acids in a reasonably well-balanced ratio. As is common with other vegetable proteins, except for soya protein, hemp protein contains a lower proportion of the essential amino acids,

leucine and lysine, than meat, egg white and soya. In contrast, it has a higher content of arginine, an amino acid, which boosts the production of nitric oxide, a compound that relaxes blood vessels, and thus may help treat angina and other cardiovascular problems. Thus, one would not want to rely on hemp seeds as the only source of protein; rather, complement it with meat, dairy or tofu.

Minerals and vitamins

Table 16.2 summarizes typical concentrations of minerals and vitamins in whole and hulled hemp seeds. It also shows the percentage of the recommended daily intake of key nutrients supplied by 30 g of whole or hulled seed. Compared to their energy content of about 8% of typical daily intake, hemp seeds and nuts provide a much higher fraction of the daily needs of phosphorus, potassium, magnesium, manganese and several B vitamins. This makes hemp seeds *nutrient dense* and a good, or even very good, source of several of these nutrients. Particularly interesting is their comparatively high magnesium content, as magnesium is deficient in the diet of many Westerners and is also not sufficiently available in commonly taken multivitamin/mineral pills. As with other nuts, hemp seeds are not a good source of Vitamins A and D. Table 16.2 indicates that hemp oil is not a particularly good source of vitamin E. But this conclusion derives from the calculation method by which tocopherols, that is, the members of the vitamin E complex, are converted into vitamin E equivalents. Hemp oil

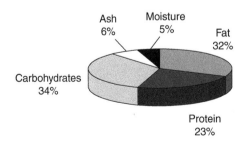

Fig. 16.5. Typical composition of whole hemp seed.

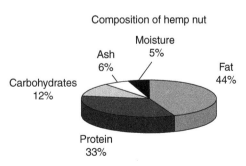

Fig. 16.6. Typical composition of hulled hemp seeds (nuts).

Table 16.2. Concentrations of vitamins and minerals in hemp seeds and nuts.

		Whole seeds		Hulled seeds (nuts)	
	RDI (mg/day)	mg/100 g	30 g seeds = RDI (%)	mg/100 g	30 g nuts = RDI (%)
Energy	2000 kCal	5.0 kCal/g	150 kCal (7.5%)	5.6 kCal/g	170 kCal (8.5%)
Phosphorus	1000	1100	33	1600	48
Potassium	2000	900	14	1100	17
Magnesium	400	450	34	670	50
Calcium	1000	150	5	80	2
Iron	18	12	20	11	18
Manganese	5	10	60	10	60
Zinc	15	7	14	11	22
Copper	2	1	15	0.5	8
Thiamine (B1)	1.5	1.3	26	1.3	26
Riboflavin (B2)	1.7	1.2	21	1.2	21
Pyroxidine (B6)	2	0.5	8	0.5	8
Vitamin C	60	1.7	1	1.7	1
Vitamin E	30	6	6	8	8

Sources: various suppliers of hemp foods; Callaway, 2004; FDA (US Food and Drug Administration), List of Reference Daily Intakes (RDI).

does not contain much alpha-tocopherol, associated with the highest vitamin E potency, but has high levels of gamma-tocopherol, a strong antioxidant, which according to recent studies has anticarcinogenic properties.

Other phytochemicals

Several other phytochemicals are known or suspected to be present in hemp seeds or oil, yet there is not much information about their typical levels and the chemical form in which they are present. These compounds include phytoestrogens, i.e. flavonoids and lignans, compounds present in soybeans and flax seeds, and the above-mentioned phytosterols. Hemp seeds also contain smaller amounts of so-called antinutrients, i.e. compounds which interfere with the digestion of protein. One such antinutrient present in hemp is phytic acid. However, its levels are comparable to those in flax seeds and soybeans, and recent research suggests that phytic acid may even have anticarcinogenic properties. Certainly, the nature and potential health benefits of phytochemicals in hemp seeds requires further research.

16.5 Research Needs and Marketing

The above summary indicates that hemp seeds and oil offer a unique nutritional composition: they have a balanced fatty acid spectrum, including several higher fatty acids which some people may not generate in sufficient quantities. Hemp protein is, as far as vegetable protein goes, reasonably complete. Finally, hemp seeds apparently contain relatively high concentrations of some vitamins and minerals. Even though very few clinical studies have so far been undertaken to demonstrate the health benefits of hemp, the available information suggests that hemp is a nutrient dense, wholesome and healthy food. Since the recent interest in hemp foods appears to be driven largely by health considerations, further growth of the food uses of hemp require that its nutritional attributes are better documented and evaluated for nutritional benefits. This is particularly important because journalists who write about nutritional health in the ever-growing number of food and health magazines now often ask for scientific validation of the claims made by manufacturers and distributors of 'health foods'.

To address that need, the Canadian Hemp Trade Alliance (CHTA) initiated in 2004 a comprehensive research programme. The CHTA represents about 60 members from all sectors of the North American hemp industry, including farmers, processors, distributors, retailers and researchers. The research programme includes the following major elements:

1. Protein characterization. This study evaluates the amino acid composition of hemp seeds from all relevant hemp varieties grown in Canada. Subsequently, the digestibility of the protein in whole seeds, nuts and protein powder is evaluated in rat feeding trials. The findings from this study will allow assessing the variability of the amino acid composition of hemp protein as a function of variety and growing conditions. It will also provide a comparison of the nutritional value of hemp protein with that of other animal and plant protein sources.
2. Minerals, vitamins and phytochemicals. This study will quantify typical levels of key vitamins, minerals and nutritionally relevant phytochemicals in hemp seeds, oil, nuts and protein powder. Again, samples will be collected from all relevant commercial hemp varieties grown in Canada.
3. Assessment of nutritional value of hemp by expert panel. To date the high cost of controlled clinical studies has prohibited such studies of the potential health benefits of hemp seeds. In order to obtain a credible assessment of the benefits expected based on the composition of hemp seeds alone, a panel of recognized experts on various aspects of nutritional health

will convene, review the results from the studies above and other evidence and then discuss the potential benefits conveyed by the nutrients present in seeds and oil.

The results from these studies will be published in scientific journals. Summaries of the results will also be made available to industry members for their use in advertising and customer education.

16.6   Future Prospects

The limited information we have today on the composition of hemp seed and its derivatives – oil, nuts and flour – indicate that they are very compatible with current trends in nutritional health. These trends favour foods that are 'nutrient dense' and offer these nutrients in a balanced form. Of course, as with all foods, they should also be tasty. Hemp seeds seem to meet these requirements par excellence.

About the Author

Gero Leson, DEnv, works as environmental researcher and consultant in Berkeley, California. He has specialized in the technical and food uses of agricultural crops, such as flax, hemp and coconut. Dr Leson initiates and coordinates research and development projects with industry groups. He serves as a scientific advisor to the Canadian Hemp Trade Alliance (CHTA) and can be reached at gl@lesonassociates.com.

17 Hemp and the Construction Industry

Laurent Arnaud,[1] Bernard Boyeux[2] and Yves Hustache[2]

[1]*ENTPE, France;* [2]*Consultant, France*

17.1 Introduction

The idea of using plant materials in the construction industry is not a recent one. It is still not uncommon to find old walls coated with lime in which hurds have been embedded. With the arrival of new mouldable materials, however, these rendering (or plastering) techniques have gradually fallen into disuse.

Recent developments in our approach to the building of houses have prompted a revival in the use of plant materials as constituents of building materials. New perspectives have been provided by technical innovations and the realization that such materials provide environmental benefits and increase market diversity.

Reference to the received wisdom we have inherited from the tale of *The three little pigs* would seem to argue strongly against this trend and [the tale denigrates] against the use of plant materials, such as straw, in the construction industry.

It would be a great shame were this subject not to be studied further. Hemp, in particular, as well as hemp wool, hurds and hemp concrete, have been shown to possess a number of advantages that have led to their gaining ground within the construction industry. Numerous laboratory investigations and field trials, both in new builds and renovation projects, have been undertaken over the past 20 years. Today, there are in excess of 100 hemp-based products on the market.

This chapter will provide a description of the raw plant material, as well as the performance results of the products it goes into making. The applications found for those products will be described, together with a brief note about the economic and environmental benefits associated with their use. Finally, ongoing standardization work and current research programmes will also be discussed.

17.2 Fibre or Hurds? Matching Materials to End Uses

With the exception of hemp seed, which has no significant use as a building material, hemp can supply two co-products: (i) the bast fibres and (ii) the woody core of the hemp stalk used to produce hurds. These co-products possess specific properties and are used to produce different construction materials (Fig. 17.1).

The fibres are used to produce insulating material in which synthetic fibres are replaced by hemp fibre. In addition to the environmental value of using plant matter, these materials benefit from the mechanical strength of the hemp fibres.

The employment of hurds in construction makes use of two important characteristics of these particles: their low density and their properties as an isolator that arise from this low

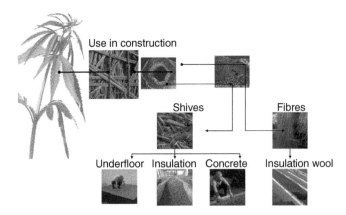

Fig. 17.1. From plant to building – hemp in the construction industry.

mass:volume ratio. Other properties, including their elasticity and permeability, are also appreciated in the construction industry. Today, particles of hurds are used primarily to produce light and superlight concretes and mortars, whose properties are very much linked to their composition. By varying the amount of binder, materials can be produced that have very different end uses, including roof insulation, paving, wall construction and rendering.

Hurds can also be used as a space-occupying insulating agent that can be poured/packed into cavities and under floors. These uses have not been developed fully due to a lack of production.

17.3   Hemp Wool

17.3.1   Mode of production

The first hemp wool products – and flax wool products – appeared in 1998. Since this date, the design of this material and all other similar woollen fibres has remained the same. The production involves the joining together of hemp fibres using heat-fusible fibres, generally of polyester. The process demands a uniform mix of the two fibre types. This is used to make a well-aerated mattress. The uniformity of the mix is essential to the reliability of the end product, as it influences the proportion of heat-fusible fibres.

The mattress is produced by a mechanical or pneumatic carding process and is then passed between two rollers and into an oven. The temperature and length of the oven, as well as the speed at which the mattress progresses, allows the fusion of the external part of the heat-fusible fibres (consisting of two fibre components) on to the hemp fibres, without the hemp fibres themselves being affected by the high temperatures. The space between the two rollers in the oven dictates the final thickness of the insulating wool produced.

On exiting the oven, the wool is cut and conditioned. The lightest wool is usually rolled and, in some cases, compacted in order to minimize transport, handling and storage costs. The denser wool (semi-rigid panels) is cut up into small panels (generally 0.6–1.2 m) and packaged.

17.3.2   Performance

The wool fibre produced is, in reality, very similar to the insulating synthetic wool commonly available today and is used in the same way. The resulting insulation is due to the very low density of this material (30–100 kg/m³) and the amount of air trapped in the fibres. The trapped air has a very low thermal conductivity and it is this that confers upon the fibres their excellent insulating properties, providing it remains immobile and is not compressed.

Insulating wool fibres are characterized by their low density and thermal conductivity λ expressed as W/m/K. The lower this value, the harder it is to transmit energy via thermal conduction and therefore the better the insulation.

When considering the performance of hemp wool fibre, the following characteristics need to be considered:

- The thermal properties of the material. Thermal insulation is the principal function of hemp wool.
- The mechanical properties of the material. The performance of the hemp wool during its conditioning and employment are a direct function of these characteristics.

Thermal properties and insulating potential of hemp wool

Let us deal first with the thermal conductivity of hemp wool. Examination of the electron micrographs of hemp fibres (Fig. 17.2) shows us the relatively regular polygonal profile of the fibre and its hollow centre. The other major feature of hemp fibres that distinguishes it from other insulating wools is its porous nature. This porosity is responsible for the very low weight and for the very low intrinsic thermal conductivity of the fibres.

The microstructure of hemp wool results in a conductivity of 0.05 W/m °C for a wool of 40 kg/m^3. This is equivalent to a thermal resistance,[1] R, of 4 W/°C for a thickness of 20 cm (R = e/λ = 0.2/0.05).

Figure 17.3 presents the thermal conductivity λ as a function of density (values measured and compared to the predictions obtained from a theoretical approach based on a self-consistent homogenization technique)[2] and shows that λ varies from 0.0038 to 0.007 W/m°C (equivalent to thermal resistances of between R = 5.26 W/m/K and R = 2.85 W/m/K for products with a constant width of 20 cm). These thermal conductivities are very similar to commercial synthetic insulating wool fibres currently in widespread use in the construction industry today.

Thermal conductivity is sensitive to and influenced by water content, a fact that is often overlooked. Water vapour in room air interacts with the material. The nature of the wool fibres and their ability to exchange water can lead to the wool taking on condensed water. The extent to which it does so is affected by the ambient humidity and temperature. Water is a very good conductor of energy (Table 17.1) and the thermal conductivity of the wool will therefore increase with increasing water content. Experimentally, increases in conductivity of the order of 30% have been measured in environments saturated with water. It is therefore essential that the installed wool fibre insulation is allowed to breathe naturally in order to regulate water exchange.

Mechanical characteristics and behaviour during installation

If we now consider the installation of the insulation material, another aspect of importance becomes clear: the mechanical characteristics and behaviour of the wool. As we have just seen, the porous microstructure is at the root of the insulating properties of this material.

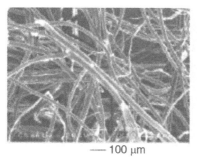

—— 100 μm —— 10 μm

Fig. 17.2. Scanning electron micrograph of hemp wool fibres.

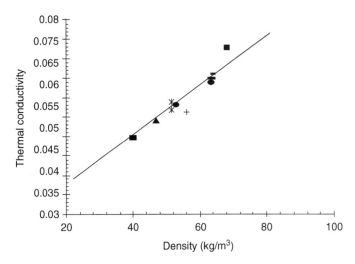

Fig. 17.3. Variation in thermal conductivity of hemp wool as a function of density. Comparison with a trial undertaken using predictions issuing from a self-consistent homogenization approach (Cerezo, 2005).

Table 17.1. Characteristic values of the materials.

	Air	Water	Hemp wool	Synthetic wool
Density ρ (kg/m³)	1.2	1000	35	35
Thermal conductivity λ (W/m/K)	0.025	0.602	0.047	0.041
Thermal resistance R = e/λ m²/K/W/)	4	0.2	2.1	2.4

They therefore must not be changed easily or modified. That said, during conditioning, and to reduce transport costs, these materials are heavily compressed. The wool must be able to withstand these compressive forces and recover its original shape and geometry in order to regain its properties. During installation, the fibres are often placed under traction. Once again, the wool must be able to resist such abuses to preserve its performance.

Laboratory tests have been undertaken to characterize the behaviour of these fibres during compression and extension, as well as to evaluate the traction resistance of hemp wool.

These experiments demonstrate that the intrinsic properties of the hemp wool (high traction resistance and mechanical 'power')

influence the mechanical performance of the wool. A very high resistance to traction is seen (Fig. 17.4): the breaking point of hemp wool with a density of 60 kg/m³ and a thickness of 45 cm is 900 kN. It would therefore be possible to hang a weight of 90 kg from a 9 cm wide strip. This means that the installation and use of these products is relatively easy and does not necessitate any particular precautions. The high degree of mechanical resistance does, however, mean that this material is very tough to cut up!

The compression and extension tests of hemp wool, as illustrated in Fig. 17.5, demonstrate their marked elastic properties with little residual distortion. It is therefore possible to reach levels of compression of 60%. It would appear, however, that above 60% the friction between plant fibres results in some significant permanent distortion. A compression of 70% will thus result in a permanent distortion of 15%. In other words, a wool with an initial thickness of 10 cm that is compressed to 3 cm during conditioning will, when installed and decompressed, have a thickness of 8.5 cm.

17.3.3   Areas of application

Among the main fields of use for hemp wool, we can list the following:

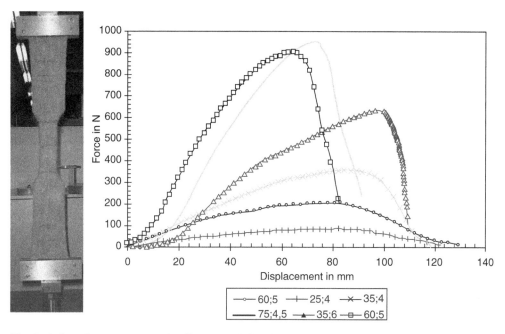

Fig. 17.4. Stretch tests on a sample of hemp wool. The photo shows the sample being tested; the graph illustrates the results for different types of hemp wool, varying in density and thickness (key: density kg/m³; thickness cm).

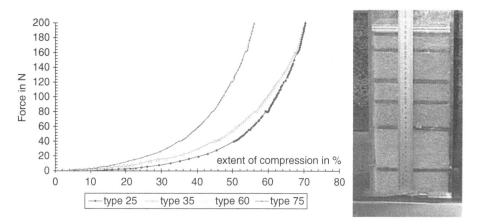

Fig. 17.5. Compression tests on hemp tiles. The photo shows the batts being tested. The graph shows the extent of compression for different types of hemp wool.

- It may be used over floorboards or instead of floorboards in roof spaces.

    Hemp wool can be applied directly to the floor (whether this be floorboards on top of joists or a solid concrete floor) in one or two layers. It can also be applied between the joists in two directions (the first across the joists, the second at right angles to them).

- It may be used underneath the floor.

    Hemp wool is installed beneath the floor (whether made of wood or masonry) on a frame.

- It may be used in loft spaces and loft conversions.

  On a traditional wooden frame, hemp wool is placed in two layers below the external covering. The first layer is introduced between the rafters; the second layer is placed over the rafters to insulate them. This second layer may be placed on or between the roofing battens that support the external covering.

  In truss roofs, hemp wool is installed in one or two layers. The first layer is placed between the trusses and the second is applied to the roof battens that support the roofing material.

  In all cases, it is important that ventilation be provided between the hemp wool and the outer shell of the roof and that the joints and extremities are weatherproofed.

  Pitched roofs for loft conversions must be equipped with an outer covering, but these may have limited permeability to water vapour. Where this is the case, a protective membrane must be placed between the hemp wool and the facing.
- Use in wall construction (both solid and partition walls).

  Hemp wool can be used in the form of semi-rigid panels:
  – between the frame of the wall, where the frame is made of wood;
  – in cavity walls or on a metal frame, in the case of walls made of bricks or breeze blocks.

17.3.4   Regulatory aspects

Hemp wool is an industrial product and is therefore subject to a number of national and European technical regulations.

Today, certain hemp wool products available on the market have been approved officially for use in roof construction.[3]

For all other uses, the manufacturer should be approached for advice and their recommendations followed with regard to installation and use.

17.3.5   Conclusions

Hemp wool products, as well as other natural materials such as flax, sheep's wool and feathers, can be substituted for other reported insulating materials. Whether they be alveolar or fibrous in construction, they all function on essentially the same principles. They also present environmental advantages – advantages that require confirmation – and benefit from the 'asbestos factor', which has shaken confidence in the safety of conventional insulating materials. These advantages provide new opportunities that may be realized if the industry can show itself to be economically competitive relative to the products it wishes to replace – products whose cost has been particularly low. In order to meet these requirements, the industry can become more efficient in a number of areas: production, automation, compression and conditioning capacity, and the mixing with other natural fibres, which may introduce complementary characteristics.

Finally, a process that uses a plant-based binding agent, rather than a synthetic one, will result in a product that meets the expectations of consumers better and is more in line with environmentally friendly objectives.

17.4   Hemp Mortars and Cement

17.4.1   Introduction

To our knowledge, hurds were first used as an aggregate for the production of light cement in 1986 by Charles Rasetti in collaboration with *Chanvrière de l'Aube* in order to renovate the *Maison de la Turque* at Nogent sur Seine, France. Contrary to popular belief, this is not an ancient/traditional technique and has been developed at all levels by French pioneers.

Certain well-informed professionals quickly recognized the technical potential of these concretes. They were confronted by a major difficulty, however: the significant capacity of hurds to absorb water, which competed with the binding agents' need for water. This could result in the mixture failing to set and the associated consequences. After exploring a number of avenues – waterproofing, various treatments, water saturation – the need to focus on the make-up of the binding agent(s) was recognized. Satisfactory results were obtained from the use of a binding agent that hardened in air (*chaux*) mixed with hydraulic and/or pozzolanic binding agents and a specially adapted adjuvant. These have led to a number of patented methods. It is, however, difficult to provide reliable recipes and the professional code of the *Construction en Chanvre* requires that manufacturers produce binding agents that correspond with, and meet the technical specifications of, the end use intended for the concrete. Different products are available on the market today.

The mixing of the binding agents with the hurds, as seen below, allows mortars and cements of very different nature to be produced for speciality markets. Hemp mortars and hemp cement are conglomerates made of a plant aggregate (hurds) and a binding agent. Variations in the ratio of hurds and binding agent allow materials with different mechanical, thermal and acoustic properties to be developed for use in flooring, wall construction, rendering and roofing.

17.4.2 Method of manufacture

The production of hemp-based mortars and concrete does not present any particular difficulties provided a few basic precautions are taken: these relate to the aggregate and its ability to absorb large quantities of water rapidly. The mixture should not contain any small clumps and should be put together in a cement mixer, vertical or horizontal paddle mixer, cement lorry, or by hand (although this is more difficult).

When using a cement mixer or mixer truck, a mixing tank containing the entire volume of water and binding agent must prepared. The hurds are then added gradually. This method is not to be used with vertical and horizontal axis mixers as the emptying traps are generally not fully sealed. The latter type of mixer does, however, produce a very uniform mixture. In the case of these machines, the binding agent and hurds will be mixed first before spraying on the water. This allows well-mixed cements to be produced with very low water content.

17.4.3 Performance

If we are interested in the performance of a particular building material, it will often be limited to a single function. Thus, for the material that assures the consolidation of mechanical forces, it is easy to determine its mechanical resistance. Current practices in the construction industry encourage the use of various products (other materials) to those providing mechanical strength, in order to provide thermal and noise insulation. This tendency to sandwich together these different materials is rapidly reaching its limitations, both in terms of function and performance. Let us, for example, consider the control of humidity in a room (or space) and its role in hygrothermic comfort. This is a much overlooked issue and requires us to factor in the performance of all the materials used in order to evaluate the behaviour of the space. This is never done. The juxtaposition of various materials also poses a number of problems related to the amount of on-site handling by different workmen, the chaotic association of materials and the need to strike compromises, all resulting in mediocre performance.

It is becoming necessary today to reconsider the role and function of building materials in order to evaluate better their overall performance in providing the shell (or envelope) for a home or other living space.

As a constitutive envelope for a building, hemp concrete can provide thermal and sound insulation while also regulating water exchange, thus conditioning the comfort of the home while contributing to the mechanical properties of the construction.

Hurds, as a plant material, produce granular particles with a highly porous structure that confers a number of 'original' properties on mortar and cement when compared with the synthetic materials normally used:

- The lightness of the particles, and therefore of the material, results in a product that is easy to install and work with and has a high resistance relative to its density.
- The compressibility/deformability of the granules plays an essential role in the mechanical behaviour of the matrix produced by the binding agent.

- The large amount of air trapped in and between the particles ensures a low thermal conductivity.
- The porous aspect of the surface ensures that these materials absorb noise well.

Finally, these same porous characteristics, including pores of variable size (micropores of the constituent materials and macropores resulting from their chaotic orientation), influence water exchange strongly (i.e. the movement of both liquid water and water vapour).

Mechanical behaviour

Hemp mortar and cement are also available as composite materials associating a rigid matrix of flexible granular particles. During the installation of such material, the chaotic arrangement of the particles relative to each other leads to the creation of pores between these same particles (Fig. 17.6).

The mechanical properties of cement and hemp are linked directly to the properties of the binding agent. The choice of binder must take into account three factors. It must first be easy to create and should coat the different constituents easily. It should run smoothly through the structure and, after setting, should provide the mechanical properties that are essential to its role as a mortar. It should be noted that the physic-chemical interactions of the different components are extremely complex, as the plant material is far from inert in this mixture.

It is strongly advised that only those binders recommended for these specific applications be used [*Règles Professionnelles de la Construction en Chanvre*; see the site on construction with hemp – http://www.construction-chanvre.asso.fr]. Many tests have been performed successfully on mixtures of air and hydraulic lime. The results presented in this chapter are from this work, ENTPE: National School of State Public Works (www.entpe.fr).

Classical trials to characterize hemp cement and mortar are available today [ENTPE and Professional Rules of the *Construction en Chanvre*]. The object here is not to provide an exhaustive account but simply to characterize the essentials and relate these to the function at the level of the envelope.

The material is characterized by an elasto-plastic behaviour: demonstrating plastic distortion with very little encouragement. It does not show obvious fragility or break easily and can be bent out of shape without snapping suddenly. Looking beyond this mechanical description, we can identify a pattern of behaviour that is interesting for several reasons. Figure 17.7 provides a comparison of the results obtained in tests to measure resistance to compression for (i) the binding agent on its own, (ii) various mixtures of hurds and the binding agent (in various proportions) and (iii) hurds alone. The binding agent on its own is a dense material and ruptures suddenly into several pieces after only a small amount of distortion. The hurds, when tested on their own, demonstrate

Fig. 17.6. Test tube of hemp cement – surface view.

a radically different behaviour. The material can undergo considerable change in shape without rupturing. For the mixtures tested, and depending on the proportion of binding agent, an intermediate range of behaviour is seen. This ability to distort without breaking confers a springiness upon the material and an ability to accommodate movement and resist any tendency to crack. It is therefore necessary that the right rendering be chosen carefully, so that this property is demonstrated in a surface that can also resist deterioration.

The significant porosity of concrete associated with a binding agent of moderate resistance results in weak resistance to compression: Rc between 0.1 and 2.5 MPa, depending on the binder used and in what proportion (Fig. 17.8). These values are not adequate to allow the construction of weight-bearing elements of standard dimensions.

If we now consider the tensile properties as measured by a flexion test and illustrated in Fig. 17.9, once again the suppleness of the material is demonstrated, allowing for considerable dis-

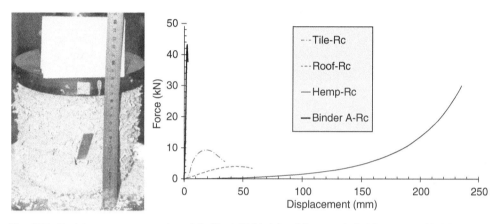

Fig. 17.7. Resistance to compression trials. The initial height of the sample is 32 cm, actual height 16 cm. Graph showing the relative changes. (Rc = resistance to compression.)

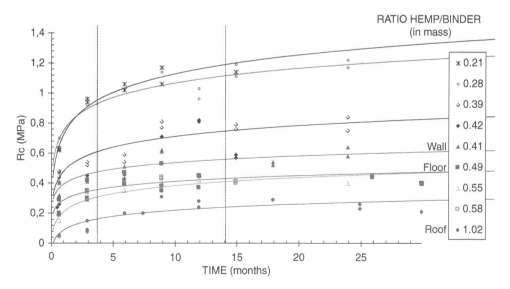

Fig. 17.8. Evolution of the resistance of hemp fibre to compression over time and as a function of the proportion of hemp to binder by mass.

ruption before breaking. The material is not fragile and flexes without breaking.

This light and supple material is distorted easily and would appear to be particularly interesting in the construction industry: the structure can be distorted without giving rise to significant constraints and without causing the material to crack. It is therefore adaptable and these properties may qualify it for use as a building material in areas prone to earthquakes.

Finally, it should be emphasized that the binder used is one that sets in air; that is to say, through a chemical reaction with CO_2 in the atmosphere. These reactions are slow to take place, as they are reliant on the diffusion of CO_2 to the interior of the material. It has been shown that resistance increases gradually over a period of 6 months, before attaining 80% of the total resistance (Fig. 17.10). This characteristic is holding back the uptake of this product by the construction industry, whose need to realize turnover imposes a faster timetable of work. Work is currently under way to produce binding agents with equally good performance but which set faster.

Thermal behaviour

From a thermal point of view, hemp cement has two advantages related to its porosity. Firstly, the density of pores allows a lot of air to be stockpiled, providing excellent insulation in much the same way as was discussed for hemp wool. In addition, the range of pore sizes – ranging from the micrometre (intra-particular porosity of hemp and the binder) to the

Fig. 17.9. Flexion tests on hemp bricks.

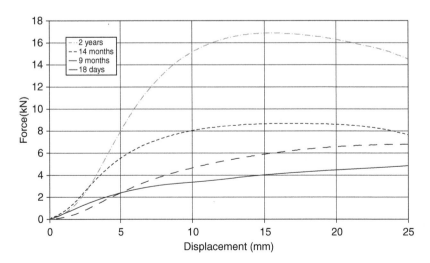

Fig. 17.10. Evolution of the resistance to compression over time.

centimetre (provided by the inter-granular pores) – allow an excellent circulation of water vapour.

These characteristics vary with the proportion of binder. Figure 17.11 illustrates the variation in thermal conductivity λ of hemp as a function of its density and therefore as a function of the proportion of binder. Even for binding agents that are relatively porous and therefore good insulators, an increase in the amount of binder will increase conductivity: λ between 0.06 and 0.18 W/m °C for dry concrete with a density ranging between 200 and 800 kg/m³.

Finally, as we have seen already for wool, the porosity of the material renders it sensitive to humidity exchange with ambient air. This phenomenon is far from insignificant, because the high concentration of water in hemp cement (density of cement 450 kg/m³) subjected to a hygrometry of 75% RH (relative humidity) will see thermal conductivity increase by 30% (Fig. 17.11).

Acoustic behaviour

From an acoustic point of view, the low density of hemp cement does not allow it to insulate in bulk. Once again, though, it is the cement's reservoir of pores left exposed on the surface that allows it to absorb the energy of sound waves.

Sound waves move through the air as a result of small oscillations of variable frequency. For humans, the range of audible waves varies between 100 and 6400 Hz. When a moving sound wave encounters a wall, the energy carried is broken up into:

• A part that is transmitted through the wall. This is particularly significant where the wall is light and permeable. In the absence of the wall, the sound can travel over long distances.

• A reflected part, which is greatest for heavy walls that are impermeable to air.

• A last part made up of energy that is absorbed by the wall. The mechanical wave is pulsed into the wall's porous interior. In doing so, it rubs along within the pores and this friction slows it down.

It should be emphasized that the acoustic comfort of a room is related directly to the capacity of the building materials to absorb noises produced within the room. Rooms that are insulated from external noise tend to have thick walls and this often results in very loud rooms, as interior noise will reverberate for some time before being absorbed.

The material's properties are measured with the help of a coefficient of acoustic absorption, α. This figure represents the percentage of energy absorbed relative to the amount

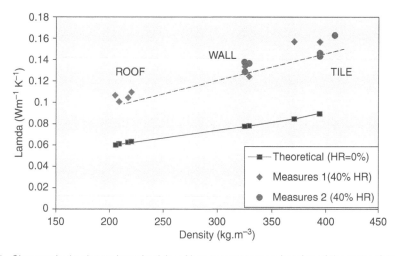

Fig. 17.11. Changes in the thermal conductivity of hemp cement as a function of the proportion of binder used and the relative ambient humidity.

arriving at the wall. The higher the coefficient, the more absorbent the material. It depends on the wave's angle of incidence, on the thickness of the wall and the frequency of the wave. Results obtained by the ENTPE are presented in Fig. 17.12 for waves arriving normally on a partition wall with a thickness of 20 cm. They are compared with results obtained under the same conditions for other commonly used building materials (Fig. 17.13). The microstructure of hemp provides these materials with excellent absorption, with coefficients ranging from 0.4 to 0.9, depending on the frequency under consideration. The intrinsic qualities of these materials are underlined further by the confidence of returning customers, who praise the acoustic comfort of rooms constructed from hemp concrete or where the walls have been lined with a hemp plaster.

It should be noted that the acoustic properties are also dependent on the quality of the connections between the different structural elements (partition wall, floor, door and window frames, etc…).

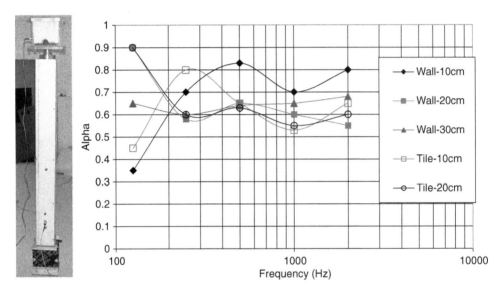

Fig. 17.12. Coefficient α for acoustic absorption for walls and tiles of different thicknesses (10, 20 and 30 cm).

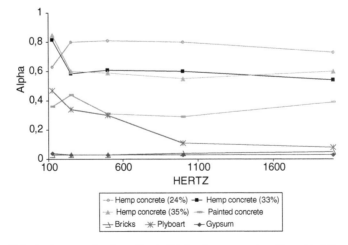

Fig. 17.13. Coefficient α for acoustic absorption in thirds of an octave for different building materials.

17.4.4 Hygrothermic performance and humiditiy regulation

It is now widely accepted that one of the essential conditions contributing to the interior comfort of buildings relates to their management of water and water vapour. For reasons of energy conservation, building materials have been developed in order to produce confined spaces. The atmosphere thus created is far from optimal in terms of sanitation. We are now presented with a promising new possibility: materials that contribute to the regulation of living spaces.

Water, in the form of water vapour, is naturally present in ambient air; the proportion will vary as a function of atmospheric conditions, temperature and pressure, together with natural or forced ventilation and exchanges taking place across the shell of the building. The exchange between ambient air and the material of which the shell is constructed is important. While the physics regulating this exchange is very complex, it is important to remember the key role played by the material's permeability, the size of its pores, the nature of its constituents and their temperature. Figure 17.14 illustrates the significant ability of hemp, the key ingredient of hemp cement, to hold water (over 30% of its mass) under different conditions, with relative humidity (RH) both increasing (sorption curve) and decreasing (desorption curve).

Trials are currently under way to evaluate various formulations of hemp concrete as part of the European Eureka Research Programme. These performances will be compared with those of other building materials.

Initial results show that the hemp cement's porous nature allows faster and easier air exchange, performing better than other materials with a similar porosity. The resulting comfort is therefore greater.

A wood-based material that does not burn

The integration of plant materials into construction leads to questions concerning the fire resistance of such products. It should first be noted that wood is an integral part of construction and has a wide range of uses. It is viewed by experts as a material that performs well with regard to fire.[4]

As we have seen already, hemp cement is made from hemp coated with a matrix of lime. This composition allows it to resist fire, because of lime's excellent fire-resistant properties and ability to preserve mechanical function. On its part, hemp burns very slowly, as the oxygen necessary for combustion is not present in significant quantity in this composite material

Conclusions

Cements and mortars that incorporate hemp bring together two materials with very different properties that can be mixed in different

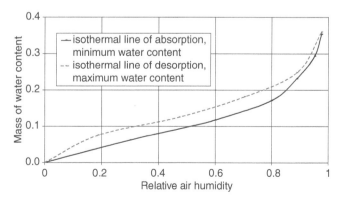

Fig. 17.14. Isothermal lines of absorption and desorption for hurds at 20°C.

proportions. It is wise, when producing a material for use in construction, to specify the mixture according to the properties one is seeking. Thus, for an interior wall in a home, the acoustic absorption of the material will be given priority. By contrast, an outer wall will benefit from materials that have been developed for their thermal resistance. In Table 17.2, the characteristics of hemp cement are detailed and compared with those of a number of other materials used in the construction industry.

17.4.5 Applications

Hemp cement, as defined previously, is used to create walls, insulating floor tiles, rendering or insulating underlay for roofing (Fig. 17.15). Each of these uses is described in the following paragraphs.

Use in wall construction

Hemp cement is used to fill the wall spaces that exist round the wall's frame or superstructure (Fig. 17.16). The frame can be covered completely in hemp cement or, alternatively, the timbers can be left visible. Generally, hemp cement with a density of approximately 420 kg/m³ is used for this purpose.

CONSTRUCTION. The hemp cement can be introduced by pouring it between two panels that act as a mould. Alternatively, it can be projected by machine on to a single panel. When constructed to a thickness in excess of 27 cm, hemp concrete guarantees a uniform insulation that respects current thermal regulations.

Use in floor construction

Hemp cement can be used to produce an insulating slab(s) of concrete sited over ventilated ballast (Fig. 17.17). Above the ground floor, it can also be applied to wooden boards (Fig. 17.18). A density of 500 kg/m³ is used for these applications.

Table 17.2. Summary of the technical characteristics of hemp cement compared to other building materials.

	Rc (MPa)	Deformation (mm)	E (GPa)	λ : Thermal Conductivity (W/(m.K))	α : Acoustic absorption
Binder A	4,8	0,01	0,600	0,214	0,23–0,42
Binder C	3	0,01	0,300		0,12–0,22
Roof (low dose of Binder A)	0,26	0,16	0,003	0,065	0,55–0,90
Wall (intermediate dose of Binder A)	0,37	0,1	0,020	0,085	0,58–0,80
Tile (intermediate dose of Binder A)	0,48	0,07	0,028	0,090	0,50–0,80
Rendering (high dose of Binder C)	0,75	0,05	0,030	0,150	0,29–0,56
Pressure sealed cellular concrete	5	25.10^{-4}	2	0,17	0,21–0,32
Panels of wood				0,058	
Panels of compressed straw				0,12	
Cement mostar	10	5.10^{-4}	20		
Hydraulic cement	20	1.10^{-4}	35	0,666	>0,1 (aggregates)

CONSTRUCTION. In both cases, the hemp cement is poured into place. It is then spread out over the surfaces to be covered and finished carefully (screed and smoothed). The thicknesses usually laid down generally exceed 7 cm.

Use of wall rendering

Hemp mortars can be used on both interior and exterior walls. They can improve the thermal and acoustic comfort of a room.

Hemp rendering is usually made up of a base coat (a mixture of lime, sand and/or

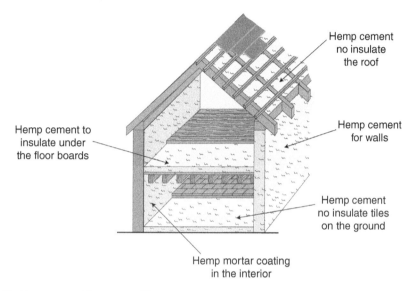

Hemp cement no insulate the roof

Hemp cement to insulate under the floor boards

Hemp cement for walls

Hemp cement no insulate tiles on the ground

Hemp mortar coating in the interior

Fig. 17.15. Hemp cement has many different uses.

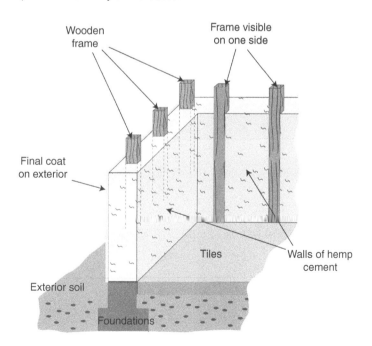

Wooden frame

Frame visible on one side

Final coat on exterior

Tiles

Walls of hemp cement

Exterior soil

Foundations

Fig. 17.16. Wall constructed with hemp cement around a wooden frame.

hemp), an undercoat (hemp mortar) and a finishing topcoat (necessary on external walls).

The hemp mortars used in these applications have a density of 950 kg/m^3.

APPLICATION (RENDERING PROCESS). Hemp rendering is applied by hand or mechanically on to the surface to be rendered. It can be applied very thickly.

Use in roofing

Hemp cement can be used as an insulating material for roofs (Fig. 17.19). It has a very low

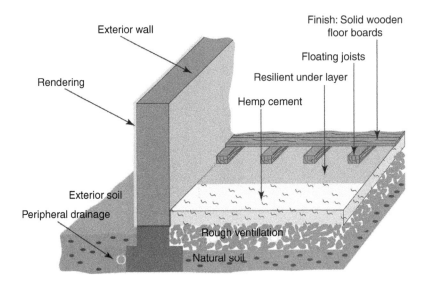

Fig. 17.17. Insulating concrete slabs at ground level.

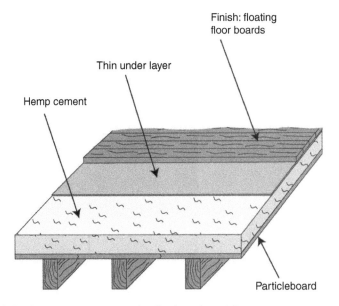

Fig. 17.18. Insulating hemp concrete on wooden floorboards upstairs.

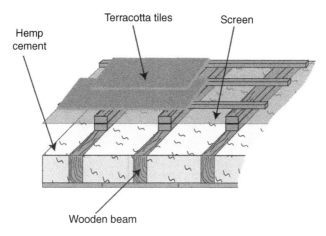

Hemp cement
Terracotta tiles
Screen
Wooden beam

Fig. 17.19. Hemp cement used to insulate a roof.

binding content and a density of the order of 250 kg/m³. As such, it is able to provide continuous and stable insulation.

INSTALLATION. The hemp cement is poured into the space in the inner aspect of the roof space.

17.4.6 Regulations

The four uses described for hemp cement in the paragraphs above (wall, floor, rendering and roof insulation) are the subject of a reference work entitled *Règles professionnelles de mise en œuvre des bétons de chanvre*. This document has been put together by a committee of experts (artisans, architects, companies and scientists) and validated by a control board.[5]

It is recommended that this type of document be consulted on all projects in countries with applicable regulations.

17.4.7 Conclusion

Hemp concrete and its use in the different roles described above has benefited from extensive use and evaluation. Today, we are no longer at the experimental stage as hemp is widely used, even though we still have not optimized all aspects fully. Ongoing work is today seeking to improve the technical and economic aspects of hemp cement on wood-frame construction systems.

17.5 Other Uses of Hemp in the Construction of Building Materials

Beyond the two uses detailed above (insulating wool and light mortars/concrete), the co-products of hemp can also be used to make other construction materials whose production today is, for various reasons, negligible.

17.5.1 Insulating lining underlay

Hurds aggregate is coated in bitumen and spread out over floor surfaces, levelled off and packed down lightly. The bitumen coating allows the components to stick together without any water. The end result has a low-density, even surface that is a good acoustic and thermal insulator.

Developed in 1963 in Germany, *Mehabit* (www.meha.de) was met with some success and is still produced, being used to renovate flooring in various ways. It can be used to level floor surfaces and provide thermal and acoustic insulation. Its qualities also include an ease of use and the fact that it does not introduce (or require) water.

It is unfortunate that this product has not been distributed more widely outside of Germany.

17.5.2 Hurds in bulk

As we have seen already, the properties of hurds make it an excellent thermal and acoustic insulator. Its use as a bulk insulator that is poured easily provides a simple and efficient method that is well suited to the insulation of floor spaces. Techniques making use of insufflation[6] have also been trialled successfully, but have not really been developed further.

The hydrophilic character of hurds does necessitate that appropriate precautions be taken. All commercial uses must be accompanied by a guarantee that this issue has been addressed.

In certain situations, a fireproofing treatment is likely to be required. Despite the utility of this product, manufacturers have ceased to produce it – or have gone out of business – so such products are currently unavailable. It is hoped that this situation will not last long.

17.5.3 Masonry blocks made of hemp cement

Contemporary construction makes extensive use of masonry blocks. They are used in 90% of individual house builds and, as early as 1990, the production of hemp blocks followed hot on the heels of the first hemp cement. Setting aside the poor fluidity of hemp cement, which can cause difficulties when transporting the freshly prepared product, the production of blocks poses relatively few problems. This is particularly the case where production machinery is available, as is the case throughout much of France.

This line of development represents an important potential for the industry to explore, providing it can respond to the following concerns. Firstly, the problem of drying times must be addressed. Hemp cement dries slowly and this is incompatible with industrial production processes. Secondly, building techniques will need to be developed that are adapted to the blocks produced, whether they be weight bearing or not. Finally, these techniques will need to take into account the regulatory climate and allow the building industry to use these products with confidence, while satisfying the needs of their insurers.

The production of building blocks made from hemp cement is marginal at present. Important French and European research programmes are under way, however, to address the questions outlined above.

17.5.4 And also...

The characteristics of the co-products of hemp fibre and hurds have been highlighted and discussed frequently. These characteristics allow us to envisage the development of numerous other construction materials provided they can attract the interest of a developer and meet the technical and economic demands of the market.

PANELS OF AGGLOMERATED HURDS.   The manufacture of chipboard panels is an established industry that makes use of all the waste from sawmills as well as entire trees.

The technology that has been perfected over the years has remained simple and low cost (attributable largely to the low cost of the raw product). A volume of wood particles (chips) is mixed with glues or polymerizing resins and compressed. This produces boards that are particularly rigid, given their thickness.

These panels [agglomerated hurds panels] have a lower density and lower thermal resistance than those made of wood. The use of hemp to produce such products is not new, but the lack of economic development afforded to this product has not encouraged hemp workers to pursue this option. Programmes under development in France and Germany are again studying this option. In order for these products to have a future, the materials produced will need to have environmental and technical specifications which give them an advantage over similar panels made from woodchip that are marketed at very low prices.

COMPOSITE MATERIALS. The use of moulded plastics has seen some important developments in the construction industry, whether that be in the manufacture of doors, windows and shutters or in the creation of terraces and swimming pools.

Decking. Both low cost and ease of use have seen hardwearing plastics used increasingly as an alternative to wood as decking material. It is used as a covering for verandas, terraces, garden furniture and anything else that is exposed to the elements. This market is well supplied by a thermoplastic composite product that is made using an extrusion process from 70% sawdust, 25% polyethylene or polypropylene and 5% additives. The European market turns over some 30,000 t, while the US market turns over 700,000 t per annum. A similar process has been developed working with powdered hemp as a raw ingredient.

Hemp-based joinery. The global market for windows produces some 6 million windows a year and PVC frames have become very popular in Europe. In most cases, these frames will, of necessity, be reinforced with steel. This market consumes 9000 ltm/year of extruded product just for France. Today, we know how to produce extruded products loaded with 30% hemp with a rigidity comparable to the equivalent steel and PVC products.

These doors and windows are lighter in weight and have better insulating characteristics, as they contain no thermal bridges. They are also recyclable and competitively priced.

Manufacturers in this sector are starting to take a serious interest in this area and in these products.

UNDERLAY. Hemp fibre that has been carded and needled to form non-woven materials has a number of uses. It can be used as underlay below floating floors or fitted carpets, as well as under floating solid floors, and as acoustic insulating material under floorboards and cement.

TEXTILE. Furnishing materials such as mats and carpets.

17.6 Building Today, Tomorrow and in the Future

This account of the possible uses of hemp in the construction industry illustrates the enormous potential of this material. This potential will only be realized if it does not fall foul of the unavoidable hurdles presented by today's economic, environmental and social demands, as well as those of tomorrow and the longer term.

17.6.1 Economic criteria

In economic terms, we need to take into account the long-term competitiveness of materials, and therefore the overall cost of buildings – their construction, upkeep and running costs. We know, for example, that 80% of the energy consumption of buildings is contributed by the running costs, the remaining 20% being contributed by the construction process. We do not currently have enough information to say that hemp-based materials will allow us to obtain a construction/upkeep/running cost that is more useful than that obtained using conventional materials. It will also be necessary to analyse the contributions made by hemp wool products and hemp cement products separately. That said, the trials under way and the testimony of owners of homes made from hemp cement, or of old homes whose walls have been rendered with hemp-based plasters, are consistent and tend to prove that these homes benefit from superior thermal comfort over conventional homes. This superior thermal comfort and the lower ambient temperatures suggest that they will result in substantial energy savings.

The cost of construction must not be overlooked as, if it proves to be excessive, this will hold back development. Here again, it is important that hemp wool and hemp cement be considered separately. We have seen already that the development of wools is linked very much to their technical and economic competitiveness. For hemp cement, the competitive objectives are currently being met. While the volumes produced do not allow for substantial economies of scale in the production of binding agents and in bringing these products to market, and while technical aspects pertaining to the use of the product have not been optimized, it is still possible to construct a house of hemp cement for the same price as for other insulating technology currently available in France and across Europe.

Using the two types of product, hemp wool and hemp cement, hemp allows us to produce building materials that are economically viable today and can be very competitive over the short term. It is regrettable that despite France having pioneered the development of this technology, it does not qualify for the sorts of simple and effective grants and financial incentives available in other European countries. If this were the case, the material would very quickly become competitive.

17.6.2   Health and environmental considerations

The interest generated by hemp materials is connected strongly with their environmental credentials. While hemp benefits from this association and the preconceptions is entails, we should still seek to confirm these ideas by conducting the appropriate building lifetime analyses (also known as life cycle assessment[7] – LCA), especially as there are now standards for this (e.g. the *Norme Français* NF 01.010). Once again, the situation with regard to wool and cement cannot be amalgamated. While evaluations still need to be undertaken for hemp wool products, walls made of hemp cement have been subject to a life cycle assessment thanks to the French Ministry of Agriculture. This assessment highlighted the environmental benefits of hemp cement according to a number of criteria and, in particular with regard to carbon stocking, as well as its ability to reduce environmental impacts still further. In respect of this last point, improvements are sought in the development of an industrial network that will increase the number of production sites (producing hemp and binding agents as well as the finished product), while reducing the impact attributable to transportation, which is currently very significant.

Here again, the dearth of financial incentives available to end users is to be lamented, as they would allow the volume of production necessary for these developments to be attained more rapidly.

These products are also thought to benefit from a positive image with regard to their status from a health and safety point of view. While there may be few studies that confirm this impression, there are a number of encouraging signs. Firstly, the production of these materials does not require the addition of any toxic ingredients that could contaminate building sites. Secondly, the ability of hemp cement to regulate the hygrothermic environment of buildings is likely to reduce the appearance of humidity, which can predispose to the development of mould. Finally, as we have mentioned earlier, the hygrothermic qualities of these materials allow rooms to reach a comfortable low ambient temperature, which is itself associated with health benefits.

17.7   Conclusions

The use of hemp-based building materials is part of the innovating and global approach to construction that is seeking to provide individuals with comfort and well-being, while paying due consideration to the environmental impact of the home and its technical and economic performance.

The following key points are to be remembered:

- Hemp-based materials demonstrate technical properties that make them suited to a range of applications and allow them to meet stringent performance requirements.
- The economic characteristics of hemp-based materials are suited to current market expectations and are likely to become increasingly popular.
- The particular characteristics of hemp-based materials allow them to contribute to the thermal, acoustic and humidity comfort of the home.
- In view of the fact that hemp-based materials are, among other things, derived from plants provides them with good environmental credentials.

Hemp-based materials, with their significant technical, economic and environmental qualities, cannot be considered without reference to the entire production process that stretches from plant cultivation to its ultimate use as a building material. These qualities could not be exploited without the network of competent professionals, including architects, entrepreneurs, distributors and others, who are able and willing to respond to the needs of end users and the proposals of manufacturers. Further improvements will require manufacturers to take into account the concerns of these professionals, while farmers will need to take into account the problems and challenges facing the manufacturers. In turn, further progress and development will require researchers to unlock new doors and move development forward. Those involved in this production line/industry are collected together in the organization *Construire en Chanvre* (Building with Hemp), which seeks to promote the use of hemp in the construction industry while providing a structure and representation for the industry (http://www.construction-chanvre. asso.fr/). The website also provides the organization's code of practice.

It would appear, therefore, that the materials produced using hemp possess a number of advantages that can benefit from further development in the construction sector. This sector is concerned increasingly with quality in the broadest sense, and therein lies a real opportunity for the industry.

Notes

1 The thermal resistance, denoted R, expresses the ability of a material to resist changes in temperature. Expressed in m^2. K/W (Kelvin per watt), the index R is calculated as the ratio of the thickness in metres to the thermal conductivity of the material. The higher the R, the higher the product is isolated. R reports the conductivity of the material to the thickness of material installed. R = thickness in metres/thermal conductivity of the material ($R = e / \lambda$).

2 See Bibliography [12]

3 Professional rules on hemp wool and/or performance of mortar and concrete structure of hemp are available on the website of the association Building with Hemp: http://www.construction-chanvre.asso.fr.

4 Source: G.C. Gosselin (1987) Protection des structures contre le feu – Méthodes de prédiction. Regard sur la science du bâtiment, CNRC.

5 Contact: http://www.construction-chanvre.asso.fr.

6 A technique of decorating a surface.

7 Definition of life cycle assessment (LCA): stroke is the extent of resources required to manufacture a product or a device for building and quantifying the environmental impacts of this production. According to the ISO, it is the 'Compilation and evaluation of energy consumption, raw material uses and releases to the environment and assessing the potential impact on the environment associated with a product, or process, or service on its entire life cycle'. The life cycle of a product, process or service brings together the phases of manufacturing, processing, use and destruction.

18 The Uses of Hemp for Domestic Animals

Pierre Bouloc
*La Chanvriere de
L' Aube (LCDA), France*

18.1 Horse Bedding

When hemp workers started working on industrial hemp again after World War II, one of their first priorities was to find a use for hurds. It is hard to forget that fibre processing yields a significant quantity of hurds, corresponding to some 60% of the weight of the plant (from a plant yield of 110–180 kg/m³).

In view of its hydrophilic properties, it was suggested that the hurds be used as bedding for horses. This use caught on and remains popular today, for it is now the primary end use of this product. In addition to its ability to absorb water, hemp straw, once on the ground, provides a stable surface that does not tend to move or slip. It is also very comfortable, for it is made up of small particles. Also, hurds are relatively dust free and are recommended as bedding for horses suffering from chronic obstructive pulmonary disease. One unfortunate inconvenience, where horses are concerned, is that the bedding is likely to be ingested where horses are left hungry with nothing else to eat. This can prove dangerous, leading to bloating when the animal drinks.

18.1.1 Recommendations for use

A clean box of some 9–10 m² will need some five bales (of 20 kg) hemp straw. These can be watered with a vinegar solution (8–10%) in order to dampen the bedding and render it unappetizing to the horse.

Horse dung should be removed with a fork 2–3 times/day. Any 'urine cakes' should be identified and removed to a depth of 4–5 cm. The bedding will then need to be evened out to a consistent depth by filling in gaps.

The deeper layers of the litter should not be disturbed. The straw should only be moved around on the surface. In this way, the maintenance of a horse's bed remains economical. Details of the cost of a hemp straw bed are given in Table 18.1. The daily work involved in maintaining the litter does not exceed 10–15 min/day. Over the course of a month, 3–4 bales will be required. The base of the bed can be kept for 20–25 weeks. Over the course of a year, the amount of hemp straw used will be approximately 45–60 bales (approximately 900–1200 kg) (Fig. 18.1).

18.1.2 Costs of using hurds as horse litter

Study of the figures in Table 18.1, derived from an economic study undertaken at a number of horse breeding yards and other establishments (studs and riding schools), show that the use of hemp straw is both economically viable and interesting.

Table 18.1. Costs of using hemp straw as horse bedding.

	Straw	Wood chips	Flax	Hemp straw
Use of the product:				
Laying the bed	100 kg	3 bales × 25 kg	5 bales × 18 kg every 6 weeks	5 bales × 20 kg every 6 months
Repair	90 kg/week	75 kg/week		
Maintenance	10 kg/day	25 kg/day	27 kg/week	60 kg/month
Total/month	660 kg	900 kg	180 kg	90 kg
Estimated price (inclusive of all taxes)	€80/t	€9.50/bale	€8.50/bale	€10.50/bale
Cost of bedding/month	€52.80	€152	€85	€47.25
Labour:				
Installation	15 min	15 min	10 min	10 min
Daily upkeep	10 min morning 10 min evening	10 min morning 5 min evening	5 min morning 5 min evening	5 min morning 5 min evening
Repair	20 min/week	20 min/week	10 min/6 months	10 min/6 months
Total	680 min or 11 h 20	530 min or 8 h 50	307 min or 5 h 07	302 min or 5 h 02
Labour cost/month	€113.30	€88.33	€51.16	€50.33
Average cost/horse/month	€166.10	€240.53[a]	€136.16[a]	€97.58[a]

Note: [a]To this must be added €10–15 of straw to compensate for any lack of intestinal fibre and to occupy the horse.

Fig. 18.1. A groom laying a bed of hemp straw.

The following parameters are used for this evaluation:

- a horse of average size (approximately 450 kg)
- leaving his stable on average 2 h/day
- boxed for 12 months of the year
- box size: 9–10 m²
- labour cost of €10/h.

18.2 Litter for Small Mammals

Small pieces of hurds are particularly suitable as bedding for hamsters, guinea pigs, chinchillas, mice, rats, gerbils, canaries, budgerigars and pigeons, as well as rabbits, snakes and even for working dogs kennelled outside (Fig. 18.2).

Fig. 18.3. Cat litter made from granules of hemp powder.

Fig. 18.2. A comfortable hemp bed.

This bedding absorbs urine and faeces and is particularly good at capturing odours. Bedding need only be changed on a weekly basis at most.

This property is relied upon for the production of small granules, which are sold as litter for cats and other small mammals.

The granular form allows the powder to absorb animal urine and excrement, as well as any associated odours. It is therefore popular as cat litter (Fig. 18.3). Cleaning of the litter tray is easy as the granules compact down after becoming soiled. As such, they can be used as a fertilizer or composted down. Only the soiled part of the litter needs to be replaced, putting an end to the daily litter duty.

18.3   Cat Litter

When examining the physical constituents of hemp powder, it was observed that its ability to absorb water was very high: approximately 350%.

18.4   Bedding for Cattle

Straw powder in its pulverised form can also be used as bedding for heavier animals. It is used in cattle sheds as an alternative to barley or wheat straw.

19 Chemical and Morphological Differences in Hemp Varieties

Gilbert Fournier

*Laboratoire de Pharmacognosie, UMR 8076 CNRS (BioCIS),
Faculté de Pharmacie, 5 rue J.-B. Clément, F - 92296 Châtenay-Malabry Cedex*

19.1 Introduction

Hash, marijuana, ganja, sinsemilla, skunk, nederwiet, pot, weed, grass…these names all evoke hemp, cannabis smoking and THC. This is not, of course, what we will be dealing with here. We will present a description of the hemp that has been anchored firmly in our society as a source of textiles and cordage up until the early 20th century. This use has witnessed a resurgence recently with the development of a wide range of new industrial applications.

19.2 History

Hemp (*Cannabis sativa* L.) is, above all else, a textile plant with cortical fibres and oil-rich seeds. It is one of the oldest non-food plants used by humans. It originates from Central Asia and its use dates back some 6000 years. Mentioned for the first time in a 4700-year-old Chinese text, it was described in detail in a botanical treatise of the 15th century BC. The use of hemp in Asia, Europe and around the Mediterranean basin stretches back into distant history. Even though its main use is not specified in ancient texts, it is certain that hemp has been considered an important plant for humans. It was widely used as a textile, for linen and for cordage, while the oil extracted

from the seeds was used as a foodstuff, as a drying oil and therapeutically.

In Southern Europe, the use of textile fibres derived from hemp in the manufacture of cordage and rough cloth materials began during the Middle Ages, and the fibres were used for finer cloth from the start of the Renaissance period. This industry continued to develop until the middle of the 19th century. Some 175,000 ha was devoted to its production at this time; its fibres were used in the production of canvas, woven material, cordage and string, while the seeds were used to produce oil for various purposes. This industry, together with other natural textile and oil industries, has subsequently fallen victim to progress. The disappearance of sailing ships, competition from other textiles (cotton, sisal and jute) and oils (groundnut) and the appearance of synthetic textiles have seen to that. This disaffection with hemp can also be attributed to the insufficient revenue that its cultivation can command and the difficulty in finding the necessary labour, especially for the onerous tasks such as harvesting and retting in water.

From a peak of 175,000 ha in 1850, the area cultivated has fallen regularly. In 1960, there were only 700 ha left devoted to hemp production in France; a trend mirrored across Europe. From 1961, hemp producers needed to undertake specific actions in order to preserve cultivation:

- The first initiative concerned harvesting, which evolved from manual methods towards an entirely mechanized cultivation, ending with fibre extraction undertaken in a workshop. This initiative was only made possible by the concurrent development of monoecious varieties that were suited for mechanical harvest.
- The second initiative consisted of developments in the use of hemp fibres for use in the manufacture of fine and resistant paper (speciality papers). Between 1970 and 1994, this industrial sector was the only one to use hemp, requiring the cultivation of only a few thousand hectares yearly. Since then, the industrial sectors using fibres have grown and we have witnessed the diversification of possible opportunities and uses and devised new openings for hemp fibres and its by-products.

Hemp currently is employed on an industrial scale and has represented approximately 10,000 ha/year in France since the start of the 21st century.

19.3 Botany

Hemp (*Cannabis sativa* L., Cannabaceae) is an annual herbaceous plant with iconic palmate-lanceolate leaves. The flowering tips are mixed with foliate bracts. The leaves, and in particular the bracts, are the only parts of the plant with secretory hairs, producing a resin whose chemical composition is complex. One of the main characteristics of hemp is its malleability: when grown under favourable conditions (soil, water, sunlight), it can attain a height of 5–6 m and a span of 2–3 m. Industrially cultivated varieties, however, are sown densely (50 kg of seed/ha) and their foliar development is therefore limited. Consequently, they rarely exceed 2–2.5 m in height.

Even though hemp is naturally a dioecious species, contemporary industrial varieties are almost entirely monoecious (plants have separate male and female flowers on each plant). This presents a number of agronomic advantages; most notably, the production of seeds on all the stems and, as has been mentioned already, the possibility of harvesting a uniform stand mechanically.

Another consequence of hemp plasticity is its variable maturation: certain varieties flower early, whereas others flower much later. This allows hemp to adapt easily to different latitudes.

Geographically, in France, hemp cultivation is limited today to a score of departments, situated principally in the west, centre, southwest and especially the east (the Aube primarily) of the country.

19.4 Chemistry

Due to its botanical, chemical and pharmacological uniqueness, hemp has been and continues to be the subject of numerous studies. During the 1970s, the United Nations substantially funded studies on hemp. As a consequence, hemp is today one of the best-known plants. This fact was demonstrated well in 1980 by the publication of a review article that provided an inventory of all the constituent chemicals that had been identified (Turner *et al.*, 1980): 421 compounds were identified, belonging to different phytochemical groups including nitrogen-containing compounds, fatty acids, steroids, terpenes (mono-, sesqui-, di- and triterpenes), flavonoids, vitamins, pigments and of course around 60 cannabinoids, the specific terpenes of hemp (Turner *et al.*, 1980; Seth and Sinha, 1991). The three most important are (Fig. 19.1):

- delta-9-tetrahydrocannabinol (Δ-9-THC), responsible for the psychoactive properties of the plant, whose concentration therefore needs to be known;
- cannabidiol (CBD), which is the main cannabinoid in hemp cultivated in France and has no psychoactive properties;
- cannabigerol (CBG), the biogenetic precursor of the previous cannabinoids, also devoid of any psychoactive properties.

It should be noted that it is due to the presence of the classified substance, Δ-9-THC, that hemp is also classified as a narcotic.

Numerous studies have demonstrated that the relative proportions of the main cannabinoids depend on genetic factors, whereas their concentration is influenced by ecological factors (Small, 1979).

Genetic factors contribute to a considerable variation in Δ-9-THC content (1 to 3-fold), whereas ecological factors exert much less influence (by a factor of approximately 1 to 2). Thus, depending on the relative proportion of Δ-9-THC and CBD, substances that seem to be related to each other can actually have several chemical varieties distinguished (these are termed chemical types or chemotypes) (Fournier, 1981, 2000). These are summarized in Fig. 19.2 and their characteristics are as follows:

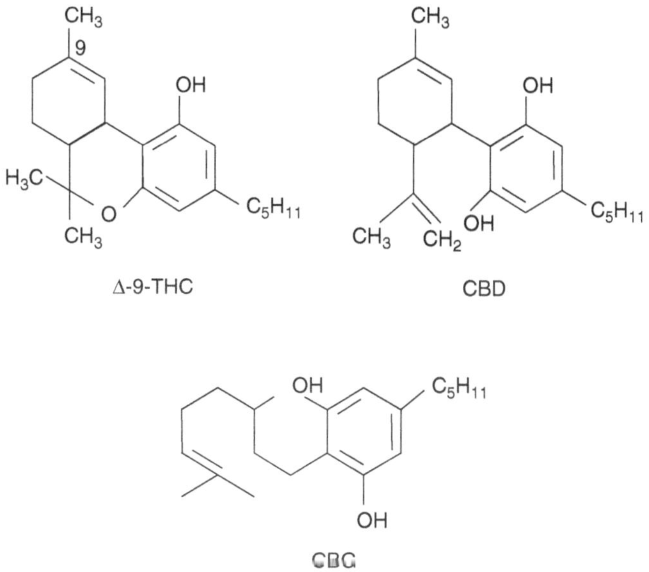

Δ-9-THC

CBD

CBG

Fig. 19.1. Chemical structures of delta-9-tetrahydrocannabinol (Δ-9-THC), cannabidiol (CBD) and cannabigerol (CBG).

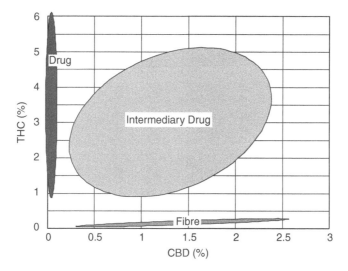

Fig. 19.2. Main chemical varieties of hemp (*Cannabis sativa*).

1. Drug varieties characterized firstly by a high Δ-9-THC content and secondly by a virtual absence of CBD. These varieties are found growing primarily in warm climates (Mexico, Afghanistan, India, Africa and Jamaica). In temperate climates, however, these varieties can be grown in greenhouses, often using hydroponic techniques (nederwiet, for example).

2. Intermediary drug varieties that possess high levels of both Δ-9-THC, conferring pharmacological properties on them, and CBD. These varieties come principally from the Mediterranean basin (Morocco and Lebanon).

3. Fibre varieties, with a particularly low Δ-9-THC content. The main cannabinoid is CBD. These are the varieties found in France and other countries with a temperate climate.

Figure 19.2 shows that the fibre varieties cultivated in France a few years ago register a Δ-9-THC concentration below 0.3%. Despite this being at the upper limit permitted by the European Community until 2000 (0.2% as from the 2001 season) (*Journal Officiel de la République Française*, 2004), a selection programme was undertaken to reduce the Δ-9-THC concentrations of French varieties. It should be noted here, however, that these levels of Δ-9-THC do not demonstrate any psychotropic effects.

Hemp selection was undertaken by the *Fédération Nationale des Producteurs de Chanvre* (FNPC), in partnership with the *Laboratoire de Pharmacognosie de la Faculté de Pharmacie de Châtenay-Malabry.*

The selection programme started by looking for a very strict operating protocol, both with regard to cultivation, the sampling method and the chemical analysis itself, which involved the testing of more than 5000 samples/year. It should be pointed out that the official texts concerning the 'EU method for the quantitative determination of Δ-9-THC in hemp varietals' (*Journal Officiel de la République Française*, 2004) has been developed from, most notably, the French experience in this field (Fournier *et al.*, 2001, 2003; Fournier, 2003a,b, 2004). This document not only describes the dosing methodology (by gas chromatography) but also specifies the procedure for sample selection (season of collection,

number of plants to be sampled, plant parts to be collected) and the treatment of the samples collected (drying and storage). This selection programme has been described elsewhere in this book.

During the selection programme, a new variety containing no Δ-9-THC (<0.005%, and therefore barely detectable by chromatographic analysis) was discovered. CBD is present in this new variety, but at a low concentration. By contrast, the major cannabinoid is CBG. This last cannabinoid, as mentioned above, is the biogenetic precursor to Δ-9-THC and CBD; this therefore suggests that the biogenesis of THC and CBD is halted early, favouring the build-up of CBG. Three new varieties presenting this cannabinoid profile have been created: Santhica 23, Santhica 27 and Santhica 70. These are second-generation fibre varieties (Fournier *et al.*, 2004).

Finally, we have discovered very recently a new chemical type characterized by the absence of cannabinoids. Biosynthesis is halted even earlier in the metabolic pathway and must occur before the assembly of cannabinoid precursors. This certainly represents an interesting and very promising discovery.

19.5 Principal Morphological and Chemical Characteristics of Hemp Cultivated in France

At this point, the following question can be asked: are there any morphological differences among these varieties and can they be distinguished from each other in the field? An unequivocal answer cannot be given.

The main differences observed concern cultivation practices and, in particular, population density. Seeds of the same variety planted at a low (1–2 kg/ha) or high (50 kg/ha) density will produce plants with very different morphologies. In the first case, they will branch considerably, whereas in the second they will be reduced to a primary stem. By contrast, in identical cultural conditions, plants with different chemotypes, as outlined above, will have similar morphological characteristics and only chromatographic analysis will distinguish them.

Nevertheless, it should be pointed out that certain plants presenting the 'drug' or the 'intermediate drug' chemotype, selected for their particularly high Δ-9-THC content, are easily recognized morphologically. This is because they present hypertrophied resin-secreting hairs on their flowering or fruiting heads.

19.6 Conclusion

In France, agroindustry cultivates hemp for fibre while carefully abiding by current French and European laws. A thorough knowledge of this cultivated plant material has allowed a selection programme to be implemented based on agronomic and chemical criteria.

The different stages in the selection process conducted in France over the past 15 years is based on chemical criteria and has allowed Δ-9-THC-free varieties (Santhica 23 and Santhica 27) to be offered to the user industries. This work is ongoing and it is reasonable to think that it will be possible to produce, in the not too distant future, new varieties that contain no cannabinoids.

Despite the positive results obtained from chemotype selection, the significant similarities that exist between the different chemical types mean that we must remain vigilant. Any crop of hemp can, after all, conceal another…

20 Hemp Production Outside the EU – North America and Eastern Europe

Pierre Bouloc[1] and Janoš Berenji[2]
[1] *La Chanvriere de L'Aube (LCDA), France;*
[2] *Institute of Field and Vegetable Crops, Novi Sad, Serbia*

20.1 Canada and the USA

In the USA, during World War II, hemp production was encouraged briefly by a government programme (1942–1945). Synthetic fibre was new and had not yet usurped hemp's maritime role. After the war, it was banned in North America and across the world and 50 years elapsed before it reappeared as a crop, and then only on Canadian soil. Hemp was legalized by the Canadian government in 1998, marking the return of the hemp industry to the continent of North America.

According to Health Canada, in 1998 the Canadian government issued 12 licences for the production of hemp on 2400 ha. The following year, the demand for licences exploded, with 578 requests made, of which 33 were for scientific purposes, the remaining 545 being for commercial production. In 1999, this translated into an area under cultivation of 14,200 ha. Over the next 2 years, the permitted area under production fell. This can be attributed to market instability and the concomitant accumulation of a large stock of raw material.

The vicissitudes of the first decade of hemp production in Canada are explained in an AgCanada report; in 2005, the area licensed for hemp production in Canada increased almost threefold to 9725 ha (24,021 acres) (Fig. 20.1). The largest increases in hemp production area were in Manitoba and Saskatchewan. The area

under hemp production increased to its highest level in 2006, at 19,458 ha (48,060 acres), almost double that in 2005. Prairie Provinces again led the country in hemp production, with almost 97% of hemp area. Manitoba had 10,705 ha (26,442 acres) of hemp, followed by Saskatchewan at 6025 ha (14,882 acres) and Alberta at 2103 ha (5194 acres). In 2007, the area under hemp production decreased by about 68%, due primarily to the lack of processing facilities for hemp fibre and stock. In 2008, the area licensed for commercial hemp production in Canada decreased further by almost 47% to 3259 ha (8050 acres). The total number of licences issued by Health Canada was 85 in 2008, a significant decrease over the past few years.

In 2009, the area licensed for hemp production increased by 72% across Canada over that in 2008, that is, from 3259 ha (8050 acres) to 5602 ha (13,837 acres). The major increases in area were again on the Prairie Provinces, led by Manitoba (145%) and Saskatchewan (34%). The area of land under hemp production in Alberta increased by 200 ha (524 acres) in 2008. The only province to report a decrease in hemp area in 2009 was Quebec, from 134 ha (331 acres) in 2008 to 92 ha (227 acres) in 2009. In British Columbia, the area for hemp production increased from 5 ha in 2008 to 84 hectares in 2009. Similarly, the area in Ontario increased from a mere 8 ha in 2008 to 132 ha in 2009. Table 20.1 below provides data on commercial hemp production in Alberta and Canada from 1998 to 2009. Tables 20.2 and 20.3 provide data on hemp production in Canada by province from 1998 to 2009 in hectares and acres.

In 2010, the area licensed for hemp production increased by almost 94% across Canada to that in 2009; that is, from 5602 ha (13,837 acres) to 10,856 ha (26,815 acres). Major increases in area were again on the Prairie Provinces, led by Saskatchewan (204%) and Manitoba (156%). The area under hemp production in Alberta increased by 1304 ha (3221 acres) in 2009, an increase of almost 267%. Similarly, the area in Ontario increased

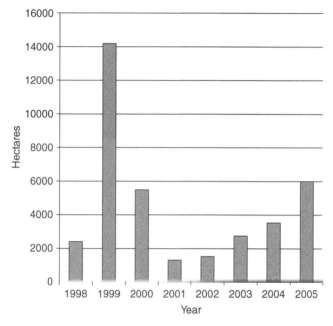

Fig. 20.1. Growth in hemp production in Canada following its authorization in 1998, based on figures from AgCanada.

Table 20.1. Hemp Seeded Acreage in Alberta and Canada, 1998–2011.

| Year | Alberta | | Canada | | |
	Hectares	Acres	Hectares	Acres	% Alberta
1998	38	94	2,400	5,927	1.58
1999	754	1,862	14,205	35,086	5.31
2000	306	756	5,485	13,549	5.58
2001	113	279	1,316	3,251	8.59
2002	123	304	1,530	3,779	8.04
2003	153	379	2,733	6,750	5.61
2004	639	1,578	3,531	8,722	18.10
2005	916	2,263	9,725	24,021	9.42
2006	2,103	5,194	19,458	48,060	10.81
2007	1,455	3,594	6,132	15,146	23.73
2008	582	1,438	3,259	8,050	17.86
2009	782	1,932	5,602	13,837	13.96
2010	2,086	5,152	10,856	26,814	19.22
2011	6,434	15,892	15,720	38,828	40.93

Table 20.2. Hemp Seeded Acreage in Canada by Province, 1998–2011 (Hectares).

Year	BC	Alberta	Sask.	Man.	Ontario	Quebec	NB	NS	PEI	Yukon	Canada
1998	72	38	263	606	1,163	24	214	19	0	0	2,400
1999	225	754	3,096	8,889	1,021	86	4	126	4	0	14,205
2000	291	306	1,426	2,902	217	239	1	102	2	0	5,485
2001	96	113	392	472	209	30	0	0	0	4	1,316
2002	200	123	449	597	142	19	0	0	0	0	1,530
2003	7	153	672	1,468	397	13	4	18	0	0	2,733
2004	18	639	1,004	1,655	183	10	4	18	0	0	3,531
2005	0	916	3,429	5,018	251	74	19	18	0	0	9,725
2006	111	2,103	6,025	10,705	398	91	8	18	0	0	19,458
2007	70	1,455	2,293	2,088	40	182	4	0	0	0	6,132
2008	5	582	1,537	993	8	134	0	0	0	0	3,259
2009	84	782	2,061	2,435	132	92	0	0	16	0	5,602
2010	64	2,086	4,214	3,799	372	321	0	0	0	0	10,856
2011	8	6,434	4,026	4,596	354	290	8	0	4	0	15,720

Table 20.3. Hemp Seeded Acreage in Canada by Province, 1998–2011 (Acres).

Year	BC	Alberta	Sask.	Man.	Ontario	Quebec	NB	NS	PEI	Yukon	Canada
1998	178	94	650	1,497	2,873	59	529	47	0	0	5,927
1999	556	1,862	7,647	21,956	2,522	212	10	311	10	0	35,086
2000	719	756	3,522	7,169	535	590	2	252	4	0	13,549
2001	237	279	968	1,166	516	74	0	0	0	10	3,251
2002	494	304	1,109	1,475	351	47	0	0	0	0	3,779
2003	18	379	1,661	3,625	981	32	10	44	0	0	6,750
2004	44	1,578	2,480	4,088	452	25	10	44	0	0	8,722
2005	0	2,263	8,470	12,394	620	183	47	44	0	0	24,021
2006	273	5,194	14,882	26,442	982	224	20	44	0	0	48,060
2007	173	3,594	5,664	5,157	99	450	10	0	0	0	15,146
2008	12	1,438	3,796	2,453	20	331	0	0	0	0	8,050
2009	207	1,932	5,091	6,014	326	227	0	0	40	0	13,837
2010	158	5,152	10,409	9,384	919	793	0	0	0	0	26,814
2011	20	15,892	9,944	11,352	874	716	20	0	10	0	38,828

from 132 ha (325 acres) in 2009 to 372 ha (919 acres) in 2010, an increase of about 283%. In Quebec, the area under commercial hemp production increased more than three-fold to 321 ha (793 acres) in 2010 over that in 2009. The only province to report a decrease in hemp area in 2010 was British Columbia, from 84 ha (207 acres) in 2009 to 64 ha (158 acres) in 2010. No hemp was grown in the Atlantic Provinces and the Yukon Territories in 2010. Table 20.1 provides data on commercial hemp production in Alberta and Canada from 1998 to 2010. Tables 20.2 and 20.3 provide hemp production in Canada by provinces from 1998 to 2010 in hectares and acres (http://www1. agric.gov.ab.ca/$department/deptdocs.nsf/all/econ9631).

Unlike the European Union, where most of the crop produced is destined for fibre production, the majority of Canadian hemp is produced to meet the growing demand for hemp as a foodstuff in the USA, where it is used as a nutraceutical, as well as in the manufacture of certain cosmetic products such as soap and shampoo.

A recent study in the USA found that the sale of hemp-based foodstuffs had increased by 60% since 2003 (Chapter 16). The net total amount of sales is actually estimated at US$12 million per annum for foodstuffs and US$50 million for cosmetic products.

This volume is still small, but its continued growth suggests that the demand for hemp seed may become as significant as the current global demand for hemp fibre. Leson (Chapter 16) addresses the American interest in this product that accounts for this expansion.

This all points to a profusion of all sorts of projects and initiatives and suggests that hemp production may continue to grow in importance over the coming years. Today, production is focused very much on hemp seed. It is not realistic to imagine that this market will be able to soak up all of the hemp produced. It will therefore be necessary to consider other end uses: in particular, the development of hemp straw as animal bedding and its use in the automotive industry are examples worth citing.

The possibility of selling hemp straw to the American equine industry is worth exploring. North American car manufacturers are already familiar with the use of natural fibres, so here the potential for the use of hemp can be explored.

In terms of prices, the purchase price of organic hemp seed is approximately US$0.85/lb, or US$1.87/kg (wholesale price during 2005). This equates to €1.55/kg in 2005. Raw fibre sold at €0.82/kg (where €1 = US$1.21).

Working from these figures, it is possible that the average gross revenue per hectare of organic hemp is of the order of €928/ha. Traditional products yield a revenue of €606/ha. A review of these figures shows that the harvesting of hemp seed alone ensures the profitability of this crop. These figures can be compared with those obtained for other hemp products produced across the world, and described in Chapter 8, 'The Agricultural Economics of Hemp'.

Finally, it is worth noting that the various interest groups participating in the development of hemp have formed themselves into an organization capable of standing up for their various professional interests with intelligence and pugnacity.

The Canadian Hemp Trade Alliance (CHTA), or *Alliance Commerciale Canadienne du Chanvre* (ACCC), represents some 80 hemp farmers, 13 processors, 15 distributors, 8 retailers, 12 researchers or consultants and

2 associations. In addition to their role in promoting hemp to the public and providing information, this association conducts studies that seek to demonstrate the various uses of hemp and its health benefits. Their work serves as an excellent example of the efficient and practical role of a professional organization.

NB In 2009/10, Canada had some 12,800 ha under hemp cultivation and anticipated that 13,000 ha would be cultivated in 2010/2011 (source: Agriculture Canada – http://www.agr.gc.ca).

20.2 Past, Present and Future of the Hemp Industry in Serbia

20.2.1 History of industrial hemp in Serbia

Hemp is a crop with a very long tradition of growing and processing worldwide, especially in the eastern part of Europe, including the territory of the former Yugoslavia and the present Serbia (Berenji, 1996b). Hemp was introduced long ago in Serbia. There is evidence of considerable hemp production in the 15th and 16th centuries. The Turks evidently taxed hemp production in Serbia. The Austro-Hungarian empress, Maria Theresa, recommended in 1765 that only high-quality hemp should be grown in this area. Therefore, the Imperial Chamber of Commerce settled colonists from northern Italy in northern Serbia (the present Vojvodina Province), who transferred high-quality Italian hemp, along with the skills of hemp cultivation and processing, from Italy to Serbia.

The production and processing of hemp in Serbia has been concentrated primarily in the Province of Voïvodine, although production was well known across the whole country. Serbia was, in fact, the biggest hemp producer in the old Yugoslavia (Makendić, 1937). Peak production was reached in 1949, with a total area under cultivation of 108,215 ha. At that time, Yugoslavia was one of the largest hemp producers in the world, the area under cultivation amounting to 25% of the total land under cultivation in Europe (excluding the USSR) and 6% of worldwide production. Despite this,

Yugoslavia and Serbia are usually overlooked in publications on the subject of hemp production in Europe.

The first sign of a crisis in the European hemp industry appeared in the 20th century, as the surface area under production went into a gradual but unstoppable decline. The crisis first affected Western Europe and was less severe in the east. Production in Yugoslavia fell by 80% in 1968, with a further decline of 80% over the next 20 years.

Serbian production of industrial hemp, essentially in the Province of Voïvodine, stabilized at the level of c.100–250 ha for fibre hemp and an additional 20–25 ha for seed hemp; that is, certified seed production (Berenji, 1996a) at the end of the 20th century. At the beginning of the 21st century, industrial hemp virtually disappeared from Serbia. Currently, only occasional farmers produce hemp seed for bird food or for oil.

20.2.2 Hemp research in Serbia

Research on hemp in Serbia began officially in 1952 with the creation of a research station in Bački Petrovac as part of the Institute of Agricultural Research (the present Institute of Field and Vegetable Crops) in Novi Sad. The founder and first director of the station, Mr Mihal Husar, was responsible for the breeding of hemp, as well as for research into production technology (Fig. 20.2). It is interesting to note that Mr Rudolf Fleischmann, the famous Hungarian breeder from Kumpolt (in Hungary), first encountered hemp in this area while staying at Ruma, a small town close to Novi Sad.

The research station conducted a great deal of research into the breeding of industrial hemp. To start with, hemp grown to produce fibre made use of Italian varieties. These were named *Futoška, Titelska, Apatinska, Vukovarska, Potočka, Osječka, Beljska, Leskovačka,* etc..., according to the name of the area in which they were mainly grown. It was interesting to note that some of the genetic stock that could be termed 'Yugoslav hemp' was sent as far away as Chile and that 34 genotypes from the old Yugoslavia, and Serbia in particular, were registered with the VIR (Vavilov Institute of Plant Industry) gene bank (Lemeshev et al., 1994).

In the 1960s, certified seed stock of the Bologna and Carmagnola cultivars was imported from Italy and was reproduced. The monoecious French cultivar 'Fibrimon' was also tried out, but with limited success. At the end of the 20th century, the Hungarian cultivars 'Kompolti', and especially 'Uniko-B' and 'Kompolti Hybrid TC', enabled a significant contribution to be made to improvements in the yield and quality of Serbian fibre-producing hemp.

The cultivar 'Novosadska konoplja' is a local dioecious cultivar developed at the

Fig. 20.2. Mihal Husar, demonstrating a fibre hemp crop density experiment in 1952.

Institute of Field and Vegetable Crops in Novi Sad (Berenji and Sikora, 1996). Currently, there are three officially registered high-yielding, low THC industrial hemp cultivars on the Serbian variety list: Marina (dioecious), Helena (monoecious) and Diana (F1 hybrid). These cultivars were developed at the Institute of Field and Vegetable Crops in Novi Sad and all were approved officially in 2002.

In the past, hemp seed traditionally was grown by planting several hemp plants at the end of cornfields, as shown in Fig. 20.3. Certified hemp seed production is now organized by the Institute of Field and Vegetable Crops in Novi Sad. Hemp seed production is contracted out with private farmers, who grow hemp in small fields (1–2 ha). Production is based almost completely on manual labour. The seed crop is sown at a row-to-row distance of 70 cm, 20–30 cm in the row. Starting from the time of pollination of dioecious cultivars, the identified male plants are removed. The crop is cut by sickles, dried in sheaves in the field and threshed by beating the tops of a few plants against an angled wooden board (Fig. 20.4). The seed is then collected, cleaned and bagged. The fibre-rich stalks left behind after threshing are usually sold as fuel or simply burned in the field.

Fig. 20.3. The traditional way of hemp seed production, growing a few plants at the end of cornfields (shown left, in the background of the group) (15 August 1960).

Fig. 20.4. Hand threshing of seed hemp.

20.2.3 Production technology for Serbian industrial hemp

Historically, hemp was produced by small farmers and virtually all the work involved in its production and processing was conducted by hand. Poverty dictated that people satisfied their own needs. Each farm was therefore able to produce its own material, as well as a range of other articles made from hemp. There virtually was not a single household without equipment for hemp processing and hemp-based products of some kind (Fig. 20.5) (Kišgeci, 1996).

The modernization of agriculture that took place during the 1930s turned hemp into an industrial crop. Between 1900 and 1960, there were over 30 factories in Serbia processing hemp (Fig. 20.6). During this period, a unique hemp 'stock exchange' existed in Odžaci, which influenced prices as far away as Italy and England (Fig. 20.7).

Hemp growers' interest in this plant can be explained by a number of advantages that hemp has over other crops. Hemp, as an annual crop sown in the spring, can be combined with other crops for the purposes of crop rotation. It is harvested relatively early in the season and leaves the soil in good condition. In view of these qualities, hemp shows itself to be an interesting crop when compared to alternatives (Drezgić et al., 1975).

Fig. 20.5. Home processing of hemp.

Нови Врбас
Ujverbász
Neuwerbass

Фабрика кудеље
Hungária kendergyár
Hanffabrik

Fig. 20.6. Hemp processing facility, popularly called hemp factory, in 1953.

Fig. 20.7. The hemp stock market in Odžaci at the beginning of the 20th century.

The vegetative period of hemp is relatively short and this allows the effects of summer droughts to be avoided, or minimized; an important consideration for crops grown on the Pannonian plain. This also explains why hemp yields over time are more reliable and less variable than those obtained for maize and other cereal crops.

The technology required for hemp cultivation is relatively simple and, except for harvesting, does not require any special equipment.

With regard to plant protection, hemp can be described as a true ecological plant, for it does not require intensive application of pesticides. In Serbia, there have been no diseases or insects causing notable economic losses on a regular basis (Čamprag *et al.*, 1996). Hemp's reaction to weeds can legitimately be compared to that of a weed, for it is able to smother and destroy them, particularly when sown at high density (Starčević, 1978), although the same phenomenon can also be seen at lower sowing densities. In recent years, Johnsongrass (*Sorghum halepense*) has become the most serious weed reported in the Vojvodina Province, and the cultivation of industrial hemp is seen as an effective ecological means of controlling and eradicating this weed.

Over the past few years, the parasitic weed, *Orobranche*, which taps into the roots of its hosts to feed, has been reported on hemp plants in a number of areas. While the robust hemp plants do not appear visibly to be affected by these attacks, this phenomenon needs to be monitored for this parasite has become a significant menace to sunflower production. Experiments aimed at determining the racial structure of *Orobanche* sp. originating from hemp plots as opposed to that from sunflower plots are under way.

The naturally growing 'weedy hemp' or 'wild hemp' occurs seemingly spontaneously across the country, even in areas where hemp has never been grown (Vrbničanin *et al.*, 2008). Roadsides, ditch banks and various undisturbed areas are the most frequent places for hemp occurrence (Fig. 20.8).

Industrial hemp production for fibre is organized by companies who contract out to farmers, large and small, with farms of 50–100 ha to only 1–2 ha.

Where hemp is grown for fibre, a sowing density of 75 kg/ha is practised, with rows spaced 12.5 cm apart.

In the production of fibre hemp, a tractor-powered mower followed by hand binding, or the Russian-made mower-binder 'Žatka', are used. The sheaves are stacked into 'fishbone' or round stacks for drying in the field. Bundles or bales pressed from them are transported to hemp-processing facilities. This production technology is rather labour-intensive and its modernization is required.

20.2.4 Industrial hemp processing technology in Serbia

The hemp straw provided by farmers to processing facilities is stored in stacks and processed in two phases.

Three operating facilities for primary hemp processing currently exist in Serbia. The primary phase of hemp processing includes water retting of bundled straw, drying and stacking of retted straw (Fig. 20.9). It is followed by decortication, performed by breakers. The baled hemp fibre is the end product of the first phase. The woody part of the stalk is sold in bulk or in the form of briquettes, in both cases as a highly demanded fuel.

The secondary phase of fibre hemp processing takes place in two hemp factories. They produce threads, twines, ropes, cordage, rough canvases and similar hemp-based textile products from the raw material obtained from the primary processors.

20.2.5 The future of the hemp industry in Serbia

The lack of available equipment adapted to the harvesting of both hemp seed and straw represents a significant limitation on the present and future production of industrial hemp in Serbia.

Fig. 20.8. Spontaneous (weedy) hemp in a sunflower field in the Vojvodina Province (2007).

Fig. 20.9. Traditional method of hemp retting in Serbia.

In order to resolve this problem, the agronomists of the Institute of Field and Vegetable Crops in Novi Sad have combined forces with mechanical engineers from the Faculty of Engineering and Science in Novi Sad and from the Faculty of Mechanical Engineering in Belgrade. Prototypes of two machines have been built and tested. The first machine is a primitive reaping and binding machine used for harvesting seed-producing hemp, but is also able to harvest straw-producing hemp. The second machine is a threshing machine that passes through the crop in advance of the first machine, thus ensuring both phases of the hemp harvesting operation. These machines need further refinement but are already producing satisfactory results. Other work to develop and improve harvesting equipment for seeds and fibre is under way (Martinov *et al.*, 1995, 1996).

In spite of the fact that industrial hemp is rapidly gaining importance due to numerous novel opportunities for utilization (Berenji, 1996b, 2004; Berenji and Sabo, 1998; Berenji *et al.*, 2001), public opinion about hemp is still frequently linked primarily to drug consumption. It is therefore important to talk and write about this crop objectively, without avoiding the topic of hemp's potential as a drug, but placing emphasis on all the advantages that make hemp an attractive crop for production, as well as on processing and utilization in the near future (Berenji, 1998).

Hemp, or more precisely the psychotrope tetrahydrocannabinol, THC, is listed among the substances (drugs) whose production, distribution and use remain illegal in Serbia. Theoretically, this means total prohibition for growing *Cannabis sativa*. In practice, presumed intention may serve as the grounds for placing one under suspicion of taking part in an illegal activity. On the other hand, no special licence is necessary for farmers to grow industrial hemp. The impression is that sentences are often passed for possessing hemp or plant parts containing THC far below the level necessary for a detectable psychoactive effect. Such sentences are based merely on the confessions of the offenders that their intention is to use hemp as a drug. From the point of view of commercial hemp production, this attitude is acceptable but obviously the legislation should be refined further in order to differentiate more precisely and more objectively between non-drug industrial hemp (low THC) and drug hemp (high THC).

The modernization of hemp production, processing techniques and technology requires considerable investments that currently are not available to the domestic hemp industry.

21 The Role of Hemp in Sustainable Development

Pierre Bouloc[1] and Hayo M.G. van der Werf[2]
[1]*La Chanvriere de L'Aube (LCDA), France;*
[2]*INRA, Rennes, France*

21.1 Introduction

This final chapter opens up new perspectives and possibilities for hemp. At a time when humans are worrying about their future and questioning the impact their behaviour has on the environment, it is reassuring to note that a number of solutions are presenting themselves which have the potential to address these concerns.

Recent studies, building on those undertaken at the end of the 1990s, have demonstrated that hemp has the potential to make an important contribution to the protection of the environment and, in particular, the campaign against the greenhouse effect.

The following serve as reminders of some of the facts presented in this book:

- 1 ha of hemp can produce nearly 12 t of dry matter (DM)/year, with a humidity level of 15%.
- Exported straw: about 8 t (or 6.8 t of DM).
- Hemp seed: 1.2 t (or 1.2 t of DM).
- Root system and leaves: over 3.5 t (or 3.5 t of DM).

NB Hemp contains approximately 45% of the atmospheric carbon taken up during photosynthesis.

Considering that calculations of the amount of carbon stored must take into account the amount that has a life expectancy in excess of the annual life cycle, we have taken into consideration only the straw destined for use in products with a long life expectancy. The hemp seed, roots and leaves are therefore not considered.

As a result, hemp straw produced by 1 ha of land can stock approximately 3.06 t of carbon.[1]

Carbon is stored for a long time where the derivatives of straw are used in a perennial way, that is to say, where the straw is used to produce materials with a long life expectancy.

The modern uses to which hemp straw and its derivatives are put do precisely that, and thus fix carbon for long periods. For example:

- Hemp wool has a life expectancy of 20 years.
- Panels made from hemp fibre used in the manufacture of car doors will last, on average, for 7 years.
- Hemp cement lasts for over 30 years.
- Plastics made using hemp fibre can be expected to last, on average, for 7 years, etc.

It is clear from this that the use of hemp (and natural fibres in general) in long-lived materials can contribute to a reduction in the greenhouse effect.

Hemp is also a crop that demonstrates its environmental credentials during cultivation, for it uses no pesticides. In an attempt to understand more fully the environmental impact of hemp products, the hemp industry has undertaken a life cycle analysis (LCA) of two products:

- a thermoplastic compound containing hemp fibres;
- the hemp cement used for cement walls laid on a wooden frame.

The results of this analysis are presented in section 21.3.

Hemp production does not result in significant pollution. The study below presented by Hayo van der Werf, although conducted in Brittany, France, demonstrates this by comparing hemp with a number of other important European crops.

Here, once again, the environmental credentials of hemp are clearly demonstrated.

21.2 The Environmental Impact of Hemp Cultivation

21.2.1 Introduction

Hemp demonstrates a number of agronomic and environmental strengths. It covers the ground quickly once it has emerged from the earth and smothers the majority of weeds, thus eliminating any requirement for herbicides. It requires only a modest amount of nitrogen fertilization, as its roots are very deep and are able to use the mineralized nitrogen deposited in the soil during the summer. In France, there are no diseases capable of provoking significant losses; as a result, there is no need for any phytosanitary interventions during the growing cycle (FNPC, 2002). Hemp is very resistant to drought and, in practice, hemp cultivated for straw does not need to be irrigated. Only those crops cultivated for straw and seed are occasionally irrigated. When harvested, hemp leaves the ground clean, relatively dry and loosened for some depth. It can therefore be said that hemp is an excellent crop to follow and can improve the yield of crops such as wheat that are planted after it (FNPC, 2002; Gorchs and Lloveras, 2003).

In 2002, the European Union financed a study called HEMP-SYS (Amaducci, 2003).

This project's objective was to promote the development of a competitive, innovating and long-lasting hemp fabric industry. This involved the development of an improved production chain to produce high-quality hemp fabric that was ecologically sustainable. This was coupled with an integrated quality control system for the stems, fibres (crude and processed) and thread founded on eco-label criteria. In order to do this, the project had to evaluate the environmental impact of the various products used in the production of hemp textiles. It was in this context that the impact of hemp production in the field was studied by means of an LCA.

The main impacts of hemp cultivation are quantified and compared to those of other annual crops. For hemp, the effects of modified farming practices and the amount of nitrate lost through washing are explored.

21.2.2 Materials and methods

The environmental impact of eight annual crops, including hemp, has been estimated by an LCA, a method that allows the potential impact of a product or function by quantifying and evaluating the resources used and the environmental emissions at each stage of the life cycle: from the point at which those resources are extracted through to the production of the materials and the product itself, to the use of the product and its ultimate disposal and recycling (Guinée *et al.*, 2002).

The environmental problems (or impact categories) that will be considered need first to be identified. In this study, we have considered: eutrophication, climate change, acidification, terrestrial pollution, energy use and the use of cultivated land. Eutrophication includes all that may result from the introduction of excessive levels of nitrogen and phosphate fertilizers into the environment. Climate change is defined as the impact of emissions on the atmosphere's ability to absorb radiated heat. Acidifying substances can have a wide range of effects on the ground, water, living organisms, ecosystems and buildings. Terrestrial pollution covers the impact of toxic substances (in this study, this is limited to heavy metal pollution) on terrestrial ecosystems. Energy use relates to the use and using up of non-renewable energy resources. The use of cultivated land concerns the temporary non-availability of cultivated land as a resource while it is being used for the production of the crop.

Subsequently, the value representing the impact indicator for each category of environmental problem is calculated by multiplying each type of resource used and each type of substance emitted by a *characterization factor* for each of the categories of problem to which the resource or substance can contribute. The impact is expressed in equivalent kg of PO_4 for eutrophication, equivalent kg of CO_2 for climate change, equivalent kg of SO_2 for acidification, equivalent kg of 1,4-dichlorobenzene for terrestrial pollution, in MJ for energy use and in $m^2/year$ for land used for cultivation. The characterization factors are quantitative

representations of the additional environmental pressure per unit of a given material emitted. The characterization factors used in this study are provided in van der Werf (2004).

This study concerned itself with hemp production from a non-combined crop (i.e. there was no production of hemp seed) and seven other annual crops in France. The inputs and yields for the eight crops are summarized in Table 21.1. Where hemp is concerned, a modification in farming practice (the use of pig slurry and no-tilling) was also studied, together with a more optimistic hypothesis regarding nitrogen losses in the form of nitrates (20 kg/ha instead of 40) (Table 21.2).

This study takes into consideration the impact resulting from field production, including harvesting, transport to the processing unit and drying of the product (this only applies to maize). The impacts resulting from the production

Table 21.1. Inputs, yields and emissions for nitrogen in the form of nitrates (all in kg/ha) for hemp and seven other annual crops produced in France.

	Hemp	Sunflower	Rape	Peas	Wheat	Maize	Potato	Sugarbeet
Inputs								
N (ammonium nitrate)	75	85	110	0	130	100	170	220
P_2O_5 (triple superphosphate)	38	32	41	46	64	51	80	101
K_2O (potassium chlorate)	113	21	30	95	90	30	293	180
CaO	333	167	167	333	333	333	0	333
Seed	55	5	2.5	200	120	20	2,000	2.5
Pesticide (active ingredient)	0	1.0	2.9	3.2	2.9	3.5	5.5	3.7
Diesel	65	79	81	87	101	91	165	137
Natural gas (for drying the seed)	0	0	0	0	0	167	0	0
Agricultural machinery	16.4	23.0	23.3	26.9	28.7	21.3	29.0	34.2
Grain yield dry matter	–	2,100	2,970	4,110	5,910	6,440	–	–
Straw yield dry matter	6,720	–	–	1,410	3,870	–	–	–
Dry matter yield	–	–	–	–	–	–	10,000	11,540
Followed by an intermediate crop (%)[a]	0	0	0	0	50	0	0	0
Following crop	Wheat	Wheat	Wheat	Wheat	Maize	Wheat	Wheat	Wheat
NO_3-N emitted	40	40	40	70	40	40	40	40

Note: [a]Indicates the percentage of cases in which we believe an intermediary crop was sown between the harvesting of the crop and the sowing of the following crop.

Table 21.2. Inputs, yield and emissions of N in the form of nitrate (all in kg/ha) for three different scenarios for hemp production in France.

	Pig slurry	No-till	Reduced washout
Inputs			
Pig slurry	20,000		
N (ammonium nitrate)	0	75	75
P_2O_5 (triple superphosphate)	0	38	38
K_2O (potassium chlorate)	51	113	113
CaO	333	333	333
Seed	55	55	55
Pesticide (active ingredient)	0	0	0
Diesel	72	39	65
Agricultural machinery	18.8	11.6	16.4
Straw yield dry matter	6,720	6,720	6,720
Followed by an intermediate crop (%)[a]	0	0	0
Following crop	Wheat	Wheat	Wheat
NO_3-N emitted	40	40	20

Note: [a]Indicates the percentage of cases in which we believe an intermediary crop was sown between the harvesting of the crop and the sowing of the following crop.

and sourcing of key inputs (agricultural machinery, diesel, fertilizer, pesticides and seeds) have been taken into consideration using the method proposed by Nemecek and Heil (2001). The data for energy contributions and transport are derived from the BUWAL 250 database (BUWAL, 1996). Buildings have not been included as their contribution to the impact of annual crops is negligible: 0–2% (van Zeijts and Reus, 1996).

For all the crops, we have assumed that good agricultural practice has been followed (*Bonnes Pratiques Agricoles* – BPA); that is to say, reasonable fertilization and integrated crop protection. For hemp, a straw yield of 8000 kg/ha with 16% humidity has been used. For the other crops, an average yield for the period 1996–2000 (AGRESTE, 2001; FAO, 2004) has been used. Details of the calculations of the field emissions are available from van der Werf (2001).

21.2.3 Results

Input requirements and impact for the eight crops

The amount of any input required varies from one crop to another (Table 21.1): from 0 (peas)

to 220 (sugarbeet) kg/ha of N; from 32 (sunflower) to 101 (sugarbeet) kg/ha of P_2O_5; from 0 (hemp) to 5.5 (potato) kg/ha for the active pesticide ingredient; and from 65 (hemp) to 165 (potato) kg/ha for diesel. Hemp and sunflower can be characterized as crops requiring little input, whereas potatoes and sugarbeet can be described as crops requiring a high input.

The impact is variable (Table 21.3). Differences are lowest for land use (10,000–10,500) and most significant for terrestrial ecotoxicity where the range is 0.1–6.7. For climate change (2300–4900), acidification (8.3–24.5) and energy use (11,400–26,300), there is a fairly sizeable variation between crops, whereas for eutrophication (20.2–34.4) this is modest.

Eutrophication is low (approximately 20 kg eq-PO_4) for hemp, sunflower and rape, but high (34kg eq-PO_4) for peas (Table 21.3). The impact on climate change is low for hemp and sunflower (2300 kg eq-CO_2) but high for potato (4120) and sugarbeet (4900). Acidification is low for peas, hemp and sunflower (8–11 kg eq-SO_2) and high for potato and sugarbeet (22–25). Terrestrial ecotoxicity is very low for peas (0.1 kg eq-1,4-DCB), low for sunflower, hemp and rape (1.8–2.5) and high for potato and sugarbeet (4.9–6.7). Energy use is low for

Table 21.3. The environmental impact arising from the production of 1 ha of hemp compared to seven other annual crops in France.

Impact category	Unit	Hemp	Sunflower	Rape	Peas	Wheat	Maize	Potato	Sugarbeet
Eutrophication	kg eq-PO_4	20.5	20.2	20.6	34.4	21.9	21.0	23.8	24.1
Climate change	kg eq-CO_2	2,330	2,300	2,700	2,890	3,370	3,280	4,120	4,900
Acidification	kg eq-SO_2	9.8	10.8	12.8	8.3	16.3	13.6	22.4	24.5
Land pollution	kg eq-1,4-DCB	2.3	1.8	2.5	0.1	4.0	3.0	4.9	6.7
Energy use	MJ	11,400	11,900	13,800	11,800	18,100	23,000	25,600	26,300
Land surface used	m²/year	10,200	10,000	10,000	10,500	10,200	10,100	10,400	10,200

Note: kg eq = kg equivalent.

hemp, peas and sunflower (11,400–11,900 MJ) and high for maize, potato and sugarbeet (23,000–26,300). The difference between the amount of land under cultivation for each crop is negligible.

For all the impact categories (with the exception of land area used), the impact is invariably low for hemp and sunflower and invariably high for potato and sugarbeet. For rape and peas, the impact is relatively low and for wheat and maize, the impact is intermediary.

Contribution made by the substances produced and resources used

The contributions to the different environmental impact categories of the substances produced and the resources used have been evaluated for hemp (characterized by low readings), wheat (intermediary readings) and sugarbeet (high readings) (Table 21.4). For these three crops, eutrophication is due primarily to NO_3 (75–89%). Climatic change is due mainly to N_2O (56–59%) and to CO_2 (40–43%). Acidification is due to NH_3, SO_2 and NO_2 emissions and the three crops contribute comparable amounts. Terrestrial ecotoxicity arises mainly from Ni (66–70%) and Cd (26–29%) in the ground. Energy use is accounted for mainly by the use of crude oil (44–46%) and natural gas (32–36%).

Even though these three crops show differing input requirements and differing levels of impact, they exhibit little difference in terms of the relative contributions of the materials and resources to the impacts.

The effect of alternative scenarios on hemp production

In certain areas, organic fertilizers are available at very low cost and their use can reduce production costs dramatically. The substitution of a mineral fertilizer with pig slurry reduces the climate change impact by −24% and energy use by −32%, but increases eutrophication (+16%), acidification (+140%) and terrestrial pollution (+1720%) (see Table 21.5). No-till farming is interesting because it reduces erosion, production costs and hours of work required. No-till farming reduces the climate change impact (−6%), acidification (−13%) and energy use (−16%).

The amount of nitrate washed out will be lower where the nitrate residue in the ground at the time of harvesting is low and when the period between harvesting and planting of the following crop is short. A reduction in washout of 40–20 kg N/ha reduces eutrophication (−43%) and climate change (−10%) (Table 21.5).

21.2.4 Discussion

This study has compared the potential environmental impact of hemp with that of seven

Table 21.4. Contributions (%) to the different environmental impact categories of the substances produced and resources used in the production of hemp, wheat and sugarbeet.

Impact category	Substance / resource	Hemp	Wheat	Sugarbeet
Eutrophication	NO_3	88.6	82.4	75.1
	NH_3	3.1	5.1	7.9
	PO_4	3.4	5.4	7.7
	NO_2	4.9	7.1	9.3
Climate change	N_2O	56.2	56.6	58.7
	CO_2	42.8	42.3	40.3
	CH_4	1.0	1.1	1.0
Acidification	NH_3	29.9	31.6	35.5
	SO_2	31.1	31.9	29.7
	NO_2	39.0	36.5	34.8
Terrestrial ecotoxicity	Zn	0	0	2.4
	Ni	70.2	67.5	65.7
	Pb	3.5	4.0	3.2
	Cd	26.3	28.5	28.7
Energy use	Crude oil	44.7	45.8	43.7
	Natural gas	32.7	32.1	36.4
	Uranium	11.4	10.6	9.0
	Coal	9.3	9.5	8.8
	Others	1.9	2.0	2.1

Table 21.5. The environmental impact arising from 1 ha of hemp production according to four different production scenarios. The reference scenario is that defined in Table 21.1, the other three are defined in Table 21.2.

Impact category	Units	Reference	Pig slurry	No-till	Reduced washout
Eutrophication	kg eq-PO_4	20.5	23.7	20.2	11.6
Climate change	kg eq-CO_2	2,330	1,770	2,200	2,090
Acidification	kg eq-SO_2	9.8	23.5	8.5	9.8
Terrestrial ecotoxicity	kg eq-1,4-DCB	2.3	41.9	2.3	2.3
Energy use	MJ	11,400	7,760	9,520	11,400
Surface used	m²/year	10,200	10,200	10,200	10,200

Note: kg eq = kg equivalent.

other crops in the context of farming practices and pedoclimatic conditions in France. Quantitative information on the environmental impact of hemp production in the field is rare. Patyk and Reinhardt (1998) present results from an approximate LCA of hemp in Germany. They found a similar energy requirement (12,300 MJ/ha) to ours (11,400). Their findings for acidification (6.6 kg eq SO_2) and for climate change (1421 kg eq CO_2) are lower than our figures (9.8 and 2330, respectively). Patyk and Reinhardt (1998) do not provide details of their methodology, making any analysis of this difference difficult.

This study has identified important differences between crops in terms of their input requirements and environmental impact. It has also shown that crops with low input requirements have a lower impact than crops with a greater input requirement.

While it is clear that the environmental impact of hemp production is less significant than that of many other crops, it is still worth considering how they could be reduced further.

The substitution of mineral fertilizers with pig slurry is of limited interest. A reduction in the impacts arising from climate change and energy use is seen, but at the cost of increases in eutrophication, and particularly acidification resulting from increased NH_3 emissions following the spreading of the slurry. Terrestrial ecotoxicity is also increased significantly, due to the presence of copper and zinc in the pig slurry. Thus, while the use of pig slurry may be of economic interest, generally speaking its effect on the environmental performance of hemp is a negative one.

By contrast, the use of no-till is interesting for it reduces energy uses, acidification and climate change. These effects result from reductions in diesel use and in the use of agricultural machinery. No-till also presents other environmental advantages, including a reduced risk of erosion and an increase in the organic material levels in the soil.

Any measures allowing the washout of nitrates to be reduced are likely to prove interesting, because a 50% reduction in nitrate washout has been shown to result in a 43% reduction in eutrophication and a 10% reduction in climate change. Whereas the reduction in eutrophication arises directly from reduced nitrate emissions, the effect on climate change is indirect and results from a lowering in N_2O emissions arising from the denitrification of NO_3. In general, optimizations in nitrogen fertilization and reductions in the time between harvesting and the planting of the next crop are the main measures recommended to reduce nitrate emissions.

21.2.5 Conclusions

This study assumed that crops were produced in France according to good agricultural practice (BPA). In France, the BPA correspond very much to those applied elsewhere in Western Europe. For this reason, the results of this study, while based on the situation in France, are applicable to the situation found elsewhere in Western Europe.

Relative to the other crops examined in this study, hemp and sunflower are the crops requiring the lowest amount of inputs and with the lowest environmental impact. This difference is particularly notable when comparisons are made with potato and sugarbeet, which can be described as crops with high-input requirements and a higher impact.

The use of low-intensity tillage (no-till) allows the environmental credentials of hemp to be improved yet further. This study has therefore allowed us to quantify the environmental impact of hemp, showing it to be relatively low compared to the other crops in this study.

Acknowledgements

These studies were conducted with a contribution from the European Union as part of the project QLK5-CT-2002-01363 'HEMP-SYS'. The data and opinions presented here are those of the author (Hayo van der Werf) and do not represent the opinions of the EU. The French corrections of Agnès van der Werf are also acknowledged.

21.3   Conclusions from the Life Cycle Analysis

21.3.1   Objectives of the study

The French hemp industry is seeking to diversify into new markets. In addition to supplying the paper industry, it is also targeting new markets in areas such as plastics/plasturgy (thermoplastic compounds containing hemp fibre) and the building trade (walls made of hemp cement on a wooden frame). Knowledge of the potential environmental impact arising from the production of these two products is required in order to identify the environmental benefits accruing as a result of the use of hemp in their manufacture. These properties have not been fully proven, despite the fact that a number of studies have been undertaken.

In order to evaluate the potential environmental impact of thermoplastics, plastics containing hemp fibre, and of walls made with hemp concrete, the FNPC and INTERCHANVRE initiated an LCA[2] of both products that would meet the ISO 14040 Standard for Life Cycle

Analysis (http://www.iso.org/iso/catalogue_detail.htm?csnumber=37456).

Description of the products studied

Two products have been studied: hemp fibre because it is used in the production of thermoplastics and hemp straw used in the production of hemp cement.

THERMOPLASTIC COMPOUNDS CONTAINING HEMP FIBRE.[3] Thermoplastics are synthetic materials derived from organic polymers that can be reversibly softened when heated and harden when cooled. These plastic materials can be mixed with fibres in order to produce composite materials. When loaded with fibres, these materials become more resistant and are termed 'reinforced thermoplastics'.

Today, the majority of thermoplastics are reinforced with glass fibre. A number of studies have shown, however, that they can be reinforced with hemp fibre and still preserve their mechanical properties.

WALLS MADE FROM HEMP CEMENT ON A WOODEN FRAME.[4] Mixed with a binder made from lime that sets in air and a hydraulic binder ready for use (Tradical 70© was used for this study), hemp straw allows the manufacture of cements that are spread mechanically over a weight-bearing wooden frame. This spreading process makes use of a panel placed behind the wooden frame, which, together with the frame, create a sort of mould. The mould is filled until the cement covers the frame. The resulting wall has a number of specific technical and performance characteristics, including high insulation, good acoustic correction and permeability to water vapour.

21.3.2 Methodology

Evaluation of the environmental impact

The objective of this section is to quantify the potential environmental impact resulting from the procedure under study. This quantification is based on a calculation of the amount of materials produced and consumed over the course of the inventory's analysis.

CHOICE OF IMPACTS TO BE CONSIDERED IN THIS STUDY. With a view to using hemp cement as a construction material, the impacts considered in this study have been selected according to the formulated recommendations of the standards that regulate the sanitary and environmental quality of construction materials (NF-P01-010). Eight different potential environmental impacts have been selected:

- the exhaustion of resources (in kg Sb eq)
- the acidification of the atmosphere (in kg SO^2 eq)
- the greenhouse effect over 100 years (in kg CO_2 eq)
- the destruction of the ozone layer (in kg CFC-11 eq)
- the creation of photochemical ozone (in kg C_2H_4 eq)
- the consumption of non-renewable energy
- the production of waste (in kg), air and water pollution (in mm^3).

NB Resource depletion (kg eq Sb). The indicator used is the abiotic depletion potential (ADP) developed by the CML of the University of Leiden (the Netherlands), 2001 edition. Its unit is kg equivalent Sb (Sb is the chemical symbol for antimony). Fuel consumption is expressed in kg and multiplied by conversion factors to take into account the importance of reserves of various elements; the reference element is antimony (coefficient 1).

MOVING FROM AN ANALYSIS OF THE INVENTORY TO AN EVALUATION OF THE IMPACT. For each material identified in the creation of the products studied here, there exists an equivalence factor for a given amount of that substance and a given environmental impact. These factors are known as characterization factors. The values calculated in the analysis of the inventory for each material are converted into impacts by means of these factors. The sum of these values is calculated for each of the potential impacts considered during the LCA.

21.3.3 Initial results are encouraging

The study was divided into three parts: the first, the agronomic part, is common to both products; then there is a thermoplastic part and a cement part.

The agronomic study

WHICH SYSTEM IS TO BE STUDIED?. The system studied in this part consists of the agricultural production (termed technical itinerary) and the production of the hemp fibres and hemp straw (termed primary transformation). The functional unit used is the kilogramme.

The system studied has not taken into account the seed production or the production of hemp seed that is sometimes associated with straw production. It has not considered either inputs with a weight less than 2% of the total mass of inputs (NF P01-010). The dust produced by the primary transformation process is considered a waste product (Chapters 11–18 on the uses of hemp). Finally, the repartition of the potential environmental impacts of hemp straw between fibre and straw was undertaken in bulk (60% hemp straw, 40% fibre) and in value (68% fibre and 32% straw).

ANALYSIS OF THE INVENTORY. In applying the ISO 14040 standard, the value of the inputs and transport distances attributed to the production of hemp straw was calculated using values supplied by the members of the partners (FNPC, LCDA, Eurochanvre and PDM Industry).

The value of the outputs attributed to straw production was calculated using formulae derived from the bibliography.

Carbon storage in the product, arising as a result of photosynthesis, has been taken into account in this study.

EVALUATION OF THE IMPACT AND INTERPRETATION OF THE LIFE CYCLE. The results show a favourable impact with regard to the greenhouse effect. In fact, the potential value of this impact is between −1.7 (allocation by mass) and −2.9 (economic allocation) kg CO_2 eq/kg of fibre and between −1 (economic allocation) and −1.9 (allocation by mass) kg CO_2 eq/kg of hemp straw. The inclusion of carbon storage in the fibre and hemp straw arising through photosynthesis explains the positive result obtained for the agronomic part. The duration of carbon storage depends on the type of end product

the material is used to produce, its use, its potential for recycling and what happens to it at the end of its life cycle.

The results also demonstrate some unfavourable potential impacts when compared with other agroindustrial activities. Nitrogen fertilization plays an important role in the environmental impact of hemp fibre and hemp straw. It is associated principally with the production of greenhouse gases (production of mineral fertilizers and transformation into fertilizers in the ground), the consumption of non-renewable energy resources and pollution of waterways with nitrates. Parameter tests applied to this stage have also shown that a reduction of 20% in the nitrogen dose can result in a reduction in potential impacts of the order of 10% (assuming a similar yield). Not included in this calculation are the production of waste, destruction of the ozone layer and pollution of water.

Looking beyond the LCA results, hemp cultivation necessitates only a limited amount of nitrogen fertilization and does not require phytosanitary products or irrigation. It can be compared favourably with other major crops, as shown in the preceding section.

Thermoplastics

LIMITATIONS OF THE STUDIES UNDERTAKEN. The compound studied here is a mix of polypropylene and hemp fibres. The proportion by mass of these two ingredients is 70% polypropylene:30% fibre. The functional unit used is the kilogramme.

The study covers the production and transportation of the primary materials, together with the compounding phase. The compound thus produced is stored in bulk ready for use.

Inputs of less than 2% of the total mass of inputs have not been included in the calculations (NF P01-010).

ANALYSIS OF THE INVENTORY. The transport distances for the primary materials and the quantity of the inputs (materials and energy) used in the reference scenario described below have been provided by AFT Plasturgie.

The analysis of the inventory for hemp fibre is presented above. The analysis of polypropylene has been obtained from the *Ecoinvent©* database.

IMPACT EVALUATION AND INTERPRETATION OF THE LIFE CYCLE. The results demonstrate that hemp fibre reduces the potential negative impact associated with polypropylene. Thus, the production of 1 kg of polypropylene containing 30% hemp consumes 67 MJ of non-renewable energy and contributes between 0.7 and 1 kg CO_2 eq to the greenhouse effect, depending on whether an economic allocation or allocation by mass is used. The energy consumption and the impact on the greenhouse effect are 20–40% lower than that which would result from the production of pure polypropylene.

Furthermore, the production of polypropylene calls on large quantities of fossil resources, including natural gas (0.73 m^3/kg), coal (63 g/kg) and uranium (5 mg/kg). In terms of eco-planning and -design, the use of materials containing renewable carbon and with the same technical properties as polypropylene has the potential to improve further the environmental credentials of compounds containing hemp fibres. The results of this comparison are presented in Table 21.6.

Construction using hemp

LIMITS OF THE SYSTEM STUDIED. The study was undertaken for a wall of hemp cement on a wooden frame, used as a weight-bearing wall. It has a surface area of 1 m^2, a thermal resistance of 2.36 m^2 K/W for 1 year. The life expectancy used for this product is 1 year.

The system studied included the production of the raw materials, the creation of the wall on site, its use[5] and disposal at the end of its life. In the absence of a known, well-identified outcome, it has been assumed that the construction waste will be stored at a landfill site (Class II (CET II)), with the transportation fragmented over the entire life cycle.

We have not included inputs whose mass is inferior to 2% of the total mass of the inputs (NF P01-010) and, in the absence of data on the breakdown of cement and wood in CET II, the end of life is restricted to the demolition waste and the emissions arising from transportation of that waste to the nearest CET II. The reuse or recycling of the hemp cement has not been factored in.

INVENTORY ANALYSIS. The organization of the production line, the distances over which the raw materials are transported and the quantities of the inputs (energy, materials) used in the reference scenario described above are supplied by BCB-Lhoist and *Construire en chanvre*.

The analysis of the inventory of hemp straw was provided in the previous section. That of Tradical 70© (lime, hydraulic binder) and the wood used for the frames are provided by the *Ecoinvent©* database.

IMPACT EVALUATION AND INTERPRETATION OF THE LIFE CYCLE. The results demonstrate a

Table 21.6. Potential environmental impacts of 1 m^3 of a thermoplastic compound containing hemp fibres compared with a thermoplastic compound containing polypropylene only and a thermoplastic compound containing glass fibre.

Environmental impact	Thermoplastic compound		
	Containing hemp fibre	Polypropylene	Containing glass fibre
Exhaustion of resources (Sb eq)	23.1	32.2	35.5
Atmospheric acidification (kg SO_2 eq)	14.6	19.7	23.8
Greenhouse effect (kg CO_2 eq)	950	1970	2680
Of which carbon storage (in kg CO_2 eq)	−543	0	0
Ozone layer (kg CFC-11 eq)	0.000227	0.000194	0.000308
Photochemical ozone (kg C_2H_4 eq)	0.483	0.662	0.813
Non-renewable energy (MJ)	63,100	82,400	93,700
Air pollution (m^3)	34,600	34,900	105,000
Water pollution (m^3)	166	151	482

potentially favourable impact vis-à-vis the greenhouse effect. During its life cycle, one m² of wall made from hemp cement, in 100 years, stocks between 14 and 35 kg of CO_2 eq per m² wall, depending on whether an economic allocation or an allocation by mass is used (Table 21.7). The storage of carbon is due largely to the hemp straw, but is also due to the wood and the lime contained within the cement (due to the recarbonation phenomenon).

As for the other largely negative impacts, and in particular the consumption of fossil fuels (between 370 and 394 MJ per m²), these need to be compared against the results obtained for other construction materials.

The production of the binder, followed by the transportation stage, are the points that account for the majority of the usage of non-renewable energy, to the greenhouse effect and photochemical ozone formation.

21.3.4 Conclusion and future perspectives

Where thermoplastic compounds containing hemp fibres are concerned, hemp contributes to the potential negative impacts in a very marginal way (through the using up of resources, acidification, production of greenhouse gases and destruction of the ozone layer), minimally (air and water pollution) and

significantly only with regard to the waste material produced.

Where walls of hemp cement are concerned, hemp contributes to the unfavourable impacts in a very marginal way (through the waste material produced, use of non-renewable energy, production of greenhouse gases and destruction of the ozone layer), minimally (using up/exhaustion of resources) and significantly only with regard to air and water pollution.

The results of this study demonstrate that, with regard to the greenhouse effect, both thermoplastic compounds containing hemp fibres and walls made of hemp cement on a wooden frame perform well. This can be attributed to the carbon produced as a result of photosynthesis and stored in the raw material, that is to say, the hemp. Furthermore, hemp cement walls are long-lived and represent a carbon store with a life expectancy of 100 years, with more carbon stored during the course of the life cycle than is emitted. As for thermoplastics, the substitution of a proportion of the polypropylene with plant fibres reduces the negative effect of thermoplastic compounds on the greenhouse effect.

The results reported are due largely to the raw product, hemp, whose production requires little fossil fuel input and fits well into agricultural production systems.

Furthermore, other environmental characteristics can also be said to be favourable. Thus, thermoplastic compounds containing hemp fibre are less dense than existing thermoplastic

Table 21.7. Potential environmental impacts, over 100 years, of a 1 m² wall of hemp cement on a wooden frame and of a 1 m² wall of cement breeze blocks covered with an insulating PSE-plaster material.

Environmental impacts	Wall made of hemp cement on a wooden frame	Wall made of breeze blocks and plaster
Exhaustion of resources (Sb eq)	1.3.10⁻1	1.8. 10⁻1
Atmospheric acidification (kg SO_2 eq)	1.0. 10⁻1	1.1. 10⁻1
Greenhouse effect (kg CO_2 eq)	−35.5	−28.8
Of which carbon storage (in kg CO_2 eq)	−75.7	0
Ozone layer (kg CFC-11 eq)	0.000227	0.000194
Photochemical ozone (kg C_2H_4 eq)	$5.4\ 10^{-3}$	$4.0\ 10^{-3}$
Non-renewable energy (MJ)	394.2	432
Air pollution (m³)	1,024	4,290
Water pollution (m³)	6.7	8.9

compounds containing glass fibre. This allows the weight of vehicles to be reduced, thus lowering their fuel consumption and the emission of greenhouse gases. These compounds are also recyclable; this allows materials such as polypropylene to be economized, further reducing greenhouse gas emissions.

Notes

[1] As compared to a growing forest that produces only 4.5 t of wet wood per annum.

[2] ACV (according to the ISO 14040 standard): compilation and evaluation of the inputs and potential environmental impacts of a production system throughout its entire life cycle.

[3] Mixed according to well-determined proportions of thermoplastics and hemp fibres.

[4] Mixed according to specified proportions of cement and hemp straw, placed between two pieces of wood and a backboard in order to create the wall.

[5] According to standard NF POI-010: stages of the life cycle of a construction material extending from the occupation of the building to the departure of the last occupants and including its maintenance and repair.

References and Further Reading

Chapter 2

Benet, S. (1936) Early diffusions and folk uses of hemp. Reprinted in: Rubin, V. (ed.) *Cannabis and Culture* (1975). Mouton, The Hague, The Netherlands.

Bennett, C. (2010) *Cannabis and the Soma Solution*. TrineDay LLC, Walterville, Oregon.

Booth, M. (2003) *Cannabis: A History*. Picador, London.

Cailly, C. (1993) *Mutations d'un espace proto industriel: le Perche au XVIII° et XIX° siecles*, 2 volumes. Fédération des amis du Perche, France.

Clarke, R.C. (1999) Botany of the Genus Cannabis. In: Ranalli, P. (ed.) *Advances in Hemp Research*. The Haworth Press, New York, pp. 1–19.

Duby, G. and Wallon, A. (1976) *Histoire rurale de la France*, 4 volumes. Editions du Seuil.

Duhamel de Monceau *L'art de la corderie*, 2nd edn 769. Desaint Editeur.

Gernet, J. (1999) *Le monde chinois*. Armand Colin, Paris.

Hillig, K.W. (2005) Genetic evidence for speciation in Cannabis (Cannabaceae). *Genetic Resources and Crop Evolution* 52, 161–180.

Hillig, K.W. and Mahlberg, P.G. (2004) A chemotaxonomic analysis of cannabinoid variation in Cannabis (Cannabaceae). *American Journal of Botany* 91(6), 966–975.

Hopkins, J.F. (1951) *A History of the Hemp Industry*. University Press of Kentucky, Lexington, Kentucky.

Kabelik, J., Krejci, Z. and Santavy, F. (1960) Cannabis as a Medicant. *Bulletin on Narcotics* 12(3), 5–23.

Renouard, M. (1909) *Fabrication des Cables*.

Rosenthal, E. (1994) *Hemp Today*. Atlantic Books, London.

Small, E. (1979) *The Species Problem in Cannabis*. Corpus, Canada.

Williams, M. (2006) *Bridport and West Bay: The Buildings of the Flax and Hemp Industry*. English Heritage, UK.

Chapter 3

Allard, H.A. (1938) Complete or partial inhibition of flowering in certain plants when days are too short or too long. *Journal of Agricultural Research* 57, 775–789.

Arnoux, M., Mathieu, G. and Castiaux, J. (1969) L'amélioration du chanvre papetier en France. Etude et sélection de la monoecie, production d'hybrides entre formes dioïques et monoïques. *Ann. Amélior. Plantes* 19, 363–389.

Bello, P.Y., Toufik, A., Gandilhon, M. and Giraudon, I. (2003) Phénomènes émergents liés aux drogues en 2003 – Cinquième rapport national du dispositif TREND. OFDT éditeur, Saint-Denis, France, pp 275.

Bercht, C.A.L., Lousberg, R.J.J., Kuppers, J.E.M. and Salemink, C.A. (1973) L(+)-isoleucine betaine in cannabis seeds. *Phytochemistry* 12, 2457–2459.

Berman, J.S., Symonds, C. and Birch, R. (2004) Efficacy of two cannabis based medicinal extracts for relief of central neuropathic pain from brachial plexus avulsion: results of a randomised controlled trial. *Pain* 112, 299–306.

Bocsa, I. and Karus, M. (1997) *The Cultivation of Hemp: Botany, Varieties, Cultivation and Harvesting.* Hemptech, Sebastopol, California, pp 184.

Bonatti, P.M., Ferrari, C., Focher, B., Grippo, C., Torri, G. and Cosentino, C. (2004) Histochemical and supramolecular studies in determining quality of hemp fibres for textile applications. *Euphytica* 140, 55–64.

Bowes, B.G. (1998) *Atlas en couleur. Structure des plantes.* Paris, pp 192.

Brett, C. and Waldron, K. (1996) *Physiology and Biochemistry of Plant Cell Walls.* Chapman and Hall, London.

Buchanan, B.B., Gruissem, W. and Jones, R.L. (2000) *Biochemistry and Molecular Biology of Plants.* American Society of Plant Physiologists, Rockville, Maryland, pp 1367.

Cappelletto, P., Brizzi, M., Mongardini, F., Barberi, B., Sannibale, M., Nenci, G., et al. (2001) Italy-grown hemp: yield, composition and cannabinoid content. *Industrial Crops and Products* 13, 101–113.

Chabbert, B., Joly, C. and Kurek, B. (2006) Les fibres végétales: retour vers le futur? In: Colonna, P. (ed.) *Chimie verte.* Tec et Doc, Paris.

Chase, M. (1998) The Angiosperm Phylogeny Group: an ordinal classification for the families of flowering plants. *Annals Missouri Botanical Garden* 85, 531–553.

Cronier, D. (2005) Caractérisation des fibres périphloémiennes au cours de la maturation du chanvre (*Cannabis sativa*). Mémoire CNAM.

Cronier, D., Monties, B. and Chabbert, B. (2005) Structure and chemical composition of bast fibers isolated from developing hemp stem. *Journal of Agricultural and Food Chemistry.*

Day, A., Ruel, K., Neutelings, G., Crônier, D., David, H. and Hawkins, S. (2005) Lignification in the flax stem: evidence for an unusual lignin in bast fibers. *Planta* 1, 3–4.

de Meijer, E.P.M., Bagatta, M., Carboni, A., Cricitti, P., Moliterni, V.M.C., Ranalli, P., et al. (2003) The inheritance of chemical phenotype in *Cannabis sativa* L. *Genetics* 163, 335–346.

del Rio, J.C., Gutierrez, A. and Martinez, A.T. (2004) Identifying acetylated lignin units in non-wood fibers using pyrolysis-gas chromatography/mass spectrometry. *Rapid Communications in Mass Spectrometry* 18, 1181–1185.

Esau, K. (1977) *Anatomy of Seed Plants.* John Wiley, London, pp 550.

Fellermeier, M. and Zenk, M.H. (1998) Prenylation of olivetolate by a hemp transferase yields cannabigerolic acid, the precursor of tetrahydrocannabinol. *FEBS Letters* 427, 283–285.

Focher, B., Palma, M.T., Canetti, M., Torri, G., Cosentino, C. and Gastaldi, G. (2001) Structural differences between non-wood plant celluloses: evidence from solid state NMR, vibrational spectroscopy and X-ray diffractometry. *Industrial Crops and Products* 13, 193–208.

Fratzl, P. (2003) Cellulose and collagen: from fibres to tissues. *Current Opinion in Colloid and Interface Science* 8, 32–39.

Garner, W.W. and Allard, H.A. (1920) Effect of the relative length of day and night and other factors of the environment on growth and reproduction in plants. *Journal of Agricultural Research* 18, 553–606.

Girault, R., Bert, F., Rihouhey, C., Jauneau, A., Morvan, C. and Jarvis, M.C. (1997) Galactans and cellulose in flax fibers: putative contributions to the tensile strengh. *International Journal of Biological Macromolecules* 21, 179–188.

Gutierrez, A. and del Rio, J.C. (2005) Chemical characterization of pitch deposits produced in the manufacturing of high-quality paper pulps from hemp fibers. *Bioresource Technology* 96, 1445–1450.

Han, J.S. and Rowell, J.S. (1997) Chemical composition of fibers. In: Rowell;, R.M., Young, R.A. and Rowell, J.K. (eds) *Paper and Composites from Agro-based Resources.* CRC Press, New York, pp 83–134.

Harborne, J.B. and Williams, C.A. (2001) Anthocyanins and other flavonoids. *Natural Products Report* 18, 310–333.

Hillig, K.W. (2004) A chemotaxonomic analysis of terpenoid variation in Cannabis. *Biochemical Systematics and Ecology* 32, 875–891.

Hills, M.J. (2004) Control of storage-product synthesis in seeds. *Current Opinion in Plant Biology* 7, 302–308.

Iversen, L. and Chapman, V. (2002) Cannabinoids: a real prospect for pain relief. *Current Opinion in Pharmacology* 2, 50–55.

Joseleau, J. (1980) Les hémicelluloses. In: Monties, B. (ed.) *Les polymères végétaux*. Bordas, Paris, pp 87–121.

Jurasek, L. (1998) Molecular modelling of fibre walls. *Journal of Pulp and Paper Science* 24, 209–212.

Kamat, J., Roy, D.N. and Goel, K. (2002) Effect of harvesting age on the chemical properties of hemp plants. *Journal of Wood Chemistry and Technology* 22, 285–293.

Kortekaas, S., Soto, M., Vicent, T., Field, J.A. and Lettinga, G. (1995) Contribution of extractives to methanogenic toxicity of hemp black liquor. *Journal of Fermentation and Bioengineering* 80, 383–388.

Kriese, U., Schumann, E., Weber, W.E., Beyer, M., Bruhl, L. and Matthaus, L. (2004) Oil content, tocopherol composition and fatty acid patterns of the seeds of 51 *Cannabis sativa* L. genotypes. *Euphytica* 137, 339–351.

Lacombe, J.P. (1978) Étude quantitative du développement végétatif du chanvre (*Cannabis sativa* L.): croissance de l'axe caulinaire. Institut National Polytechnique.

Leizer, C., Ribnicky, D., Poulev, A., Dushenkov, S. and Raskin, I. (2000) The composition of hemp seed oil and its potential as an important source of nutrition. *J. Nutraceu. Function. Med. Food* 2, 35–53.

Linger, P., Mussig, J., Fischer, H. and Kobert, J. (2002) Industrial hemp (*Cannabis sativa* L.) growing on heavy metal contaminated soil: fibre quality and phytoremediation potential. *Industrial Crops and Products* 16, 33–42.

Lucchese, C., Venturi, G., Amaducci, M.T. and Lovato, A. (2001) Electrophoretic polymorphism of *Cannabis sativa* L. cultivars: characterisation and geographical classification. *Seed Science and Technology* 29, 239–248.

Makriyannis, A., Mechoulam, R. and Piomelli, D. (2005) Therapeutic opportunities through modulation of the endocannabinoid system. *Neuropharmacology* 48, 1068–1071.

Mandolino, G. and Carboni, A. (2004) Potential of marker-assisted selection in hemp genetic improvement. *Euphytica* 140, 107–120.

Mautthaus, B. (1997) Antinutritive compounds in different oil seeds. *Fett/Lipid* 99, 170–174.

Mechoulam, R. and Hanus, L. (2000) A historical overview of chemical research on cannabinoids. *Chemistry and Physics of Lipids* 108, 1–13.

Mechoulam, R. and Hanus, L. (2002) Cannabidiol: an overview of some chemical and pharmacological aspects. Part I: chemical aspects. *Chemistry and Physics of Lipids* 121, 35–43.

Mediavilla, V., Leupin, M. and Keller, A. (2001) Influence of the growth stage of industrial hemp on the yield formation in relation to certain fibre quality traits. *Industrial Crops and Products* 13, 49–56.

Molleken, H. and Theimer, R. (1997) Survey of minor fatty acids in *Cannabis sativa* L. fruits of various origins. *Journal of the International Hemp Association* 4, 13–17.

Mustafa, A.F., McKinnon, J.J. and Christensen, D.A. (1999) The nutritive value of hemp meal for ruminant. *Canadian Journal of Animal Science* 79, 91–95.

Novak, J., Zitter-Eglseer, K., Deans, S.G. and Franz, C.M. (2001) Essential oils of different cultivars of *Cannabis sativa* and their anti-microbial activity. *Flavour and Fragrance* 16, 259–262.

Odani, S. and Odani, S. (1998) Isolation and primary structure of a methionine- and cystine-rich seed protein of *Cannabis sativa*. *Bioscience, Biotechnology and Biochemistry* 62, 650–654.

Oomah, B.D., Busson, M., Godfrey, D.V. and Drover, J.C.G. (2002) Characteristics of hemp (*Cannabis sativa* L.) seed oil. *Food Chemistry* 76, 33–43.

Patel, S. (1994) Crystalllographic characterization. *Journal of Molecular Biology* 235, 361–363.

Pichersky, E. and Gang, D.R. (2000) Genetics and biochemistry of secondary metabolites in plants: an evolutionary perspective. *Trends in Plant Science* 5, 439–445.

Rahman, M. (1979) Morphology of the fibres of jute, flax, and hemp as seen under a scanning electron microscope. *Indian Journal of Agricultural Sciences* 49, 483–487.

Ridley, B.L., O'Neill, M.A. and Mohnen, D. (2001) Pectins: structure, biosynthesis, and oligogalacturonide-related signaling. *Phytochemistry* 57, 929–967.

Robert, D. and Catesson, A.M. (2000) *Biologie Végétale: caractéristiques et stratégie évolutive des plantes. Organisation vegetative*. Doin, Paris, pp 356.

Roland, J.C., Mosiniak, M. and Roland, D. (1995) Dynamique du positionnement de la cellulose dans les parois des fibres textiles de lin (*Linum usitatissimum*). *Acta Botanica Gallica* 142, 463–484.

Ross, S.A. and ElSohly, M.A. (1996) The volatile oil composition of fresh and air dried buds of *Cannabis sativa*. *Journal of Natural Products* 59, 49–51.

Ross, S.A., ElSohly, M.A., Sultana, G.N.N., Mehmedic, Z., Hossain, C.F. and Chandra, S. (2005) Flavonoids, glycosides and cannabinoides from the pollen of *Cannabis sativa*, L. *Phytochemical Analysis* 16, 45–48.

Schofield, J.E. and Waller, M.P. (2005) A pollen analytical record for hemp retting from Dungeness Foreland, UK. *Journal of Archaeological Science* 32, 715–726.

Shewry, P.R., Beaudoin, F., Jenkins, J., Griffiths-Jones, S. and Mills, E.N.C. (2001) Plant protein families and their relationships to food allergy. *Biochemical Society Transactions* 30, 906–910.

Stott, C.G. and Guy, G.W. (2004) Cannabinoids for the pharmaceutical industry. *Euphytica* 140, 83–93.

Thygesen, L.G. and Hoffmeyer, P. (2005) Image analysis for the quantification of dislocations in hemp fibres. *Industrial Crops and Products* 21, 173–184.

Tournois, J. (1912) Influence de la lumière sur la floraison du houblon japonais et du chanvre déterminée par des semis hâtifs. *Comptes Rendus Hebdomadaires des Seances de l'Academie des Sciences (Paris)* 155, 297–300.

Turner, S.R. (1980) Constituents of *Cannabis sativa* L. XVII. A review of the natural constituents. *Journal of Natural Products* 43, 169–234.

van der Werf, H.M.G., Harsveld van der Veen, J.E., Bouma, A.T.M. and ten Cate, M. (1994) Quality of hemp (*Cannabis sativa* L.) stems as a raw material for paper. *Industrial Crops and Products* 2, 219–227.

Vanhoenacker, G., Van Rompaey, P., D., d. K.; Sandra, P. (2002) Chemotaxonomic features associated with flavonoids of cannabinoid-free cannabis in relation to hops. *Journal of Natural Products* 16, 57–63.

Vignon, M.R. and Garcia-Jaldon, C. (1996) Structural features of the pectic polysaccharides isolated from retted hemp bast fibres. *Carbohydrate Research* 296, 249–260.

Vignon, M.R., Garcia-Jaldon, C. and Dupoyro, D. (1995) Steam explosion of woody hemp chenevotte. *International Journal of Biological Macromolecules* 17, 395–404.

Wingerchuk, D. (2004) Cannabis for medical purposes: cultivating science, weeding out the fiction. *The Lancet* 364, 315–316.

Chapter 4

Beherec, O. (2000) FNPC hemp breeding and CCPSC seeds production. Bioresources Hemp 2000, Wolfsburg, Germany, pp. 13–16.

Berenji, J. (1992) Konoplja. *Bilten za hmelj, sirak i lekovito bilje* 23–24(64–65), 79–85.

Berenji, J. (1996) Present status and perspectives of hemp in Yugoslavia. *Agricultural Engineering* 2(1–2), 1–11.

Berenji, J. (1998) Istine i zablude o konoplji. *Zbornik radova Instituta za ratarstvo i povrtarstvo Novi Sad* 30, 271–281.

Berenji, J. (2004) Perspectives and future of hemp in Europe. Proceedings of the Global Workshop (General Consultation) of the FAO/ESCORENA European Cooperative Research Network on Flax and other Bast Plants 'Bast Fibrous Plants for Healthy Life'. Banja Luka (Bosnia and Herzegovina/Republika Srpska), proceedings on CD.

Berenji, J. and Sabo, A. (1998) Konoplja (*Cannabis* sp.) kao lekovita biljka. *Medicinal Plant Report* 5(5), 43–57.

Berenji, J. and Sikora, V. (1996) Oplemenjivanje i semenarstvo konoplje. *Zbornik radova Instituta za ratarstvo i povrtarstvo Novi Sad* 26, 9–38.

Berenji, J. and Sikora, V. (2000) Selekcija konoplje na povećani sadržaj vlakna. Zbornik izvoda 'Treći Jugoslovenski naučno-stručni simpozijum iz selekcije i semenarstva – III JUSEM', Zlatibor, p. 39.

Berenji, J., Kišgeci, J. and Sikora, V. (1997) Genetički resursi konoplje. *Savremena poljoprivreda* 47(5–6), 89–98.

Bócsa, I. (1954) Kender heterózis-nemesítési eredmények. *Növénytermelés* 3(4), 301–316.

Bócsa, I. (1967) Kender fajtahibrid előállításához szükséges unisexuális (hímmentes) anyafajta nemesítése. *Rostnövények* 3–7.

Bócsa, I. (1999) Genetic improvement: conventional approaches. In: Ranalli, P. (ed.) *Advances in Hemp Research*. Food Products Press, New York, pp. 133–152.

Bócsa, I. and Karus, M. (1998) *The Cultivation of Hemp. Botany, Varieties, Cultivation and Harvesting*. Hemptech, Sebastopol, California.

Bócsa, I., Pummer, L. and Finta-Korpel'ová, Z. (1999) *Előzetes eredmények a fajták közötti és fajtán belüli olaj- és gamma-linolénsav tartalom variabilitásáról kendernél*. Növénynemesítési Tudományos Napok, Budapest, p. 42.

Bócsa, I., Finta-Korpel'ová, Z. and Máthé, P. (2005) Preliminary results of selection for seed oil content in hemp (*Cannabis sativa* L.). *Journal of Industrial Hemp* 10(1), 5–15.

Bredemann, G. (1942) Züchtung auf hohen Fasergehalt bei Hanf (*Cannabis sativa* L.). *Züchter* 14, 201–213.

Callaway, J.C. (2002) Hemp as food at high latitudes. *Journal of Industrial Hemp* 7(1), 105–117.

de Meijer, E.P.M. (1995) Fibre hemp cultivars: a survey of origin, ancestry, availability and brief agronomic characteristics. *Journal of the International Hemp Association* 2(2), 66–75.

de Meijer, E.P.M. (1999) Cannabis germplasm resources. In: Ranalli, P. (ed.) *Advances in Hemp Research*. Food Products Press, New York, pp. 133–152.

de Meijer, E.P.M. and van Soest, L.J.M. (1992) The CPRO *Cannabis* germplasm collection. *Euphytica* 62, 201–211.

Dempsey, J.M. (1975) *Fiber Crops*. University of Florida Press, Gainesville, Florida, pp. 46–89.

Feeney, M. and Punja, Z.K. (2003) Tissue culture and *Agrobacterium*-mediated transformation of hemp (*Cannabis sativa* L.). *In Vitro Cellular and Developmental Biology* 39(6), 578–585.

Finta-Korpel'ová, Z. (2006) *Az olajtartalom növelésének következtében a kenderolaj összetételében bekövetkezett változások*. XII. Növénynemesítési Tudományos Napok, Budapest, p. 91.

Finta-Korpel'ová, Z. and Berenji, J. (2007) Trends and achievements in industrial hemp (*Cannabis sativa* L.) breeding. *Bilten za hmelj, sirak i lekovito bilje* 39(80), 63–75.

Fleischmann, R. (1938) Der Einfluss der Tageslänge auf den entwickelungsrhytmus von Hanf und Ramie. *Faserforschung* 13(2), 93–99.

Hennik, S. (1994) Optimisation of breeding for agronomic traits in fibre hemp by study of parent–offspring relationships. *Euphytica* 78, 69–76.

Hoffmann, W. (1938) Das Geschletsproblem des Hanfes in der Züchtung. *Zeitschrift für Pflanzenzüchtung* 22, 453–467.

Hoffmann, W. (1961) Hanf, *Cannabis sativa*. In: Kappert, H. and Rudolf, W. (eds) *Handbuch der Pflanzenzüchtung*. Vol. V. Paul Parey, Berlin and Hamburg, pp. 204–261.

Hoffmann, W., Mudra, A. and Plarre, W. (1970) *Lehrbuch der Züchtung landwirtschaftlicher Kulturpflanzen*. Vol. 2, Verlag Paul Parey, Berlin and Hamburg, pp. 415–430.

Kutuzova, S., Rumyantseva, L., Grigoryev, S. and Clarke, R.C. (1997) Maintenance of Cannabisgermplasm in the Vavilov Research Institute gene bank – 1996. *Journal of the International Hemp Association* 4(1), 17–21.

Laakkonen, T.T. and Callaway, J.C. (1998) Update on FIN-314. *Journal of the International Hemp Association* 5(1), 34–35.

Mandolino, G. and Ranalli, P. (2002) The applications of molecular markers in genetics and breeding of hemp. *Journal of Industrial Hemp* 7, 7–23.

Menzel, M.Y. (1964) Meiotic chromosomes of monoecious Kentucky hemp (*Cannabis sativa*). *Bulletin of the Torrey Botanical Club* 91(3), 193–205.

Müssig, J. (2003) Quality aspects in hemp fiber production – influence of cultivation, harvesting and retting. *Journal of Industrial Hemp* 8(1), 11–32.

Neuer, H. and von Sengbusch, R. (1943) Die Geschlechtsvererbung bei Hanf und die Züchtung eines monö-zichschen Hanfes. *Züchter* 15, 49–62.

Ranalli, P. (2004) Current status and future scenarios of hemp breeding. *Euphytica* 140, 121–131.

Scheifele, G. (2000) Studies on factors affecting hemp oil extraction from hemp grain grown in northern Ontario (www.gov.on.ca/omafra).

Shao, H. and Song, S.J. (2003) Female-associated DNA polymorphisms of hemp (*Cannabis sativa* L.). *Journal of Industrial Hemp* 8(1), 5–9.

Törjék, O., Bucherna, N., Kiss, E., Homoki, H., Finta-Korpelová, Z., Bócsa, I., *et al.* (2002) Novel male-specific molecular markers (MADC5, MADC6) in hemp. *Euphytica* 127, 209–218.

van der Werf, H.M.G., Haasken, H.J. and Wijlhuizen, M. (1994) The effect of day length on yield and quality of fibre hemp (*Cannabis sativa* L.). *European Journal of Agronomy* 3(2), 117–123.

Watson, D.P. and Clarke, R.C. (1997) Genetic future of hemp. *Journal of the International Hemp Association* 4(1), 32–36.

Watson, J.D., Baker, T.A., Bell, S.P., Gann, A., Levine, M. and Losick, R. (2004) *Molecular Biology of the Gene*. 5th edition. Benjamin Cummings and Cold Spring Harbor Laboratory Press, San Francisco, California.

Chapter 6

Beherec, O. (2009) FNPC's Hemp Breeding and CCPSC's Hemp Seeds Production, EIHA Conference, Wesseling, Germany.

Dempsey, J.M. (1975) *Fiber Crops*. University of Florida Press, Gainesville, Florida, pp. 46–89.

van der Werf, H.M.G., Haasken, H.J. and Wijlhuizen, M. (1994) The effect of day length on yield and quality of fibre hemp (*Cannabis sativa* L.) *European Journal of Agronomy* 3(2), 117–123.

Chapter 9

Kessler, R.W. and Kohler, R. (1996) New strategies for exploiting flax and hemp. *Chemtech* 26(10), 34–43.

Renouard, M. (1909) *Fabrication des Cables*.

Chapter 10

Bassetti P., Mediavilla, V., Spiess, E., Ammann, H., Strasser, H. and Mosimann, E. (1998) *Hanfanbau in der Schweiz – Geschichte, aktuelle Situation, Sorten, Anbau- und Erntetechnik, wirtschaftliche Aspekte und Perspektiven*. FAT-Berichte 1998 No 516, Ed. Eidg. Forschungsanstalt für Agrarwirtschaft und Landtechnik (FAT), Tänikon, Switzerland [in German].

Batra, S.K. (1998) Other long vegetable fibers. In: Lewin, M. and Pearce, E. (eds) *Handbook of Fiber Chemistry*, 2nd edn, revised and expanded, International Fiber Science and Technology Series 15. Marcel Dekker, New York, pp. 505–575.

Baxter, B.P., Brims, M.A. and Taylor, T.B. (1992) Description and performance of the optical fibre diameter analyser (OFDA). *Journal of the Textile Institutes* 83(4), 507–526.

Bobeth, W. (ed.) (1993) *Textile Faserstoffe – Beschaffenheit und Eigenschaft*. Springer-Verlag, Berlin [in German].

Bos, H.L., Müssig, J. and van den Oever, M. (2006) Properties of short-flax-fibre reinforced compounds. *Composites Part A* 37, 1591–1604.

Cescutti, G. and Müssig, J. (2005) Industrial quality management – natural fibres. *Kunststoffe plast europe* 1, 1–4.

Cichocki, F.R. and Thomason, J.-L. (2002) Thermoelastic anisotropy of a natural fiber. *Composites Science and Technology* 62(5), 669–678.

Dreyer, J., Müssig, J., Koschke, N., Ibenthal, W.-D. and Harig, H. (2002) Comparison of enzymatically separated hemp and nettle fibre to chemically separated and steam exploded hemp fibre. *Journal of Industrial Hemp* 7(1), 43–59.

Drieling, A., Bäumer, R., Müssig, J. and Harig, H. (1999) Möglichkeiten zur Charakterisierung von Festigkeit, Feinheit und Länge von Bastfasern. *Technische Textilien* 42(4), 261–262 (and E66) [in German].

FIBRE (1994) *Bremer Baumwoll-Rundtest 1994/1 – Auswertung der Testergebnisse – Evaluation of the Test Results.* Bremer Baumwollbörse (ed.). Faserinstitut Bremen e.V. – FIBRE –, Bremen, Germany, pp. 1–18.

Fischer, H., Müssig, J. and Bluhm, C. (2004a) *Enzymatic Modification of Hemp Fibres for Sustainable Production of High Quality Materials.* Program and Abstracts, Lecture No 13. INTB 04 – 3rd International Conference on Textile Biotechnology, 13–16 June 2004. Graz University of Technology, Graz, Austria.

Fischer, H., Müssig, J., Geppert, N. and Bluhm, C. (2004b) *Beurteilung des Geruchspotenzials von Naturfasern für den Einsatz im Automobilbereich.* In: Conference Proceedings/Manual. (AVK-TV) Internationale AVK-TV Tagung für verstärkte Kunststoffe und duroplastische Formmassen, Baden-Baden, 28–29 September 2004. Arbeitsgemeinschaft Verstärkte Kunststoffe – Technische Vereinigung e.V., Frankfurt/Main, Germany, pp. B11-1–B11-7 [in German].

Fischer, H., Müssig, J. and Bluhm, C. (2006) Enzymatic modification of hemp fibres for sustainable production of high quality materials: influence of processing parameters. *Journal of Natural Fibers* 3(2/3), 39–53.

Flemming, M. and Roth, S. (2003) *Faserverbundbauweisen: Eigenschaften; mechanische, konstruktive, thermische, elektrische, ökologische, wirtschaftliche Aspekte.* Springer Verlag, Berlin, pp. 3–4 [in German].

Fröter, K. and Zienkiewicz, H. (1952) *Bericht über erhaltene Resultate bei laboratoriums- und fabrikmäßiger Untersuchung bestimmter Qualitätseigenschaften der schwedischen Flachsfaser der Produktion von 1946 bis 1951.* (Flachs 6) Lin 6 1952, p. 125. Cited in Simor (1965) [in German].

Gassan, J. (2003) Lightweight construction: natural fibres in automotive interiors. *Kunststoffe international* 2003/08, 31–34.

Griffith, A.A. (1920) *Philosophical Transactions of the Royal Society* 221(A), 163–198.

Grignet, J. (1981) Microprocessor improves wool fiber-length measurements and extends the application – Part I. General description of the system and a review of its applications. *Textile Research Journal* March, 174–181.

Harig, H. and Müssig, J. (1999) *Heimische Pflanzenfasern für das Automobil.* In: Harig, H. and Langenbach, C.J. (eds) *Neue Materialien für innovative Produkte – Entwicklungstrends und gesellschaftliche Relevanz.* (Wissenschaftsethik und Technik und Technikfolgenbeurteilung Bd.3). Springer Verlag, Berlin, pp. 235–251 [in German].

Haudek, H.W. and Viti, E. (1980) *Textilfasern: Herkunft, Herstellung, Aufbau, Eigenschaften, Verwendung.* Verlag Johann L. Bondi & Sohn, Wien-Perchtoldsdorf [in German].

Hearle, J.W.S. and Peters, R.H. (1963) *Fibre Structure.* Butterworth, Manchester, UK, pp. 667.

Joseph, P.V., Mathew, G., Joseph, K., Thomas, S. and Pradeep, P. (2003) Mechanical properties of short sisal fiber-reinforced polypropylene composites: comparison of experimental data with theoretical predictions. *Journal of Applied Polymer Science* 88(3), 602–611.

Karus, M., Ortmann, S., Gahle, C. and Pendarovski, C. (2006) *Use of Natural Fibres in Composites for the German Automotive Production from 1999 till 2005.* nova-Institut, Hürth, Germany, December 2006 (http://www.nova-institut.de/pdf/06-12_nova_NF-CompositesAutomotive.pdf).

Kaw, A.K. (1997) *Mechanics of Composite Materials.* Mechanical Engineering Series 6. CRC Press LLC, Boca Raton, Florida, pp. 12–13.

Koch, P.A. (1997) Lyocell-Fasern. Faserstoff-Tabellen. *Melliand Textilberichte* 9/97, pp. 575–581 [in German].

Krässig, H., Schurz, J., Steadman, R.G., Schliefer, K. and Albrecht, W. (1989) *Cellulose.* In: Elvers, B., Hawkins, S., Ravenscroft, M. and Schulz, D. (eds) *Ullmann's Encyclopedia of Industrial Chemistry*, Vol A28. Verlag Chemie VCH, Weinheim, Germany, pp. 375–419.

Madsen, B. and Lilholt, H. (2003) Physical and mechanical properties of unidirectional plant fibre composites – an evaluation of the influence of porosity. *Composites Science and Technology* 63, 1265–1272.

Müssig, J. (2001a) *Untersuchung der Eignung heimischer Pflanzenfasern für die Herstellung von naturfaserverstärkten Duroplasten – vom Anbau zum Verbundwerkstoff.* (Fortschritt-Bericht VDI, Reihe

5, Grund- und Werkstoffe/Kunststoffe, No. 630). VDI Verlag GmbH, manuscript, 214 pages [in German].

Müssig, J. (2001b) *Neue technische Textilien aus heimischen Pflanzenfasern (Hanf und Nessel)* (AiF 11918N). Faserinstitut Bremen e.V. – FIBRE –, Bremen, Germany, manuscript, 98 pages [in German].

Müssig, J. and Harig, H. (2000) Fahrzeugbauteile vom Acker. *Landwirtschaftsblatt Weser-Ems* 147(13), 11–14 [in German].

Müssig, J. and Martens, R. (2003) Quality aspects in hemp fibre production – influence of cultivation, harvesting and retting. *Journal of Industrial Hemp* 8(1), 11–32.

Müssig, J. and Schmid, H.G. (2004) Quality control of fibers along the value added chain by using scanning technique – from fibers to the final product. In: Anderson, I.M., Price, R., Clark, E. and McKernan, S. (eds) *Microscopy and Microanalysis 2004*. Proceedings to the Conference – Microscopy and Microanalysis, Volume 10, Supplement 2, 2004, 1–5 August 2004, Savannah, Georgia, USA. Press Syndicate of the University of Cambridge, Cambridge, New York, Melbourne, pp. 1332CD–1333CD.

Müssig, J., Karus, M. and Franck, R.R. (2005) Bast and leaf fibre composite materials. In: Franck, R.R. (ed.) *Bast and Other Plant Fibres*. Woodhead Publishing, Cambridge, UK, pp. 345–376.

Müssig, J., Rau, S. and Herrmann, A.S. (2006) Influence of fineness, stiffness and load-displacement characteristic of natural fibres on the properties of natural fibre-reinforced polymers. *Journal of Natural Fibers* 3(1), 59–80.

Nechwatal, A., Mieck, K.-P. and Reueszettmann, T. (2003) Developments in the characterization of natural fibre properties and in the use of natural fibres for composites. *Composites Science and Technology* 63, 1273–1279.

Ruys, D., Crosky, A. and Evans, W.J. (2002) Natural bast fibre structure. *International Journal of Materials and Product Technology* 17(1/2), 2–10.

Schnegelsberg, G. (1999) *Handbuch der Faser – Theorie und Systematik der Faser*. (Theorien und Systeme in Technik und Ökonomie; (Vol. 1.) Deutscher Fachverlag, Frankfurt, Germany [in German].

Schwill, R., Ortmann, S., Karus, M. and Vogt, D. (2004) *Überblick über die PP-NF-Spritzguss-Technologie und ihre Eigenschaften*. nova-Institut GmbH, Arbeitsgemeinschaft Verstärkte Kunststoffe – Technische Vereinigung e.V. (AVK-TV) (Hrsg.). Hürth: Eigenverlag, 2004 (2004-08-31) Marktstudie in Auftrag für die Arbeitsgemeinschaft Verstärkte Kunststoffe – Technische Vereinigung e.V. (AVK-TV), Frankfurt, Germany, p. 17 [in German].

Simor, P. (1965) Faserfeinheits- und Spaltbarkeitsmessung an Flachs mit dem Micronaire-Apparat. *Spinner Weber Textilveredelung* 1, 29–33 [in German].

Suh, M.W., Cui, X. and Sasser, P.E. (1994) New understanding on HVI tensile data based on mantis single fiber test results. In: *Proceedings*, Vol 3 of 3, Beltwide Cotton Conferences, 5–8 January 1994, San Diego, California, pp. 1400–1403.

Warrier, J.K.S. and Munshi, V.G. (1982) Relationship between strength-elongation characteristics of single fibres and fibre bundles of cotton *Indian Journal of Textile Research* 7, 42–44.

Chapter 13

Felton, A. (1976) Nonwood stock preparation – a system concept. *TAPPI* 59(1).

Jeyasingam, J.T. (1993) *Industrial Experience in the Manufacture of Cigarette Tissue Using Hemp Pulp*. Pulping Conference 1993.

Kovacs, I. (1992) Hemp as a possible raw material for the paper industry. *Cellulose Chemistry and Technology*.

Marpiliero, P. (1977) *20 Years' Experience in Pulping Non-homogeneous Plants*. Tappi Press Progress Report No 8.

Chapter 15

Aragona, M., Onesti, E., Tomassini, V., Conte, A., Gupta, S., Gilio, F., *et al.* (2008) Psychopathological and cognitive effects of therapeutic cannabinoids in multiple sclerosis: a double-blind, placebo controlled, crossover study. *Clinical Neuropharmacology* 23 October (electronic publication).

Aviello, G., Romano, B. and Izzo, A.A. (2008) Cannabinoids and gastrointestinal motility: animal and human studies. *European Review for Medical and Pharmacological Sciences* 12(Supp. 1), 81–93.

BMA [British Medical Association] (1997) *Therapeutic Uses of Cannabis*, Harwood Academic Publishers, 142 pp.

Brady, C.M., DasGupta, R., Dalton, C., Wiseman, O.J., Berkley, K.J. and Fowler, C.J. (2004) An open-label pilot study of cannabis-based extracts for bladder dysfunction in advanced multiple sclerosis. *Multiple Sclerosis* 10(4), 425–433.

Caballero, F. and Bisiou, Y. (2000) *Droit de la drogue* (2nd edition). Dalloz, Paris, 827 pp.

Campbell, F.A., Tramer, M.R., Carroll, D., Reynolds, D.J., Moore, R.A. and McQuay, H.J. (2001) Are cannabinoids an effective and safe treatment option in the management of pain? A qualitative systematic review. *British Medical Journal* 323(7303), 13–16.

Carroll, C.B., Bain, P.G., Teare, L., Liu, X., Joint, C., Wroath, C., *et al.* (2004) Cannabis for dyskinesia in Parkinson disease: a randomised double-blind crossover study. *Neurology* 63(7), 1245–1250.

Clapper, J.R., Mangieri, R.A. and Piomelli, D. (2009) The endocannabinoid system as a target for the treatment of cannabis dependence. *Neuropharmacology* 56(Supp. 1), 235–243.

Correa, F., Mestre, L., Molina-Holgado, E., Arévalo-Martin, A., Docagne, F., Romero, E., *et al.* (2005) The role of cannabinoid system on immune modulation: therapeutic implications on CNS inflammation. *Mini Reviews in Medicinal Chemistry* 5(7), 671–675.

Croxford, J.L. and Miller, S.D. (2004) Towards cannabis and cannabinoid treatment of multiple sclerosis. *Drugs Today* 40(8), 663–676.

Davis, M.P. (2008) Oral nabilone capsules in the treatment of chemotherapy-induced nausea and vomiting and pain. *Expert Opinion on Investigational Drugs* 17(1), 85–95.

Degenhardt, L. and Hall, W.D. (2008) The adverse effects of cannabinoids: implications for use of medical marijuana. *CMAJ* 178(13), 1685–1686.

De Jong, B.C., Prentiss, D., McFarland, W., Machekano, R. and Israelski, D.M. (2005) Marijuana use and its association with adherence to antiretroviral therapy among HIV-infected persons with moderate to severe nausea. *Journal of Acquired Immune Deficiency Syndrome* 38(1), 43–46.

Elkhasef, A., Vocci, F., Huestis, M., Haney, M., Budney, A., Gruber, A., *et al.* (2008) Marijuana neurobiology and treatment. *Substance Abuse* 29(3), 17–29.

EMCDDA [European Monitoring Centre for Drugs and Drug Addiction] (2008) *A Cannabis Reader: Global Issues and Local Experiences*. Monographie No 8 (2 volumes), 746 pp (www.emcdda.europa.eu/publications/monographs).

Engels, F.K., de Jong, F.A., Mathijssen, R.H., Erkens, J.A., Herings, R.M. and Verweij, J. (2007) Medicinal cannabis in oncology. *European Journal of Cancer* 43(18), 2638–2644.

Grinspoon, L. and Bakalar, J. (1993) *Marihuana, The Forbidden Medicine*. Yale University Press, New Haven, Connecticut.

Gross, D.W., Hamm, J., Ashworth, N.L. and Quigley, D. (2004) Marijuana use and epilepsy: prevalence in patients of a tertiary care epilepsy center. *Neurology* 62(11), 1924–1925.

Hanus, L.O. (2009) Pharmacological and therapeutics secrets of plant and brain (endo) cannabinoids. *Medicinal Research Reviews* 29(2), 213–271.

Hazekamp, A., Ruhaak, P., Zuurman, L., van Gerven, J. and Verpoorte, R. (2006) Evaluation of a vaporizing device (Volcano™) for the pulmonary administration of tetrahydrocannabinol. *Journal of Pharmacological Sciences* 95(6), 1308–1317.

Hosking, R.D. and Zajicek, J.P. (2008) Therapeutic potential of cannabis in pain medicine. *British Journal of Anaethesia* 101, 59–68.

Iuvone, T., Esposito, G., de Fillippis, D., Scuderi, C. and Steardo, L. (2009) Cannabidiol: a promising drug for neurodegenerative disorders? *CNS Neuroscience and Therapeutics* 15(1), 65–75.

Killestein, J. and Polman, C. (2004) The therapeutic value of cannabinoids in MS: real or imaginary? *Multiple Sclerosis* 10(4), 339–340.

Killestein, J., Hoogervorst, E.I.J., Reif, M., Kalkers, N.F., van Loenen, A.C., Staats, P.G.M., *et al.* (2002) Safety, tolerability and efficacy or orally administered cannabinoids in MS. *Neurology* 58(9), 1404–1407.

Klein, T.W. (2005) Cannabinoid-based drugs as anti-inflammatory therapeutics. *Nature Reviews Immunology* 5, 400–411.

Liang, Y.C., Huang, C.C. and Hsu, K.S. (2004) Therapeutic potential of cannabinoids in trigeminal neuralgia. *Current Drug Targets. CNS and Neurological Disorders* 3(6), 507–514.

Mach, F., Montecucco, F. and Steffens, S. (2008) Cannabinoid receptors in acute and chronic complications of atherosclerosis. *British Journal of Pharmacology* 153, 290–298.

Machado Rocha, F.C., Stefano, S.C., de Cassia Haiek, R., Rosa Oliveira, L.M. and da Silveira, D.X. (2008) Therapeutic use of *Cannabis sativa* on chemotherapy-induced nausea and vomiting among cancer patients: systematic review and meta-analysis. *European Journal of Cancer Care* 17(5), 431–443.

Manzanares, J., Julian, M. and Carrascosa, A. (2006) Role of the cannabinoid system in pain control and therapeutic implications for the management of acute and chronic pain episodes. *Current Neuropharmacology* 4(3), 239–257.

Martin, B.R. and Wiley, J.L. (2004) Mechanism of action of cannabinoids: how it may lead to treatment of cachexia, emesis and pain. *Journal of Supportive Oncology* 2(4), 305–314.

Martin Fontelles, M.I. and Goicoechea Garcia, C. (2008) Role of cannabinoids in the management of neuropathic pain. *CNS Drugs* 22(8), 645–653.

Mendizabal, V.E. and Adler-Graschinsky, E. (2007) Cannabinoids as therapeutic agents in cardiovascular disease: a tale of passions and illusions. *British Journal of Pharmacology* 151(4), 427–440.

Michalon, M. (2005) Rôles du cannabis et des cannabinoïdes en médecine. Aspects cliniques et scientifiques. *Le Courrier des Addictions* 7(Suppl. 1), 21–25.

Michka, Cervantes, J., Clarke, R.C., Conrad, C., *et al.* (2009) *Cannabis médical: du chanvre indien au THC de synthèse*. Mama éditions, Paris, 267 pp.

Radbruch, L. and Elsner, F. (2005) Emerging analgesics in cancer pain management. *Expert Opinion on Emerging Drugs* 10(1), 151–171.

Ramirez, B.G., Blazquez, C., Gomez del Pulgar, T., Guzman, M. and de Ceballos, M.L. (2005) Prevention of Alzheimer's disease pathology by cannabinoids: neuroprotection mediated by blockade of microglial activation. *Journal of Neuroscience* 25(8), 1904–1913.

Ramos, J.A., Gonzalez, S., Sagredo, O., Gomez-Ruiz, M. and Fernandez-Ruiz, J. (2005) Therapeutic potential of the endocannabinoid system in the brain. *Mini Reviews in Medicinal Chemistry* 5(7), 609–617.

Ribuot, C., Lamontagne, D. and Godin-Ribuot, D. (2005) Cardiac and vascular effects of cannabinoids: toward a therapeutic use? *Annales de Cardiologie et d'Angéiologie* 54(2), 89–96.

Richard, D. (2009) *Le cannabis et sa consommation*. Armand Colin, Paris, 128 pp.

Richard, D., Senon, J.-L. and Valleur, M. (2004) *Dictionnaire des drogues et des dépendances*. Larousse, 626 pp.

Rosenthal, E., Mikuriya, T. and Gieringer, D. (1997) *Marijuana Medical Handbook. A Guide to Therapeutic Use*. Quick American Archives, Oakland, California.

Russo, E.B. (2007) History of cannabis and its preparations in saga, science, and sobriquet. *Chemistry and Biodiversity* 4(8), 1614–1648.

Sacerdote, P., Martucci, C., Vaccani, A., Bariselli, F., Panerai, A.E., Colombo, A., *et al.* (2005) The non-psychoactive component of marijuana cannabidiol modulates chemotaxis and IL-10 and IL-12 production of murine macrophages both in vivo and in vitro. *Journal of Neuroimmunology* 159, 97–105.

Sarfaraz, S., Afaq, F., Adhami, V.M. and Mukhtar, H. (2005) Cannabinoid receptor as a novel target for the treatment of prostate cancer. *Cancer Research* 65(5), 1635–1641.

Scuderi, C., Fillippis, D.D., Iuvone, T., Blasio, A., Steardo, A. and Esposito, G. (2008) Cannabidiol in medicine: a review of its therapeutic potential in CNS disorders. *Phytotherapy Research* 9 October (electronic publication).

Sieradzan, K.A., Fox, S.H., Hill, M., Dick, J.P.R., Crossman, A.R. and Brotchie, J.M. (2001) Cannabinoids reduce levodopa-induced dyskinesia in Parkinson's disease: a pilot study. *Neurology* 57, 2108–2111.

Smith, P.F. (2004) Medicinal cannabis extracts for the treatment of multiple sclerosis. *Current Opinion in Investigational Drugs* 5(7), 727–730.

Teixeira-Clerc, F., Julien, B., Grenard, P., Tran van Nhieu, J., Deveaux, V., Hezode, C., *et al.* (2008) Le système endocannabinoïde, une nouvelle cible pour le traitement de la fibrose hépatique [The endocannabinoids system as a novel target for the treatment of liver fibrosis]. *Pathologie Biologie* 56(1), 36–38.

Tramer, M.R., Carroll, D., Campbell, F.A., Reynolds, D.J., Moore, R.A. and McQuay, H.J. (2001) Cannabinoids for control of chemotherapy induced nausea and vomiting: quantitative systematic review. *British Medical Journal* 323(7303), 16–21.

Vaney, C., Heinzel-Gutenbrunner, M., Jobin, P., Tschopp, F., Gattlen, B., Hagen, U., *et al.* (2004) Efficacy, safety and tolerability of an orally administered cannabis extract in the treatment of spasticity in patients with multiple sclerosis: a randomized, double-blind, placebo-controlled, crossover study. *Multiple Sclerosis* 10(4), 417–424.

Wang, T., Collet, J.P., Shapiro, S. and Ware, M.A. (2008) Adverse effects of medical cannabinoids: a systematic review. *CMAJ* 178(13), 1669–1678.

Watson, S.J., Benson, J.A. Jr and Joy, J.E. (2000) Marijuana and medicine: assessing the science base. A summary of the 1999 Institute of Medicine Report. *Archives of General Psychiatry* 57, 547–552.

Widmer, M., Hanemann, C.O. and Zajicek, J. (2008) High concentrations of cannabinoids activate apoptosis in human U373MG glioma cells. *Journal of Neuroscience Research* 86(14), 3212–3220.

Yazulla, S. (2008) Endocannabinoids in the retina: from marijuana to neuroprotection. *Progress in Retinal and Eye Research* 27(5), 501–526.

Zajicek, J. (2004) The cannabinoids in Multiple Sclerosis Study. Final results from 12 months follow-up. *Multiple Sclerosis* 10(Suppl. 2), 115.

Zajicek, J., Fox, P., Sanders, H., Wright, D., Vickery, J., Nunn, A., *et al.* (2003) Cannabinoids for treatment of spasticity and other symptoms related to multiple sclerosis (CAMS study): multicentre randomised placebo-controlled trial. *The Lancet* 362(9395), 1517–1526.

Chapter 16

Callaway, J.C. (2004) Hempseed as a nutritional resource: an overview. *Euphytica* 140, 65–72.

Leson, G., Pless, P., Grotenhermen, F., Kalant, H. and ElSohly, M.A. (2001) Evaluating the impact of hemp food consumption on workplace drug tests. *Journal of Analytical Toxicology* 25, 691–698.

Matthäus, B. (1997) Antinutritive compounds in different oilseeds. *Fett/Lipid* 99, 170–174.

Chapter 17

Arnaud, L. (1999) Qualification physique des matériaux de construction à base de chanvre. Progress report June 1998 to June 1999. ENTPE (Laboratoire de Géomatériaux), France, 73 pp (www.entpe.fr).

Arnaud, L. (1999) Les bétons de chènevottes: approches mécaniques et thermiques. La journée scientifique du chanvre, 21 August 1999, AFLAM, Montjean sur Loire, France.

Arnaud, L. (2000) Qualification physique des matériaux de construction à base de chanvre. Progress report June 1999 to June 2000. ENTPE (Laboratoire de Géomatériaux), France, 73 pp.

Arnaud, L. (2000) Les propriétés des mortiers à base de chanvre. Conférence au CRAB-Info Batir de Lyon, 26 October 2000, France.

Arnaud, L. (2000) Properties of Hemp Concrete. 3rd Internationnal Symposium Bioresource Hemp 2000, Hanover, Germany.

Arnaud, L., Monnet, H. and Sallet, F. (2000) Modélisation par homogénéisation auto cohérente de la conductivité thermique des bétons et laines de chanvre. In: Lallemand, A. and Leone, J.F. (eds) *Congrès français de thermique*, 2000. Elsevier, pp. 543–548.

Arnaud, L., Monnet, H. and Sallet, F. (2000) Mechanical and thermal properties of vegetal concrete. In: Barbosa, N.P., Swamy, R.N. and Lynsdale, C. (eds) *Sustainable Construction Into the Next Millenium: Environmentally Friendly and Innovation Cement-based Materials*, November 2000. Elsevier, pp. 302–311.

Avis technique 20/04-50 (2005) Produit d'isolation thermique de toiture, Isonat végétal, Florapan Plus ou Flararol Plus, April 2005 (http://www.cstb.fr).

Cerezo, V. (2005) Mechanical, thermal and acoustic properties of a material based on plant particles: experimental approach and theoretical modelling. Doctoral thesis, NHI and ENTPE, Lyon, France, 242 pp.

Cerezo, V. and Arnaud, L. (2001) Propriétés des bétons et des laines de chanvre. Conférence au 2nd Assises du Chanvre, March 2001, Bar sur Aube, France.

Cordier, C. (1999) Caractérisation thermique et mécanique des bétons de chanvre. Travail de Fin d'Etudes. ENTPE (Laboratoire de Géomatériaux), France, 56 pp.

Couedel, I. (1998) Le béton de chanvre comme matériau de construction, première approche mécanique du béton de chanvre. DEA:ENTPE (Laboratoire de Géomatériaux), France, 69 pp.

Garnier, P. (2000) Le séchage des matériaux poreux. Approche expérimentale et approche théorique par homogénéisation des structures périodiques. DEA:ENTPE (Laboratoire de Géomatériaux), France, 100 pp.

Lienard, P. and Francois, P. (1972) Acoustique Industrielle: éléments fondamentaux et métrologie. Monographies d'Acoustiques du Groupement des Acousticiens de Langue Française, France, 283 pp.

Mainguy, M., Coussy, O. and Eymard, R. (1995) Modélisation des transferts hydriques isothermes en milieu poreux phénomènes de séchage. Application au séchage des matériaux à base de ciment. Études et recherches du LCPC – OA32.

Menguy, G., Laurent, M., Société Weber and Broutin, Moutarda, A. and Leveau, J. (1986) Cellule de mesure des caractéristiques thermophysiques des matériaux E1700. Technical Bulletin. Deltalab, 16 pp.

Monnet, H. (1999) Caractérisation mécanique et thermique de la laine de chanvre. Travail de Fin d'Etudes. ENTPE (Laboratoire de Géomatériaux), France, 49 pp.

Règles professionnelles de Construction en Chanvre CenC FFB, FNPC, MEL, MACP (http://www.construction-chanvre.asso.fr).

Chapter 19

Fournier, G. (1981) Les chimiotypes du Chanvre (*Cannabis sativa* L.). Intérêt pour un programme de sélection. *Agronomie* 1(8), 679–688.

Fournier, G. (2000) La sélection du chanvre à fibres (*Cannabis sativa* L.) en France. Chanvre et THC. *C.R. Acad. Agric.* 86, 209–217.

Fournier, G. (2003a) À propos de la teneur en Δ-9-THC dans les variétés de chanvre à fibres cultivées en France. *Annales de Toxicologie Analytique* 15, 30–34.

Fournier, G. (2003b) Le chanvre (*Cannabis sativa* L.) et sa réglementation Européenne en 2003. Teneur en Δ-9-THC dans les variétés cultivées en France. *Annales de Toxicologie Analytique* 15, 190–196.

Fournier, G. (2004) Évolution de la réglementation française relative au Cannabis. *Annales de Toxicologie Analytique* 16, 221–222.

Fournier, G. (2005) Une nouvelle souche de chanvre à fibres sans cannabinoïdes. *Annales de Toxicologie Analytique* 17, 109–111.

Fournier, G., Beherec, O. and Bertucelli, S. (2003) Intérêt du rapport Δ-9-THC/CBD dans le contrôle des cultures de chanvre industriel. *Annales de Toxicologie Analytique* 15, 250–259.

Seth, R. and Sinha, S. (1991) Chemistry and pharmacology of Cannabis. *Progress in Drug Research* 36, 71–115.

Small, E. (1979) *Cannabis. The Species Problem in Science and Semantics.* Corpus Ed., Toronto.

Turner, C.E., Elsohly, M.A. and Boeren, E.G. (1980) Constituents of *Cannabis sativa* L. XVII: a review of the natural constituents. *Journal of Natural Products* 43(2), 169–234.

Chapter 20

Berenji, J. (1996a) Konoplja. *Bilten za hmelj, sirak i lekovito bilje* 23/24 (64–65), 79–85.

Berenji, J. (1996b) Present status and perspectives of hemp in Yugoslavia. *Agricultural Engineering* 2(1–2), 1–11.

Berenji, J. (1998) Istine i zablude o konoplji. *Zbornik radova Instituta za ratarstvo i povrtarstvo Novi Sad* 30, 271–281.

Berenji, J. (2004) Perspectives and future of hemp in Europe. Proceedings of the Global Workshop (General Consulation) of the FAO/ESCORENA European Cooperative Research Network on Flax and other Bast Plants 'Bast Fibrous Plants for Healthy Life', Banja Luka (Bosnia and Herzegovina/Republika Srpska), proceedings on CD.

Berenji, J. and Sabo, A. (1998) Konoplja (*Cannabis* sp.) kao lekovita biljka. *Medicinal Plant Report* 5(5), 43–57.

Berenji, J. and Sikora, V. (1996) Oplemenjivanje i semenarstvo konoplje. *Zbornik radova naučnog institiuta za ratarstvo i povrtarstvo Novi Sad* 27, 19–38.

Berenji, J., Martinov, M. and Sikora, V. (2001) Perspektive konoplje. *Savremena poljoprivredna tehnika* 27(3–4), 131–138.

Čamprag, D., Jovanić, M. and Sekulić, R. (1996) Štetočine konoplje i integralne mere suzbijanja. *Zbornik radova naučnog institiuta za ratarstvo i povrtarstvo Novi Sad* 27, 55–68.

Drezgić, P., Stanaćev, S. and Starčević, Lj. (1975) *Posebno ratarstvo – drugi deo.* Univerzitet u Novom Sadu, Poljoprivredni fakultet, Novi Sad, Serbia.

Kišgeci, J. (1996) Praise to hemp. Nolit Beograd & NIP Novi Sad. Presentation of this book. *Journal of the International Hemp Association* 3(2), 88.

Lemeshev, N., Rumyantseva, L. and Clarke, R.C. (1994) Maintenance of Cannabis germplasm in the Vavilov Research Institute Gene Bank – 1993. *Journal of the International Hemp Association* 1(1), 1–5.

Makendić, V. (1937) Gnojidba konoplje. *Poljoprivredni glasnik* 17(5), 5–7.

Martinov, M., Veselinov, V. and Berenji, J. (1995) Žetva semenske konoplje. *Revija agronomska saznanja* 2, 71–74.

Martinov, M., Marković, D., Tešić, M. and Grozdanić, N. (1996) Hemp harvesting mechanization. *Agricultural Engineering* 2(1–2), 23–38.

Starčević, Lj. (1966) *Savremena tehnologija proizvodnje konoplje.* Privredna komora Vojvodine, Novi Sad, Serbia.

Starčević, Lj. (1978) *Gustina sklopa kao činilac formiranja prinosa i kvaliteta konoplje za vlakno.* Zbornik radova institiuta za ratarstvo i povrtarstvo, Novi Sad, Serbia, pp. 357–364.

Vrbničanin, S., Malidža, G., Stefanović, L., Elezović, I., Stanković-Kalezić, R., Marisavljević, D., *et al.* (2008) Distribucija nekih ekonomski štetnih, invazivnih i karantinskih korovskih vrsta na području Srbije. II deo: prostorna distribucija i zastupljenost devet korovskih vrsta. *Biljni lekar* 34(6), 408–418.

Chapter 21

AGRESTE (2001) Tableaux de l'Agriculture Bretonne 2001. Direction Régionale de l'Agriculture et de la Forêt, Service Régional de Statistique Agricole, Rennes, France.

Amaducci, S. (2003) HEMP-SYS: Design, development and up scaling of a sustainable production system for hemp textiles – an integrated quality systems approach. *Journal of Industrial Hemp* 8(2), 79–83.

BUWAL (1996) Ökoinventare für Verpackungen. Schriftenreihe Umwelt Nr. 250/1+2, Bundesamt für Umwelt, Wald und Landschaft, Bern, Switzerland.

FAO (2004) FAOSTAT Agriculture Data (http://faostat.fao.org/).

FNPC (2002) *La culture du chanvre. Les techniques culturales, le contexte économique.* Fédération Nationale des Producteurs de Chanvre, Le Mans, France.

Gorchs, G. and Lloveras, J. (2003) Current status of hemp production and transformation in Spain. *Journal of Industrial Hemp* 8(1), 45–64.

Guinée, J.B., Gorrée, M., Heijungs, R., Huppes, G., Kleijn, R., Koning, A. de, *et al.* (2002) *Handbook on Life Cycle Assessment. An Operational Guide to the ISO Standards.* Kluwer Academic Publishers, Dordrecht, Netherlands.

Nemecek, T. and Heil, A. (2001) SALCA – Swiss Agricultural Life Cycle Assessment Database. Version 012, December 2001. FAL, Swiss Federal Research Station for Agroecology and Agriculture, Zurich, Switzerland.

Patyk, A. and Reinhardt, G.A. (1998) Life cycle assessment of hemp products. In: Ceuterick, D. (ed.) *International Conference on Life Cycle Assessment in Agriculture, Agro-industry and Forestry* Proceedings. Brussels, Belgium, pp. 39–44.

Van der Werf, H.M.G. (2004) Life cycle analysis of field production of fibre hemp, the effect of production practices on environmental impacts. *Euphytica* 140, 13–23.

Van Zeijts, H. and Reus, J.A.W.A. (1996) *Toepassing van LCA voor Agrarische Produkten. 4a. Ervaringen met de methodiek in de case akkerbouw.* LEI-DLO, The Hague, Netherlands.

Index